T0212558

Lecture Notes in Artificial Intelligence   10147

Subseries of Lecture Notes in Computer Science

More information about this series at http://www.springer.com/series/1244

Alfredo Petrosino · Vincenzo Loia
Witold Pedrycz (Eds.)

Fuzzy Logic and Soft Computing Applications

11th International Workshop, WILF 2016
Naples, Italy, December 19–21, 2016
Revised Selected Papers

 Springer

Editors
Alfredo Petrosino
University of Naples "Parthenope"
Naples
Italy

Witold Pedrycz
University of Alberta
Edmonton, AB
Canada

Vincenzo Loia
University of Salerno
Fisciano, (Salerno)
Italy

ISSN 0302-9743   ISSN 1611-3349   (electronic)
Lecture Notes in Artificial Intelligence
ISBN 978-3-319-52961-5   ISBN 978-3-319-52962-2   (eBook)
DOI 10.1007/978-3-319-52962-2

Library of Congress Control Number: 2016963602

LNCS Sublibrary: SL7 – Artificial Intelligence

Printed on acid-free paper

This Springer imprint is published by Springer Nature
The registered company is Springer International Publishing AG
The registered company address is: Gewerbestrasse 11, 6330 Cham, Switzerland

Preface

The 11th International Workshop on Fuzzy Logic and Applications, WILF 2016, held in Naples (Italy) during December 19–21, 2016, covered all topics in theoretical, experimental, and application areas of fuzzy, rough, and soft computing in general, with the aim of bringing together researchers from academia and industry to report on the latest advances in their fields of interest. A major objective of WILF in the present rich data era is the presentation of the consolidated results of fuzzy, rough, and soft computing and of their potential applications to the analysis of big data and computer vision tasks and the potential impact on deep learning as mechanisms to capture hidden information from data.

This event represents the pursuance of an established tradition of biannual inter-disciplinary meetings. WILF returned to Naples for the third time, after the first edition in 1995, when it was formerly established, and after the edition of 2003 that consolidated the international validity of the workshop. The previous editions of WILF have been held, with an increasing number of participants, in Naples (1995), Bari (1997), Genoa (1999), Milan (2001), Naples (2003), Crema (2005), Camogli (2007), Palermo (2009), Trani (2011), and Genoa (2013). Each event has focused on distinct main thematic areas of fuzzy logic and related applications. From this perspective, one of the main goals of the WILF workshop series is to bring together researchers and developers from both academia and high-tech companies and foster multidisciplinary research.

WILF 2016 certainly achieved the goal. This volume consists of 22 selected peer-reviewed papers, discussed at WILF 2016 as oral contributions. Two invited speakers provided useful links between logic and granular computing and applications:

- Hani Hagras (University of Essex, UK) "General Type-2 Fuzzy Logic Systems For Real World Applications"
- Witold Pedrycz (University of Alberta, Canada) "Algorithmic Developments of Information Granules of Higher Type and Higher Order and Their Applications"

A tutorial by Francesco Masulli in a happy moment of his life gave insight into the role of computational intelligence in big data with an emphasis on health and well-being applications:

- Francesco Masulli (University of Genoa, Italy) "Computational Intelligence and Big Data in Health and Well-Being"

WILF 2016 was also an occasion to fully recognize the achievements of Antonio Di Nola, who, as honorary chair, pointed out how fuzzy logic may be seen as a logic itself:

- Antonio Di Nola (University of Salerno, Italy) "Fuzzy Logic as a Logic"

In addition, awards made available by the Italian Group of Pattern Recognition Researchers (GIRPR), affiliated to the International Association of Pattern Recognition (IAPR), the European Society for Fuzzy Logic and Technology (EUSFLAT), were handed to PhD students who reported their achievements and research plans in a successful PhD Forum, as well as young researchers who were authors of WILF 2016 papers. All of the award recipients were invited to submit their papers to the *Information Sciences* journal as encouragement of their valuable work in the field.

Thanks are due to the Program Committee members for their commitment to provide high-quality, constructive reviews, to the keynote speakers and the tutorial presenters, and to the local Organizing Committee for the support in the organization of the workshop events. Special thanks to all the CVPRLab staff and specifically Francesco, Alessandro, Mario, Gianmaria, and Vincenzo, for their continuous support and help.

December 2016   Alfredo Petrosino
                                                          Vincenzo Loia
                                                          Witold Pedrycz

Organization

WILF 2016 was jointly organized by the Department of Science and Technology, University of Naples Parthenope, Italy, the EUSFLAT, European Society for Fuzzy Logic and Technology, the IEEE, Computational Intelligence Society, Italian Chapter, and the GIRPR, Group of Italian Researchers in Pattern Recognition.

Conference Chairs

Alfredo Petrosino	University of Naples Parthenope, Italy
Vincenzo Loia	University of Salerno, Italy
Witold Pedrycz	University of Alberta, Canada

Program Committee

Andrzej Bargiela	University of Nottingham, UK
Isabelle Bloch	CNRS, LTCI, Université Paris-Saclay, France
Gloria Bordogna	CNR IREA, Italy
Humberto Bustince	Universidad Publica de Navarra, Spain
Giovanna Castellano	University of Bari, Italy
Oscar Castillo	Tijuana Institute of Technology, Mexico
Ashish Ghosh	Indian Statistical Institute, India
Fernando Gomide	University of Campinas, Brazil
Hani Hagras	University of Essex, UK
Enrique Herrera-Viedma	University of Granada, Spain
Tzung-Pei Hong	National University of Kaohsiung, Taiwan
Ronald Yager	Iona College, USA
Javier Montero	Universidad Complutense de Madrid, Spain
Janusz Kacprzyk	Polish Academy of Sciences, Poland
Nikola Kasabov	Auckland University of Technology, New Zealand
Etienne Kerre	Ghent University, Belgium
László Kóczy	Budapest University of Technology and Economics, Hungary
Vladik Kreinovich	University of Texas at El Paso, USA
Sankar Pal	Indian Statistical Institute, Kolkata, India
Marek Reformat	University of Edmonton, Canada
Stefano Rovetta	University of Genova, Italy

Organizing Committee

Francesco Camastra	University of Naples Parthenope, Italy
Giosuè Lo Bosco	University of Palermo, Italy
Antonino Staiano	University of Naples Parthenope, Italy

WILF Steering Committee

Antonio Di Nola	University of Salerno, Italy
Francesco Masulli	University of Genoa, Italy
Gabriella Pasi	University of Milano Bicocca, Italy
Alfredo Petrosino	University of Naples Parthenope, Italy

Scientific Secretary

Federica Andreoli	University of Naples Parthenope, Italy

Financing Institutions

DIBRIS, University of Genoa, Italy
EUSFLAT, European Society for Fuzzy Logic and Technology
GIRPR, Group of Italian Researchers in Pattern Recognition

Contents

Invited Speakers

Towards a Framework for Singleton General Forms of Interval Type-2 Fuzzy Systems

Gonzalo Ruiz-García[1], Hani Hagras[2(✉)], Ignacio Rojas[1],
and Hector Pomares[1]

[1] Department of Computer Architecture and Computer Technologies,
Universidad de Granada, Granada, Spain
[2] The Computational Intelligence Centre, School of Computer Science
and Electronic Engineering, University of Essex, Colchester, UK
hani@essex.ac.uk

Abstract. Recently, it has been shown that interval type-2 fuzzy sets (IT2FSs) are more general than interval-valued fuzzy sets (IVFSs), and some of these IT2FSs can actually be non-convex. Although these IT2FSs could be considered within the general type-2 fuzzy sets' (GT2FSs) scope, this latter have always been studied and developed under certain conditions considering the convexity and normality of their secondary grades. In recent works the operations of intersection and union for GT2FSs have been extended to include non-convex secondary grades. Hence, there is a need to develop the theory for those general forms of interval type-2 fuzzy logic systems (gfIT2FLSs) which use IT2FSs that are not equivalent to IVFSs and can have non-convex secondary grades. In this chapter, we will present the mathematical tools to define the inference engine for singleton gfIT2FLSs. This work aims to introduce the basic structure of such singleton gfIT2FLSs, paying special attention to those blocks presenting significant differences with the already well known type-2 FLSs which employ IT2FSs which are equivalent to IVFSs (we will term IVFLSs).

1 Introduction

It is a well-known fact that both Type-1 (T1) and Type-2 (T2) fuzzy logic systems (FLSs) have received significant attention from the research community, and both have been successfully used in many real world applications, such as robotics [5, 6, 29], control [3, 25], image processing [33], network traffic control [9], function approximation [24], pattern recognition [8] and many others.

Type-2 fuzzy logic and systems are usually divided in the literature between Interval Type-2 (IT2) and General Type-2 (GT2) fuzzy logic and systems. Some authors argue that the term "General Type-2 Fuzzy Logic" should not be used, as there is no formal definition for it, so "Type-2 Fuzzy Logic" should be the proper name to use; nonetheless, the term is widely used to make explicit difference with the IT2 case, and we will do so in this Chapter. Although GT2FLSs were defined as soon as 1999, their practical application has been limited due to their higher computational complexity, favouring the simpler version of IT2FLSs. More recently, some authors have proposed some approximations of GT2FLSs based on several IT2FLSs working in

© Springer International Publishing AG 2017
A. Petrosino et al. (Eds.): WILF 2016, LNAI 10147, pp. 3–26, 2017.
DOI: 10.1007/978-3-319-52962-2_1

parallel (α-planes representation in [20] or the zSlices based GT2FLS in [30]); however, these systems require the secondary grades to be convex type-1 fuzzy sets.

Traditionally IT2FSs have been considered to be equivalent to IVFSs [16]. Nevertheless, Bustince et al. showed recently [2] that IT2FSs are more general than IVFSs. In addition, some of these IT2FSs have secondary grades which are non-convex T1FSs. Hence, although there is a big literature in both IT2 and GT2FLSs, most of the existing work focuses either on IVFSs or GT2FSs with convex and normal secondary grades. In this Chapter, we consider those IT2FSs whose secondary grades can be non-convex, and thus are not equivalent to IVFSs. Those sets will be referred to in this text as *general forms of interval type-2 fuzzy sets* (gfIT2FSs).

These new gfIT2FSs can easily capture the faced uncertainty without introducing unneeded and unrealistic uncertainty to the IT2FS. For instance, a method frequently used in the literature to obtain IVFSs [19] for the antecedents or consequents involve a survey from different people (who know about fuzzy logic) to provide the *person type-1 fuzzy set Membership Function* (MF) which represents a given concept from that person's point of view. Some of these MFs are depicted in Fig. 1(a). These type-1 fuzzy sets are then aggregated in order to obtain upper and lower MFs to define an IVFS. The resulting set can actually embed many type-1 fuzzy sets obtained within the resulting footprint of uncertainty (FOU) as shown in Fig. 1(b).

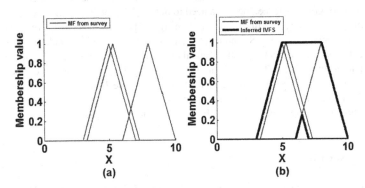

Fig. 1. (a) Type-1 MFs obtained from different people in a survey. (b) IVFS obtained aggregating the T1 MFs.

However, if the different T1 MFs provided are sparse (which might happen frequently, as shown in Fig. 1(a)), a very wide footprint of uncertainty (FOU) can be obtained, which implies that a higher level of uncertainty is included in the set and modelled than what might actually be present in real world situations. Besides, the obtained FOU in Fig. 1(b) might include huge number of emerging non triangular embedded sets which might not represent the surveyed population opinion.

This problem can be naturally solved by using a specific class of the gfIT2FSs, which is called *multi-singleton IT2FSs*, whose secondary membership is comprised of several singletons at each point within the X-domain (as depicted in Fig. 2). These sets can easily

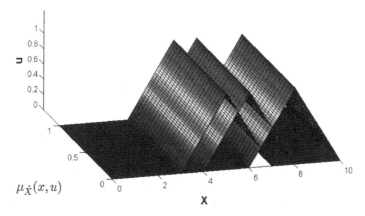

$\mu_{\tilde{X}}(x,u)$

Fig. 2. A multi-singleton IT2FS to model the uncertainty amongst the type-1 fuzzy sets in Fig. 1(a).

represent all sets gathered in the survey to model the faced uncertainty without adding extra unneeded and/or unrealistic uncertainty to the final type-2 fuzzy set.

In [26] the theory for the join and meet operations on GT2FSs with non-convex/ arbitrary secondary grades was presented; in addition, special attention was drawn to the case of non-convex IT2FSs which are gfIT2FSs. Once these set theoretic operations are available, the fuzzy inference engine for the gfIT2FLSs can be defined. In this Chapter we aim to present the structure of a singleton gfIT2FLSs. We will specially focus on those parts having significant differences, as many of the elements are analogous to those in other kinds of FLS (i.e. IVFLSs and GT2FLSs).

This chapter is organised as follows: Sect. 2 presents an overview on the theoretical background of the gfIT2FSs. Section 3 presents the join and meet operations for GT2FSs; subsequently, those results will be particularised to gfIT2FSs, in order to obtain specific equations for the inference engine within a gfIT2FLS. Section 4 presents the structure of a singleton gfIT2FLS, including a worked example. Finally, Sect. 5 presents applications and future work.

2 Theoretical Background on General Forms of IT2FSs (gfIT2FSs)

In this section, we will review the theoretical background of the gfIT2FSs. In order to do so, we will revisit the initial notion of T2FSs, their initial formal definitions and, finally, their generalisation.

Based on Zadeh's initial notion of a type-2 fuzzy set in [32], Mizumoto and Tanaka [23] presented the mathematical definition of a type-2 fuzzy set. After their work, many authors have used different approaches and notations to define these sets, which have been defined by Liu and Mendel as follows [18]:

Definition 1: A type-2 fuzzy set $\tilde{A}$ in a non-empty universe of discourse X is given by:

$$\tilde{A} = \{((x, u), \mu_{\tilde{A}}(x, u)) | \text{ for all } x \in X, \text{ for all } u \in J_x \subseteq [0, 1] | 0 \leq \mu_{\tilde{A}}(x, u) \leq 1\} \quad (1)$$

Definition 2: Using Definition 1, in [18] IT2FSs are defined as follows:
   If all $\mu_{\tilde{A}}(x, u) = 1$, then $\tilde{A}$ is an IT2FSs.

It is important to note that, from Definition 1, $J_x \subseteq [0, 1]$ does not necessarily imply that J_x is a continuous and closed interval contained in or equal to [0, 1], but rather a subset within that interval. This fact was further discussed in [2] and, hence, IT2FSs as defined in Definition 2 are not necessarily interval-valued fuzzy sets. The authors of [2] offered detailed itemisation of all kinds of IT2FSs (shown in Fig. 3), which is summarised subsequently:

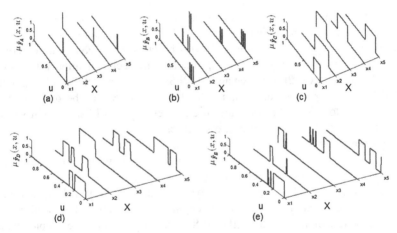

Fig. 3. Examples of all types of IT2FSs. (a) T1FS. (b) Multi-singleton IT2FS. (c) IVFS. (d) Multi-interval IT2FS. (e) gfIT2FS containing both singletons and intervals [16].

- Case (A) is when the subset $J_x \subseteq [0, 1]$ is a single element, i.e., $J_x = \{u_x\} \in [0, 1]$. This condition is equivalent to $\mu_{\tilde{A}}(x) = u_x \in [0, 1]$; which is equivalent to type-1 fuzzy sets.
- Case (B) is when the subset $J_x \subseteq [0, 1] = \{u_x^1, u_x^2, \ldots, u_x^n\} \subseteq [0, 1]$ which is a set of different singletons, such that $u_x^i \neq u_x^j, \forall i \neq j$. This is referred to as the *multi-singleton IT2FS*.
- Case (C) is when the subset $J_x \subseteq [0, 1] = [\underline{A_x}, \overline{A_x}] \subseteq [0, 1]$ is a connected and closed interval contained in [0, 1]. This is analogous to IVFSs.

- Case (D) is when the subset $J_x \subseteq [0,1] = \left[\underline{A_x^1}, \overline{A_x^1}\right] \cup \left[\underline{A_x^2}, \overline{A_x^2}\right] \cup \ldots \cup \left[\underline{A_x^n}, \overline{A_x^n}\right]$, where each $\left[\underline{A_x^i}, \overline{A_x^i}\right] \subseteq [0,1]$ and $\left[\underline{A_x^i}, \overline{A_x^i}\right]$ and $\left[\underline{A_x^j}, \overline{A_x^j}\right]$ are closed and disjointed intervals if $i \neq j$. We will usually refer to this case as *multi-interval IT2FSs*.

- Case (E) is a particular case of (D), when some of the intervals $\left[\underline{A_x^i}, \overline{A_x^i}\right]$ degenerate in singletons, i.e., $\underline{A_x^i} = \overline{A_x^i}$ for one or more i.

All cases of IT2FSs, from A to E, can be represented as the union of a finite number of closed intervals, say $N_{\tilde{X}}(x)$. It is worthwhile to highlight that $N_{\tilde{X}}$ depends on the primary variable x, as not all the secondary grades may have the same number of disjointed singletons/intervals. Hence, a gfIT2FS $\tilde{X}$ whose membership function is $\mu_{\tilde{X}}(x)$ which is defined over a universe of discourse X (with $x \in X$), can be written as follows:

$$\mu_{\tilde{X}}(x) = \bigcup_{i=1}^{N_{\tilde{X}}(x)} \mu_{\tilde{X}}^i(x) = \bigcup_{i=1}^{N_{\tilde{X}}(x)} 1 / \left[l_{\tilde{X}}^i(x), r_{\tilde{X}}^i(x)\right] \tag{2}$$

Where each of the closed intervals $\mu_{\tilde{X}}^i(x) = \left[l_{\tilde{X}}^i(x), r_{\tilde{X}}^i(x)\right]$ comprising $\mu_{\tilde{X}}(x)$ will be referred to as *the subintervals of* $\mu_{\tilde{X}}(x)$. As we are dealing with gfIT2FSs, all the non-zero secondary membership values will be equal to one, hence the 1 in Eq. (2) is redundant and sometimes we will drop it from the notation for the sake of simplicity. Moreover, in Eq. (2) we have explicitly indicated the dependency of $N_{\tilde{X}}$, $l_{\tilde{X}}^i$, and $r_{\tilde{X}}^i$ with x, also for the sake of simplicity, we may not always indicate so. Hence, Eq. (2) can be simplified as:

$$\mu_{\tilde{X}}(x) = \bigcup_{i=1}^{N_{\tilde{X}}} \mu_{\tilde{X}}^i(x) = \bigcup_{i=1}^{N_{\tilde{X}}} \left[l_{\tilde{X}}^i, r_{\tilde{X}}^i\right] \tag{3}$$

At this point it would be a good idea to revise the concepts of *lower membership function* (LMF) and *upper membership function* (UMF). In [14] Liang and Mendel defined the LMF and UMF of an IVFS as the bounds for the footprint of uncertainty (FOU); however, in their definition the FOU was a closed and connected region. When dealing with gfIT2FSs, the FOU may have several boundaries; thus, to avoid any kind of ambiguity, we will redefine the LMF and UMF of a gfIT2FS:

Definition 3: Let $\tilde{X}$ be a gfIT2FS and let $\mu_{\tilde{X}}(x)$ be a vertical slice placed at each x, $\mu_{\tilde{X}}(x) = f(x, \theta)$, $\theta \in [0,1]$. For each $f(x, \theta)$, let $v_1 = v_1(x)$ be the infimum of the support of $f(x, \theta)$. Hence, the *lower membership function (LMF)* of $\mu_{\tilde{X}}(x)$ is:

$$LMF(\tilde{X}) = LMF(\mu_{\tilde{X}}(x)) = \underline{\mu_{\tilde{X}}}(x) = \{v_1(x) | x \in X\} \tag{4}$$

Definition 4: Let $\tilde{X}$ be a gfIT2FS and let $\mu_{\tilde{X}}(x)$ be a vertical slice placed at each x, $\mu_{\tilde{X}}(x) = f(x, \theta)$, $\theta \in [0, 1]$. For each $f(x, \theta)$, let $v_{end} = v_{end}(x)$ be the supremum of the support of $f(x, \theta)$. Hence, the *upper membership function (UMF)* of $\tilde{X}$ is:

$$UMF(\tilde{X}) = UMF(\mu_{\tilde{X}}(x)) = \bar{\mu}_{\tilde{X}}(x) = \{v_{end}(x)|x \in X\} \tag{5}$$

Definitions 3 and 4 can also be applied to every single subinterval $\mu_{\tilde{X}}^{i}(x)$ comprising a gfIT2FS. Hence, we will distinguish between the LMF/UMF of a gfIT2FS $\mu_{\tilde{X}}(x)$ and the LMF/UMF of a subinterval $\mu_{\tilde{X}}^{i}(x)$ within a gfIT2FS, which will be denoted as $LMF\left(\mu_{\tilde{X}}^{i}(x)\right) = \underline{\mu}_{\tilde{X}}^{i}(x)$ and $UMF\left(\mu_{\tilde{X}}^{i}(x)\right) = \bar{\mu}_{\tilde{X}}^{i}(x)$, respectively, with $i = 1, \ldots, N_{\tilde{X}}$. Once the LMF and UMF for subintervals have been defined, Eq. (3) could be rewritten as follows:

$$\mu_{\tilde{X}}(x) = \bigcup_{i=1}^{N_{\tilde{X}}} \mu_{\tilde{X}}^{i}(x) = \bigcup_{i=1}^{N_{\tilde{X}}} \left[\underline{\mu}_{\tilde{X}}^{i}(x), \bar{\mu}_{\tilde{X}}^{i}(x)\right] \tag{6}$$

It is worthwhile to highlight that for the sake of simplicity, we will drop (x) from the notation of $\underline{\mu}_{\tilde{X}}^{i}(x)$ and $\bar{\mu}_{\tilde{X}}^{i}(x)$ and they will be written as $\underline{\mu}_{\tilde{X}}^{i}$ and $\bar{\mu}_{\tilde{X}}^{i}$.

3 Operations on gfIT2FSs

In order to define the inference engine operations of the gfIT2FLSs, we will examine the operations of *intersection* and *union* on gfIT2FSs. To do so, we firstly focus our attention in those operations on GT2FSs.

Although these operations were defined in 1975 by Zadeh using the Extension Principle [32], this method was relatively computationally expensive, which motivated researchers to look for some simpler methods or closed formulas to compute them. In [16], Mendel proved the set theoretic operations for IVFSs, whereas in [12] Karnik and Mendel proved the mathematical formulas for these operations on GT2FSs where the secondary grades are normal and convex T1FSs. This work was later generalised by Coupland [4] to incorporate non-normal sets. Later works [21, 22, 27] studied the geometrical properties of some GT2FSs to find closed formulas or approximations for the join and meet operations in some specific cases. However, none of these results are applicable in the present work, as the involved secondary grades can be non-convex type-1 sets. In [26], two theorems were presented for the join and meet operations of GT2FSs (including gfIT2FSs) with arbitrary secondary grades which we will review in this section.

3.1   The Join Operation on GT2FSs

Definition 5: Let $\tilde{F}_1$ and $\tilde{F}_2$ be two type-2 fuzzy sets in a universe of discourse X. Let $\mu_{\tilde{F}_1}(x)$ and $\mu_{\tilde{F}_2}(x)$ denote the membership grades of $\tilde{F}_1$ and $\tilde{F}_2$, respectively, at $x \in X$.

Then, for each $x \in X$, using minimum t-norm and maximum t-conorm, the union set $\tilde{F}_1 \cup \tilde{F}_2$, which is characterised by its membership grade $\mu_{\tilde{F}_1 \cup \tilde{F}_2}(x, \theta)$, is given by the *join* operation on $\mu_{\tilde{F}_1}(x)$ and $\mu_{\tilde{F}_2}(x)$, and is as follows[1]:

$$
\begin{aligned}
\mu_{\tilde{F}_1 \cup \tilde{F}_2}(x, \theta) &= \left(\mu_{\tilde{F}_1}(x) \sqcup \mu_{\tilde{F}_2}(x) \right)(\theta) \\
&= sup_{v \in [0,\theta]} \{f_1(v)\} \wedge (f_1(\theta) \vee f_2(\theta)) \wedge sup_{w \in [0,\theta]} \{f_2(w)\}
\end{aligned}
\tag{7}
$$

Such that $(v \vee w) = \theta$.

Theorem 1: The union operation on two type-2 fuzzy sets defined in [12] using minimum t-norm is equivalent to the union defined in Eq. (7).

Proof of Theorem 1: Let the join operation be performed on two type-2 fuzzy sets, denoted $\tilde{F}_1$ and $\tilde{F}_2$, in a universe of discourse X. The membership grades at $x \in X$ of $\tilde{F}_1$ and $\tilde{F}_2$ are denoted as $\mu_{\tilde{F}_1}(x)$ and $\mu_{\tilde{F}_2}(x)$, respectively, which are fuzzy sets defined in $V, W \subseteq [0, 1]$. According to [3], the union of two type-2 fuzzy sets, denoted as $\tilde{F}_1 \cup \tilde{F}_2$, is given by the join operation between $\tilde{F}_1$ and $\tilde{F}_2$ as follows:

$$
\begin{aligned}
\tilde{F}_1 \cup \tilde{F}_2 \leftrightarrow \mu_{\tilde{F}_1 \cup \tilde{F}_2}(x) &= \mu_{\tilde{F}_1}(x) \sqcup \mu_{\tilde{F}_2}(x) \\
&= \int_{v \in V} \int_{w \in W} (f_1(v) * f_2(w))/(v \vee w) \\
& x \in X
\end{aligned}
\tag{8}
$$

Where $*$ indicates the minimum t-norm (hence $*$ will be replaced by $\wedge$ the rest of the chapter) and $\vee$ indicates the maximum t-conorm. Thus, any element $\theta = (v \vee w)$ in the primary membership of $\tilde{F}_1 \cup \tilde{F}_2$, can be obtained by any of the following two cases:

(1) **Case 1:** if v is any value between 0 and θ, and $w = \theta$; i.e., $\{(v, w)|v \leq \theta \, and \, w = \theta\} \to (v \vee w) = (v \vee \theta) = \theta$. This condition is equivalent to state that $v \in [0, \theta]$ and $w = \theta$.
(2) **Case 2:** if w is any value between 0 and θ, and $v = \theta$; i.e., $\{(v, w)|w \leq \theta \, and \, v = \theta\} \to (v \vee w) = (\theta \vee w) = \theta$. This condition is equivalent to state that $w \in [0, \theta]$ and $v = \theta$.

The membership value associated with θ can be obtained by applying the minimum t-norm on the secondary grades $f_1(v)$ and $f_2(w)$ where v and w are as described in Cases 1 or 2; hence, $\mu_{\tilde{F}_1 \cup \tilde{F}_2}(x, \theta) = f_1(v) \wedge f_2(w)$.

---

[1] It should be noted that Eq. (7) has some similarity to Eq. (10) in [31] (see also [7]) as both equations refer to the join operation of general type-2 fuzzy sets. However, the representation of Eq. (7) is quite different to simplify the computations and analysis.

It is important to note that if more than one pair $\{v, w\}$ result in the same $\theta = (v \vee w)$ but with different membership grade $\mu_{\tilde{F}_1 \cup \tilde{F}_2}(x, \theta) = f_1(v) \wedge f_2(w)$, then we keep the maximum membership grade obtained from all $\{v \vee w\}$ pairs. Hence, $\mu_{\tilde{F}_1 \cup \tilde{F}_2}(x, \theta)$ is obtained by the following steps:

- **Step 1**: calculate $\Phi_1(\theta)$, where:

$$\Phi_1(\theta) = \sup_{v \in [0, \theta]} \{f_1(v) \wedge f_2(\theta)\} \tag{9}$$

According to the notation used in [31], Eq. (9) would be $f_1^L(\theta) \wedge f_2(\theta)$. See [31] for this notation.

- **Step 2**: calculate $\Phi_2(\theta)$, where:

$$\Phi_2(\theta) = \sup_{w \in [0, \theta]} \{f_1(\theta) \wedge f_2(w)\} \tag{10}$$

According to the notation used in [31], Eqs. (3–4) would be $f_1(\theta) \wedge f_2^L(\theta)$.

- **Step 3**: calculate $\mu_{\tilde{F}_1 \cup \tilde{F}_2}(x, \theta)$ where:

$$\mu_{\tilde{F}_1 \cup \tilde{F}_2}(x, \theta) = \Phi_1(\theta) \vee \Phi_2(\theta) \tag{11}$$

$f_1(\theta)$ and $f_2(\theta)$ are fixed as θ is fixed. Hence, $f_1(\theta)$ and $f_2(\theta)$ will not be considered for the suprema calculation. Consequently, we can rewrite Eqs. (9) and (10) as Eqs. (12) and (13), respectively, and combine them in Eq. (14).

$$\Phi_1(\theta) = \sup_{v \in [0, \theta]} \{f_1(v)\} \wedge f_2(\theta) \tag{12}$$

$$\Phi_2(\theta) = f_1(\theta) \wedge \sup_{w \in [0, \theta]} \{f_2(w)\} \tag{13}$$

$$\mu_{\tilde{F}_1 \cup \tilde{F}_2}(x, \theta) = \left(\sup_{v \in [0, \theta]} \{f_1(v)\} \wedge f_2(\theta)\right) \vee \left(f_1(\theta) \wedge \sup_{w \in [0, \theta]} \{f_2(w)\}\right) \tag{14}$$

Using four labels denoted as A_1, B_1, C_1 and D_1 as illustrated in Eq. (15), Eq. (14) can be rewritten as shown in Eq. (16).

$$\mu_{\tilde{F}_1 \cup \tilde{F}_2}(x, \theta) = \left(\underbrace{\sup_{v \in [0, \theta]} \{f_1(v)\}}_{A_1} \wedge \underbrace{f_2(\theta)}_{B_1}\right) \vee \left(\underbrace{f_1(\theta)}_{C_1} \wedge \underbrace{\sup_{w \in [0, \theta]} \{f_2(w)\}}_{D_1}\right) \tag{15}$$

$$\mu_{\tilde{F}_1 \cup \tilde{F}_2}(x, \theta) = (A_1 \wedge B_1) \vee (C_1 \wedge D_1) \tag{16}$$

The distributive property of minimum and maximum operations allows us to re-write the right hand side of Eq. (16) as follows:

$$(A_1 \wedge B_1) \vee (C_1 \wedge D_1) = (A_1 \vee C_1) \wedge (A_1 \vee D_1) \wedge (B_1 \vee C_1) \wedge (B_1 \vee D_1) \qquad (17)$$

By substituting Eq. (17) in Eq. (16) and replacing the labels denoted as A_1, B_1, C_1 and D_1, Eq. (17) can be written as below:

$$\mu_{\tilde{F}_1 \cup \tilde{F}_2}(x, \theta) = \left(sup_{v \in [0,\theta]}\{f_1(v)\} \vee f_1(\theta) \right)$$
$$\wedge \left(sup_{v \in [0,\theta]}\{f_1(v)\} \vee sup_{w \in [0,\theta]}\{f_2(w)\} \right) \qquad (18)$$
$$\wedge (f_2(\theta) \vee f_1(\theta)) \wedge (f_2(\theta) \vee sup_{w \in [0,\theta]}\{f_2(w)\})$$

Using four labels denoted as A_2, B_2, C_2 and D_2 as illustrated in Eq. (19), Eq. (18) can be re-written as shown in Eq. (20).

$$\mu_{\tilde{F}_1 \cup \tilde{F}_2}(x, \theta) = \underbrace{\left(sup_{v \in [0,\theta]}\{f_1(v)\} \vee f_1(\theta) \right)}_{A_2}$$
$$\wedge \underbrace{\left(sup_{v \in [0,\theta]}\{f_1(v)\} \vee sup_{w \in [0,\theta]}\{f_2(w)\} \right)}_{B_2} \qquad (19)$$
$$\wedge \underbrace{(f_2(\theta) \vee f_1(\theta))}_{C_2} \wedge \underbrace{(f_2(\theta) \vee sup_{w \in [0,\theta]}\{f_2(w)\})}_{D_2}$$

$$\mu_{\tilde{F}_1 \cup \tilde{F}_2}(x, \theta) = A_2 \wedge B_2 \wedge C_2 \wedge D_2 \qquad (20)$$

It is worthwhile to analyse two of the terms in Eq. (20), which are A_2 and D_2, separately:

$$A_2 = sup_{v \in [0,\theta]}\{f_1(v)\} \vee f_1(\theta) \qquad (21)$$

In the term A_2 shown in Eq. (21), it is important to note that the value $f_1(\theta)$ is *included* in the value $sup_{v \in [0,\theta]}\{f_1(v)\}$, as the value $v = \theta$ belongs to the interval $v \in [0, \theta]$. Hence, the maximum $f_1(\theta) \vee sup_{v \in [0,\theta]}\{f_1(v)\}$ will always be represented in the value $sup_{v \in [0,\theta]}\{f_1(v)\}$, regardless of the value of θ and the shape of the function $f_1(v)$. Consequently, term A_2 in Eq. (21) can be written as A_2' as shown in Eq. (22):

$$A_2 = sup_{v \in [0,\theta]}\{f_1(v)\} \vee f_1(\theta) = sup_{v \in [0,\theta]}\{f_1(v)\} = A_2'$$
$$\rightarrow f_1(\theta) \in \{f_1(v)|v \in [0, \theta]\} \rightarrow f_1(\theta) \leq sup_{v \in [0,\theta]}\{f_1(v)\} \qquad (22)$$

Similarly, we will use the abovementioned approach for the term D_2 in Eq. (19):

$$D_2 = f_2(\theta) \vee sup_{w \in [0,\theta]}\{f_2(w)\} \qquad (23)$$

Analogously, D_2 is equivalent to D_2':

$$D_2 = sup_{w \in [0,\theta]}\{f_2(w)\} \vee f_2(\theta) = sup_{w \in [0,\theta]}\{f_2(w)\} = D_2' \rightarrow$$
$$\rightarrow f_2(\theta) \in \{f_2(w)|w \in [0,\theta]\} \rightarrow f_2(\theta) \leq sup_{w \in [0,\theta]}\{f_2(w)\} \tag{24}$$

By using A_2' instead of A_2, and using D_2' instead of D_2 in Eq. (20), we have Eq. (25) as follows:

$$\mu_{\tilde{F}_1 \cup \tilde{F}_2}(x, \theta) = A_2' \wedge B_2 \wedge C_2 \wedge D_2' \tag{25}$$

Substituting each label A_2', B_2, C_2 and D_2' with their corresponding contents, we obtain Eq. (26):

$$\mu_{\tilde{F}_1 \cup \tilde{F}_2}(x, \theta) = sup_{v \in [0,\theta]}\{f_1(v)\} \wedge \left(sup_{v \in [0,\theta]}\{f_1(v)\} \vee sup_{w \in [0,\theta]}\{f_2(w)\}\right)$$
$$\wedge (f_2(\theta) \vee f_1(\theta)) \wedge sup_{w \in [0,\theta]}\{f_2(w)\} \tag{26}$$

In order to simplify the notations in the equation we will again label each term in Eq. (26) separately as shown below:

$$\mu_{\tilde{F}_1 \cup \tilde{F}_2}(x, \theta) = \underbrace{sup_{v \in [0,\theta]}\{f_1(v)\}}_{A_3} \wedge \left(\underbrace{sup_{v \in [0,\theta]}\{f_1(v)\}}_{A_3} \vee \underbrace{sup_{w \in [0,\theta]}\{f_2(w)\}}_{B_3}\right)$$
$$\wedge \underbrace{(f_2(\theta) \vee f_1(\theta))}_{C_3} \wedge \underbrace{sup_{w \in [0,\theta]}\{f_2(w)\}}_{B_3} \tag{27}$$

Using three labels denoted as A_3, B_3 and C_3 as illustrated in Eq. (27), Eq. (26) can be expressed as shown in Eq. (28).

$$\mu_{\tilde{F}_1 \cup \tilde{F}_2}(x, \theta) = A_3 \wedge (A_3 \vee B_3) \wedge C_3 \wedge B_3 \tag{28}$$

We will focus on the partial expression $A_3 \wedge (A_3 \vee B_3)$ in Eq. (28). Using the fact that $a \wedge (a \vee b) = a$ for any real numbers a and b, then $A_3 \wedge (A_3 \vee B_3) = A_3$, and Eq. (28) becomes Eq. (29). Substituting each label by its content, we have Eq. (30).

$$\mu_{\tilde{F}_1 \cup \tilde{F}_2}(x, \theta) = A_3 \wedge (A_3 \vee B_3) \wedge C_3 \wedge B_3 = A_3 \wedge C_3 \wedge B_3 \tag{29}$$

$$\mu_{\tilde{F}_1 \cup \tilde{F}_2}(x, \theta) = sup_{v \in [0,\theta]}\{f_1(v)\} \wedge (f_1(\theta) \vee f_2(\theta)) \wedge sup_{w \in [0,\theta]}\{f_2(w)\} \tag{30}$$
$$x \in X$$

Equation (30) is the same as Eq. (7) and this concludes the proof of Theorem 1. This equation is the final result for the join operation performed on two type-2 fuzzy sets, $\tilde{F}_1$ and $\tilde{F}_2$, for each $x \in X$. It is important to note that this result is obtained without any assumption regarding the normality or convexity of the secondary grades, denoted $f_1(v)$ and $f_2(w)$, that belong to the fuzzy sets $\tilde{F}_1$ and $\tilde{F}_2$, respectively.

3.2 The Meet Operation on GT2FSs

Definition 2: Let $\tilde{F}_1$ and $\tilde{F}_2$ be two type-2 fuzzy sets in a universe of discourse X. Let $\mu_{\tilde{F}_1}(x)$ and $\mu_{\tilde{F}_2}(x)$ denote the membership grades of $\tilde{F}_1$ and $\tilde{F}_2$, respectively, at $x \in X$. Then, using minimum t-norm, the intersection set $\tilde{F}_1 \cap \tilde{F}_2$, which is characterised by its membership function $\mu_{\tilde{F}_1 \cap \tilde{F}_2}(x, \theta)$, is given by the *meet* operation on $\mu_{\tilde{F}_1}(x)$ and $\mu_{\tilde{F}_2}(x)$, and is as follows[2]:

$$\mu_{\tilde{F}_1 \cap \tilde{F}_2}(x, \theta) = \left(\mu_{\tilde{F}_1}(x) \sqcap \mu_{\tilde{F}_2}(x) \right)(\theta)$$
$$= sup_{v \in [\theta,1]}\{f_1(v)\} \wedge (f_1(\theta) \vee f_2(\theta)) \wedge sup_{w \in [\theta,1]}\{f_2(w)\} \qquad (31)$$
$$x \in X$$

Theorem 2: The intersection operation on two type-2 fuzzy sets defined in [12] using minimum t-norm is equivalent to the intersection defined in Eq. (31).

Proof of Theorem 2: Proof of Theorem 2 is very similar to the proof of Theorem 1. In this case, any element θ in the primary membership of $\tilde{F}_1 \cap \tilde{F}_2$ is of the form $\theta = (v \wedge w)$, and can be obtained by any of the following two cases:

(1) **Case 1:** $v \in [\theta, 1]$ and $w = \theta$;
(2) **Case 2**: $w \in [\theta, 1]$ and $v = \theta$;

The rest of the proof is exactly the same as the one for the join operation, but changing the intervals $v \in [0, \theta]$ and $w \in [0, \theta]$ by $v \in [\theta, 1]$ and $w \in [\theta, 1]$, respectively. The final result will be as in Eq. (31).

3.3 Join and Meet Operations on gfIT2FSs

In this section, we will focus on the particular case where $f_1(v)$ and $f_2(w)$ *are either 0 or 1 and their supports are non-empty closed sets*. In other words, we will focus on gfIT2FSs, as presented in [2]. We will obtain specific versions of Eqs. (7) and (31) when sets are gfIT2FSs. It is important to note that all examples in [2] satisfy that the supports of $f_1(v)$ and $f_2(w)$ are non-empty closed sets.

Let $g_1(\theta) = sup_{v \in [0,\theta]}\{f_1(v)\}$. For a given value of θ, $g_1(\theta)$ is the maximum value that the function $f_1(v)$ has attained for all values of v lower than or equal to θ, i.e., $\forall v \leq \theta$. Let v_1 be the infimum of the support of f_1. Hence, for all $\theta < v_1$:

$$g_1(\theta) = sup_{v \in [0,\theta]}\{f_1(v)\} = sup_{v \in [0,\theta]}\{0\} = 0 \; \forall \theta < v_1 \qquad (32)$$

---

[2] It should be noted that Eq. (31) has some similarity to Eq. (11) in [31] as both equations refer to the meet operation of general type-2 fuzzy sets. However, the representation of Eq. (31) is different to simplify the computations and analysis.

For values $\theta \geq v_1$, as $f_1(v_1) = 1$, the following stands:

$$g_1(\theta) = sup_{v \in [0,\theta]}\{f_1(v)\} = f_1(v_1) \vee sup_{\substack{v \in [0,\theta] \\ v \neq v_1}}\{f_1(v)\}$$

$$= 1 \vee sup_{\substack{v \in [0,\theta] \\ v \neq v_1}}\{f_1(v)\} = 1 \quad \forall \theta \geq v_1 \tag{33}$$

Hence, combining Eqs. (32) and (33):

$$g_1(\theta) = sup_{v \in [0,\theta]}\{f_1(v)\} = \begin{cases} 0 & \forall \theta < v_1 \\ 1 & \forall \theta \geq v_1 \end{cases} \tag{34}$$

Analogously, let $g_2(\theta) = sup_{w \in [0,\theta]}\{f_2(w)\}$ and let w_1 be the infimum of the support of f_2. Hence:

$$g_2(\theta) = sup_{w \in [0,\theta]}\{f_2(w)\} = \begin{cases} 0 & \forall \theta < w_1 \\ 1 & \forall \theta \geq w_1 \end{cases} \tag{35}$$

Let $g(\theta) = g_1(\theta) \wedge g_2(\theta)$. Combining Eqs. (34) and (35):

$$g(\theta) = \begin{cases} 0, & \theta < max(v_1, w_1) \\ 1, & \theta \geq max(v_1, w_1) \end{cases} \tag{36}$$

Considering the definition of $g(\theta)$, we can rewrite Eq. (7) as:

$$\mu_{\tilde{F}_1 \cup \tilde{F}_2}(x, \theta) = g(\theta) \wedge (f_1(\theta) \vee f_2(\theta)) \tag{37}$$

Combining Eqs. (36) and (37):

$$\mu_{\tilde{F}_1 \cup \tilde{F}_2}(x, \theta) = \begin{cases} 0, & \theta < max(v_1, w_1) \\ f_1(\theta) \vee f_2(\theta), & \theta \geq max(v_1, w_1) \end{cases} \tag{38}$$

Let v_{end} and w_{end} be the supremum of the supports of f_1 and f_2, respectively. Hence, $f_1(v) = 0 \, \forall v > v_{end}$ and $f_2(w) = 0 \, \forall w > w_{end}$. Therefore, the term $f_1(\theta) \vee f_2(\theta)$ will be $0 \, \forall \theta > max(v_{end}, w_{end})$. Consequently we can rewrite Eq. (38) as:

$$\mu_{\tilde{F}_1 \cup \tilde{F}_2}(x, \theta) = \begin{cases} f_1(\theta) \vee f_2(\theta), & \theta \in [max(v_1, w_1), max(v_{end}, w_{end})] \\ 0, & elsewhere \end{cases} \tag{39}$$

Now let's consider the case of the meet operation. In this case, let $g_1(\theta) = sup_{v \in [\theta,1]}\{f_1(v)\}$. Given a value of $\theta > v_{end}$:

$$g_1(\theta) = sup_{v \in [\theta,1]}\{f_1(v)\} = sup_{v \in [\theta,1]}\{0\} = 0 \quad \forall \theta > v_{end} \tag{40}$$

For values $\theta \leq v_{end}$, as $f_1(v_{end}) = 1$, the following stands:

$$g_1(\theta) = sup_{v\in[0,1]}\{f_1(v)\} = f_1(v_{end}) \vee \underset{\substack{v \in [0,1] \\ v \neq v_{end}}}{sup} \{f_1(v)\}$$

$$= 1 \vee \underset{\substack{v \in [0,1] \\ v \neq v_{end}}}{sup} \{f_1(v)\} = 1 \quad \forall \theta \geq v_{end} \tag{41}$$

Hence, combining Eqs. (40) and (41):

$$g_1(\theta) = sup_{v\in[0,1]}\{f_1(v)\} = \begin{cases} 1 & \forall \theta \leq v_{end} \\ 0 & \forall \theta > v_{end} \end{cases} \tag{42}$$

A similar expression can be found for $g_2(\theta) = sup_{w\in[0,1]}\{f_2(w)\}$.

$$g_2(\theta) = sup_{w\in[0,1]}\{f_2(w)\} = \begin{cases} 1 & \forall \theta \leq w_{end} \\ 0 & \forall \theta > w_{end} \end{cases} \tag{43}$$

Let $g(\theta) = g_1(\theta) \wedge g_2(\theta)$. We can write:

$$\mu_{\tilde{F}_1 \cap \tilde{F}_2}(x, \theta) = g(\theta) \wedge (f_1(\theta) \vee f_2(\theta)) \tag{44}$$

Considering $g(\theta) = g_1(\theta) \wedge g_2(\theta)$ and using Eqs. (42) and (43), we can rewrite Eq. (44) as follows:

$$\mu_{\tilde{F}_1 \cap \tilde{F}_2}(x, \theta) = \begin{cases} f_1(\theta) \vee f_2(\theta), & \theta \leq min(v_{end}, w_{end}) \\ 0, & \theta > min(v_{end}, w_{end}) \end{cases} \tag{45}$$

By definition of v_1 and w_1, $f_1(v) = 0 \,\forall v < v_1$ and $f_2(w) = 0 \,\forall w < w_1$. Therefore, the term $f_1(\theta) \vee f_2(\theta)$ will be $0 \,\forall \theta < min(v_1, w_1)$. Consequently we can rewrite Eq. (45) as:

$$\mu_{\tilde{F}_1 \cap \tilde{F}_2}(x, \theta) = \begin{cases} f_1(\theta) \vee f_2(\theta), & \theta \in [min(v_1, w_1), min(v_{end}, w_{end})] \\ 0, & elsewhere \end{cases} \tag{46}$$

It is worthwhile to highlight that Eqs. (39) and (46) lead to the well-known results of the join and meet when the involved sets are type-1 sets or IVFSs. For the join in type-1 sets, as $v_1 = v_{end}$ and $w_1 = w_{end}$, then θ is non zero only when $\theta = max(v_1, w_1)$, so $f_1(\theta) \vee f_2(\theta)$ is a singleton placed a this value $\theta = max(v_1, w_1)$. An analogous reasoning for the meet operation, given $\theta = min(v_1, w_1)$, leads to a singleton placed at this $\theta = min(v_1, w_1)$.

For the case of IVFSs, as f_1 and f_2 have continuous supports, then $f_1(\theta) \vee f_2(\theta)$ will also be continuous in $\theta \in [max(v_1, w_1), max(v_{end}, w_{end})]$ for the join, and $\theta \in [min(v_1, w_1), min(v_{end}, w_{end})]$ for the meet, regardless of the relative positions of v_1, w_1, v_{end} and w_{end}, thus leading to the well-known equations for IVFSs:

$$\mu_{\tilde{F}_1 \cup \tilde{F}_2}(x, \theta) = \begin{cases} 1, & \theta \in [max(v_1, w_1), max(v_{end}, w_{end})] \\ 0, & elsewhere \end{cases} \qquad (47)$$

$$\mu_{\tilde{F}_1 \cap \tilde{F}_2}(x, \theta) = \begin{cases} 1, & \theta \in [min(v_1, w_1), min(v_{end}, w_{end})] \\ 0, & elsewhere \end{cases} \qquad (48)$$

An example of both meet and join operations on gfIT2FSs, as in Case E, is depicted in Fig. 4.

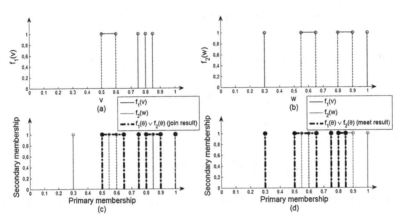

Fig. 4. (a) $f_1(v)$. (b) $f_2(w)$. (c) Sets and join result. (d) Sets and meet result.

4   General Forms of Singleton IT2FLSs

The structure of a singleton gfIT2FLS is represented in Fig. 5, and is analogous to a GT2FLSs [10, 22] and IT2FLSs [14, 16], although some of its blocks may perform operations in a different way.

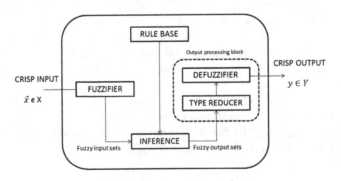

Fig. 5. Structure of a gfIT2FLS.

In the following subsections we will briefly discuss each of the parts of the system, paying special attention to the parts having remarkable differences, as the type reduction block.

4.1   The Fuzzifier

In a multiple-input-single-output (MISO) gfIT2FLS having p inputs, the fuzzifier blocks maps the crisp input values to gfIT2FSs. Although, a priori, the fuzzifier can map inputs into arbitrary fuzzy sets, we will consider only *singleton fuzzification*, which maps each crisp input value x_i into a gfIT2 singleton fuzzy set; this set has non-zero primary membership just in one point $x = x_i$, and the primary membership value will be $J_x = \{1\} \subseteq [0, 1]$. The secondary grade at this point $x = x_i$ will also be equal to 1, i.e., $\mu_{\tilde{X}_i}(x_i, 1) = 1$, and 0 otherwise.

4.2   The Rule Base

As in the T1, IV and GT2 cases, the rule base comprising a gfIT2FLS is a set of M IF-THEN rules, each rule relating the input antecedents using the AND or OR operators, with the rule consequent.

Although fuzzy rules can take any form as an IF-THEN statement, it is customary in the FLS literature [16] to express the rules within the rule-base relating the antecedents just with the AND operator. Thus, according to this, the l-th rule of the rule base, $l = 1, \ldots, M$, is as in Eq. (49):

$$R^l : IF x_1 is \tilde{F}_1^l \ AND \ x_2 \ is \ \tilde{F}_2^l \ AND \ldots AND \ x_p \ is \ \tilde{F}_p^l \ THEN \ y \ is \ \tilde{Y}^l \tag{49}$$

Where x_is are the inputs to the system, $\tilde{F}_i^l$ is the antecedent of the i-th input $(i = 1, \ldots, p)$ of the l-th rule $(l = 1, \ldots, M)$, y is the output of the system and $\tilde{Y}^l$ is the consequent set for that rule. Although the rule presented in Eq. (49) is valid for any GT2FLS, we expect to have all involved fuzzy sets (both antecedents and consequents) to be gfIT2FSs, as this is the framework under study. For a given input vector $\vec{x} = (x_1, \ldots, x_p)$, an integer number of N rules will be fired, $N \leq M$; each of these fired rules will combine the antecedent sets for the given input vector with the consequent set $\tilde{Y}^l$. How to combine the antecedents with their consequents, and how to combine the fired rules are the tasks of the inference engine, which will be explained subsequently.

4.3   Inference Engine

After each component of the input vector, which is a crisp number, has been fuzzified into a gfIT2FS, the fuzzifier block provides the type-2 input fuzzy sets to the inference engine. This block is in charge of *meeting* the input fuzzy sets (using the meet operation) $\mu_{\tilde{X}_i}(x_i)$ with their corresponding antecedents $\mu_{\tilde{F}_i^l}(x_i)$, and combines all the resulting sets within the rule with the consequent set $\mu_{\tilde{Y}^l}(y)$ to provide the rule output set, $\mu_{\tilde{B}^l}(y)$, as stated in Eq. (50):

$$\mu_{\tilde{B}^l}(y) = \left[\sqcap_{i=1}^{p} \left(\sqcup_{x_i \in X_i} \mu_{\tilde{X}_i}(x_i) \sqcap \mu_{\tilde{F}_i^l}(x_i) \right) \right] \sqcap \mu_{\tilde{Y}^l}(y) \tag{50}$$

Equation (50) is valid for the general non-singleton fuzzification case; however, when using singleton fuzzification, the input fuzzy sets $\mu_{\tilde{X}_i}(x_i)$, with $i = 1,\ldots,p$, are type-2 singletons (as explained in Sect. 4.1), and the term $\mu_{\tilde{X}_i}(x_i) \sqcap \mu_{\tilde{F}_i^l}(x_i)$ reduces to $\mu_{\tilde{F}_i^l}(x_i)$; this simplified term is the same as evaluating the given antecedent $\mu_{\tilde{F}_i^l}(x)$ at $x = x_i$, or equivalent to taking the vertical slice from $\mu_{\tilde{F}_i^l}(x)$ placed an such $x = x_i$. An example of such result is depicted in Fig. 6(Left) and Fig. 6(Right); the former shows a gfIT2FS, which represents an example of an arbitrary antecedent $\mu_{\tilde{F}_i^l}(x)$. The latter depicts the vertical slice at $x = x_i$, which will be a gfIT1FS.

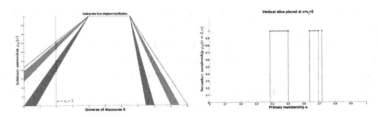

Fig. 6. (Left) Arbitrary antecedent $\mu_{\tilde{F}_i^l}(x)$ evaluated at $x = x_i = 2$. (Right) Vertical slice placed at $x = x_i = 2$.

Thus, Eq. (50) can be significantly simplified and becomes Eq. (51):

$$\mu_{\tilde{B}^l}(y) = \left[\sqcap_{i=1}^{p} \mu_{\tilde{F}_i^l}(x_i) \right] \sqcap \mu_{\tilde{Y}^l}(y) \tag{51}$$

Usually, the term $\left[\sqcap_{i=1}^{p} \mu_{\tilde{F}_i^l}(x_i) \right]$ in Eq. (51) is referred to as the *rule firing strength*. As we are considering gfIT2FLSs, it is expected that all antecedent and consequents sets will be gfIT2FSs, and that *every term* $\mu_{\tilde{F}_i^l}(x_i)$ is a gfIT1FS, i.e., a *vertical slice* [16] from $\mu_{\tilde{F}_i^l}(x)$ at $x = x_i$. On the other hand, $\mu_{\tilde{Y}^l}(y)$ will be, in general, a gfIT2FS and consequently, $\mu_{\tilde{B}^l}(y)$ will also be a gfIT2FS; however, as we will see in the following subsection, some type-reduction strategies replace $\mu_{\tilde{Y}^l}(y)$ with a T1FS; in this latter situation $\mu_{\tilde{B}^l}(y)$ will be a T1FS.

After obtaining each rule output set, the classical inference engine combines them using the join operation to obtain the system output fuzzy set, as in Eq. (52):

$$\mu_{\tilde{B}}(y) = \sqcup_{l=1}^{M} \mu_{\tilde{B}^l}(y) \tag{52}$$

The resulting gfIT2FS $\mu_{\tilde{B}}(y)$ is the output of the inference engine. However, depending on the chosen type-reduction strategy, the system may require the inference engine to provide as its output all rule firing strengths and consequents for each rule.

4.4 Type Reduction

In GT2FLSs, the type-reduction (TR) block receives as its input a GT2FS, and will provide as its output a T1FS which is representative of the whole GT2FS; this T1FS will later be defuzzified to provide a crisp number as the system output. Actually, type-reduction can be considered as an extension, via the Extension Principle [32], of the defuzzification process. Thus, the *centroid type-reducer*, which is the extended version of the centroid defuzzifier, is the most widely used method in the literature when discussing theoretical frameworks [10, 11, 13], and is as follows:

$$C_{\tilde{A}} = \int\limits_{\theta_1 \in J_{x_1}} \cdots \int\limits_{\theta_N \in J_{x_N}} [f_{x_1}(\theta_1) \star \ldots \star f_{x_N}(\theta_N)] / \frac{\sum_{i=1}^{N} x_i \theta_i}{\sum_{i=1}^{N} \theta_i} \tag{53}$$

Where the domain X has been discretised in N values, and each of the primary memberships J_{x_i}, $i = 1, \ldots, N$, has been discretised in, say, M_j values. Hence, this scheme requires enumerating all embedded type-1 fuzzy sets [17] within the type-2 fuzzy sets, which sum a total of $\prod_{j=1}^{N} M_j$; this number can be huge, even for small values of N and M_j, due to the so *called curse of dimensionality*.

Thus, its high computational complexity has motivated researchers to define other simpler type-reduction operations, as the *centre-of-sets* (COS) type-reducer [16]. This method requires the inference engine to provide as its output all firing strengths and consequents from each rule.

In the gfIT2FLS framework presented in this work, the classical centroid type-reduction as defined in [16] can be used. Besides, a modified version of the COS TR, which is computationally simpler, is presented.

4.4.1 Approximating a Finite Set of Disjointed Intervals

The proposal presented in this subsection is based on some definitions in [28] and some intermediate results in [1]. In this latter work, the authors propose a method to approximate fuzzy quantities (i.e., non-convex and/or non-normal fuzzy sets) by fuzzy numbers (normal and convex fuzzy sets verifying some other conditions). Across their work, they used a method to approximate a finite set of disjointed intervals by one single interval, which would be representative of all of them. To do so, it is first needed to define a *distance* between two intervals $B = [b_l, b_r]$ and $D = [d_l, d_r]$. If we denote $mid(B) = (b_r + b_l)/2$ and $spr(B) = (b_r - b_l)/2$, then the distance between B and D would be:

$$d_{\bar{\theta}}(B, D) = \sqrt{(mid(B) - mid(D))^2 + \bar{\theta} \cdot (spr(B) - spr(D))^2} \tag{54}$$

Where $\bar{\theta} \in (0, 1]$ is a real-valued parameter to weight the importance of spreads with respect to the midpoints. Once a distance between two intervals is defined, the *distance* between a given interval C and a finite set of N disjointed intervals $A = \bigcup_{i=1}^{N} A_i = \bigcup_{i=1}^{N} [a_L^i, a_R^i]$ can be defined, as in Eq. (55):

$$J(C; A, \bar{\theta}) = \sum_{i=1}^{N} d_{\bar{\theta}}^2(C, A_i) \tag{55}$$

Hence, the interval $C^* = [c_L^*, c_R^*]$ that best approximates a given set of N disjointed intervals A is the one that minimises J in Eq. (55). In [1] it is proven that such interval C^* is given by Eq. (56):

$$C^* = [c_L^*, c_R^*] = \left[\frac{1}{N}\sum_{i=1}^{N} a_L^i, \frac{1}{N}\sum_{i=1}^{N} a_R^i\right] \tag{56}$$

It is important to note that the solution in Eq. (56) does not depend on the parameter $\bar{\theta}$.

As an example, we will approximate the vertical slice presented in Fig. 6(Left), which is comprised by a finite set of continuous, closed and disjointed intervals, by the interval C^* that minimises function given in Eq. (56), i.e., the interval that best approximates them. First of all, we have to express the given vertical slice in Fig. 7 (Left):

$$\mu_{\tilde{X}}(x = 2) = \bigcup_{i=1}^{3} \mu_{\tilde{X}}^i(x = 2) = \bigcup_{i=1}^{3} [l_{\tilde{X}}^i, r_{\tilde{X}}^i] \tag{59}$$
$$= [0.3846, 0.5] \cup [0.6364, 0.6923] \cup [0.7143, 0.7143]$$

Where:

$$\mu_{\tilde{X}}^1(x = 2) = [l_{\tilde{X}}^1, r_{\tilde{X}}^1] = [0.3846, 0.5]$$
$$\mu_{\tilde{X}}^2(x = 2) = [l_{\tilde{X}}^2, r_{\tilde{X}}^2] = [0.6364, 0.6923] \tag{60}$$
$$\mu_{\tilde{X}}^3(x = 2) = [l_{\tilde{X}}^3, r_{\tilde{X}}^3] = [0.7143, 0.7143] = 0.7143$$

Thus, applying the solution given in Eq. (59), we get the following result:

$$
\begin{aligned}
C^* = \left[c_L^*, c_R^*\right] &= \left[\frac{1}{3}\sum_{i=1}^{3}l_{\tilde{X}}^i, \frac{1}{3}\sum_{i=1}^{3}r_{\tilde{X}}^i\right] \\
&= \frac{1}{3}\left[(0.3846 + 0.6364 + 0.7143), (0.5 + 0.6923 + 0.7143)\right] \\
&= \frac{1}{3}\left[1.7353, 0.5 + 0.6923 + 0.7143\right] = [0.5784, 0.6355]
\end{aligned}
\tag{61}
$$

4.4.2 Type-Reduction Strategy

The strategy adopted in the type-reduction stage for gfIT2FLSs requires the gfIT2FS $\tilde{Y}^l$ to be replaced by a T1FS Y^l that is representative of it. In our case, there are several options:

- Obtain Y^l as the centroid of $\tilde{Y}^l$. This option is only valid for theoretical purposes, as the type-2 centroid operation is extremely computationally demanding.
- Directly define the consequent as a gfIT1FS.
- Obtain an interval T1FS from a gfIT2FS consequent set. This option is the most general, and is the one to be considered.

Hence, to approximate a gfIT2FS by an interval T1FS, we shall proceed as follows:

1. From a gfIT2FS (as in Fig. 6(Left)), perform the approximation *slice by slice*. This step will transform the gfIT2FS into an IVFS.
2. Once we have an IVFS, we can apply the well-known KM algorithm [16] to obtain its centroid interval T1FS.

These operations can be done ahead of time, and will replace each consequent set $\tilde{Y}^l$ by an interval; if we consider the gfIT2FS depicted in Fig. 6(Left) is a rule consequent $\tilde{Y}^l$ instead of an antecedent, then Fig. 7 represents the IVFS that best approximates it, and Fig. 8 represents the interval obtained after applying the KM algorithm to that IVFS; this latter interval is the one we will use in our *modified centre-of-sets type-reduction operation*, which can be summarised as follows:

1. For each consequent set $\tilde{Y}^l$, obtain the interval $\left[y_L^l, y_R^l\right]$ that best approximates it using the steps previously specified.
2. For each rule firing strength $F^l = \left[\prod_{i=1}^{p}\mu_{\tilde{F}_i^l}(x_i)\right]$, which is a gfIT1FS, obtain the interval $\left[f_L^l, f_R^l\right]$ that best approximates it. This process has to be done for each input value.
3. Once we have one interval per firing strength $\left[f_L^l, f_R^l\right]$ and consequent $\left[y_L^l, y_R^l\right]$ for each rule, the classical COS type-reduction operation [11] can be applied.

The result will be a single T1FS comprised by an interval, $Y = [y_L, y_R]$, which is the output of the type-reduction block. This interval will be the input to the final defuzzification stage.

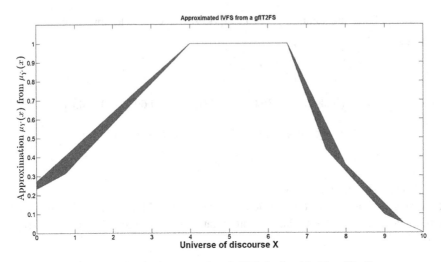

Fig. 7. IVFS approximating the gfIT2FS depicted in Fig. 6(Left).

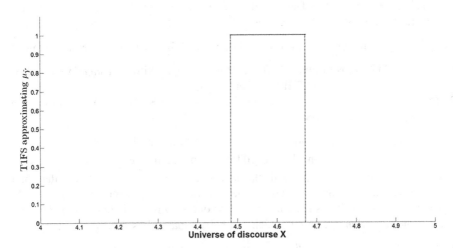

Fig. 8. Centroid of the IVFS depicted in Fig. 7.

4.5 Defuzzification

When the output of the type-reduction block is a T1FS consisting in a single interval, the most usual defuzzification process [16] is to provide its midpoint as the output of the system:

$$y = \frac{y_L + y_R}{2} \tag{62}$$

4.6   Summary of gfIT2FLSs and Its Relation with IVFLSs

It is important to highlight that the classical IVFLSs can be considered as a specific case of gfIT2FLSs: in every block within the system, if we replace the gfIT2FSs by IVFSs, the resulting FLS will be an IVFLS; i.e., the inference engine using the meet and join operations reduces to the inference engine for IVFLSs as presented in [16]. The same happens with the type-reduction block: if all involved sets are IVFSs (which are a particular case of gfIT2FSs), then the interval that best approximates a given interval is that interval itself, as the distance J in Eq. (55) will reduce to 0. Hence, the modified COS TR reduces to the classic COS operation.

As an example, we will consider a 2-input-1-output gfIT2FLS. The system will control the wheel speed of an autonomous mobile robot, whose main task is to follow a wall at a specific distance. The inputs will be the distance measured from two sonar sensors, one placed at the front and another placed at the rear of the robot. Then, the whole FLS would operate as follows:

1. The fuzzifier block receives the input vector $\vec{x} = (x_1, x_2)$, which is comprised of crisp numbers, and maps each value into a gfIT2 singleton (as we are only considering singleton fuzzification). These input fuzzy sets are depicted in Fig. 9(a) and (b), respectively.
2. For each rule within the rule base, the inference engine meets the gfIT2 input singletons with their corresponding antecedent (depicted in Fig. 9(c); this is equivalent to evaluating the antecedent sets $\mu_{\tilde{F}_i^l}(x)$ at $x = x_i$) and combines all the antecedents within the rule to obtain the *rule firing strength* $\left[\sqcap_{i=1}^{p} \mu_{\tilde{F}_i^l}(x_i)\right]$. An example is plotted in Fig. 9(d).
3. The inference engine provides as its outputs the firing strength and the rule consequent for each of the M rules. These gfIT1FSs will be the inputs to the type-reduction block.

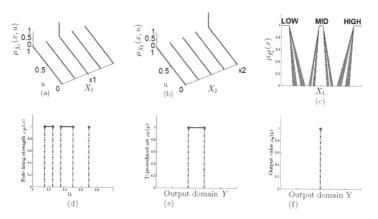

Fig. 9. (a) Singleton fuzzy set for input $x1$. (b) Singleton fuzzy set for input $x2$. (c) Example of antecedents. (d) An example of a rule firing strength. (e) A type-reduced set. (f) System output.

4. The type reduction block will apply the modified COS type-reduction operation, as explained in Sect. 4.4. The output of this block will be an interval $Y = [y_L, y_R]$; an example is illustrated in Fig. 9(e).
5. Finally, the defuzzification block obtains the midpoint of the type-reduced interval and provides it as the system output. This singleton value is depicted in Fig. 9(f).

5 Applications and Future Work

In this chapter we have presented the basic framework of the general forms of IT2FLSs, considering singleton fuzzification and modified COS type-reduction. Work in progress by the authors intends to present two new methodologies applying these new types of sets/systems: the first one will use multi-interval gfIT2FSs to model sensor input noise, extending the present framework using non-singleton fuzzification; the new gfIT2FSs will allow representing noise and uncertainty in a way the classic IVFSs cannot. The second methodology will use the classical survey method [15] to obtain multi-interval gfIT2FSs to be used as the antecedents within the inference engine.

References

1. Anzilli, L., Facchinetti, G., Mastroleo, G.: A parametric approach to evaluate fuzzy quantities. Fuzzy Sets and Syst. **250**, 110–133 (2014)
2. Bustince, H., Fernandez, J., Hagras, H., Herrera, F., Pagola, M., Barrenechea, E.: Interval type-2 fuzzy sets are generalization of interval-valued fuzzy sets: towards a wider vie won their relationship. IEEE Trans. Fuzzy Syst. **23**(5), 1876–1882 (2015)
3. Cara, A.B., Rojas, I., Pomares, H., Wagner, C., Hagras, H.: On comparing non-singleton type-2 and singleton type-2 fuzzy controllers for a nonlinear servo system. In: Proceedings IEEE Symposium on Advances in Type-2 Fuzzy Logic Systems, pp. 126–133, April 2011
4. Coupland, S., John, R.: Geometric type-1 and type-2 fuzzy logic systems. IEEE Trans. Fuzzy Syst. **15**, 3–15 (2007)
5. Figueroa, J., Posada, J., Soriano, J., Melgarejo, M., Rojas, S.: A type-2 fuzzy controller for delta parallel robot. IEEE Trans. Ind. Inf. **7**(4), 661–670 (2011)
6. Hagras, H.: A hierarchical type-2 fuzzy logic control architecture for autonomous mobile robots. IEEE Trans. Fuzzy Syst. **12**(4), 524–539 (2004)
7. Hernandez, P., Cubillo, S., Torres-Blanc, C.: On T-norms for type-2 fuzzy sets. IEEE Trans. Fuzzy Syst. **23**(4), 1155–1163 (2014)
8. Hosseini, R., Qanadli, S.D., Barman, S., Mazinari, M., Ellis, T., Dehmeshki, J.: An automatic approach for learning and tuning Gaussian interval type-2 fuzzy membership functions applied to lung CAD classification system. IEEE Trans. Fuzzy Syst. **20**(2), 224–234 (2012)
9. Jammeh, E.A., Fleury, M., Wagner, C., Hagras, H., Ghanbari, M.: Interval type-2 fuzzy logic congestion control for video streaming across IP networks. IEEE Trans. Fuzzy Syst. **17**(5), 1123–1142 (2009)

10. Karnik, N.N., Mendel, J.M.: Type-2 fuzzy logic systems: type-reduction. In: Proceedings of the IEEE Conference on Systems, Man and Cybernetics, San Diego, CA, pp. 2046–2051, October 1998
11. Karnik, N.N., Mendel, J.M.: Type-2 fuzzy logic systems. IEEE Trans. Fuzzy Syst. **7**(6), 643–658 (1999)
12. Karnik, N.N., Mendel, J.M.: Operations on type-2 fuzzy sets. Fuzzy Sets Syst. **122**, 327–348 (2001)
13. Karnik, N.N., Mendel, J.M.: Centroid of a type-2 fuzzy set. Inf. Sci. **132**(1), 195–220 (2001)
14. Liang, Q., Mendel, J.M.: Interval type-2 fuzzy logic systems: theory and design. IEEE Trans. Fuzzy Syst. **8**(5), 535–550 (2000)
15. Liu, F., Mendel, J.M.: An interval approach to fuzzistics for interval type-2 fuzzy sets. In: IEEE International Fuzzy Systems Conference, FUZZ-IEEE 2007, London, pp. 1–6 (2007)
16. Mendel, J.M.: Uncertain Rule-Based Fuzzy Logic Systems: Introduction and New Directions. Prentice-Hall, Upper-Saddle River (2001)
17. Mendel, J.M., Bob John, R.I.: Type-2 fuzzy sets made simple. IEEE Trans. Fuzzy Syst. **10** (2), 117–127 (2002)
18. Mendel, J.M., John, R.I., Liu, F.: Interval type-2 fuzzy logic systems made simple. IEEE Trans. Fuzzy Syst. **14**(6), 808–821 (2006)
19. Mendel, J.M.: Computing with words and its relationships with fuzzistics. Inf. Sci. **177**(4), 988–1006 (2007)
20. Mendel, J.M., Liu, F., Zhai, D.: Alpha-plane representation for type-2 fuzzy sets: theory and applications. IEEE Trans. Fuzzy Syst. **17**(5), 1189–1207 (2009)
21. Mendel, J.: On the geometry of join and meet calculations for general type-2 fuzzy sets. In: Proceedings of 2011 IEEE International Conference on Fuzzy Systems, Taiwan, pp. 2407–2413, June 2011
22. Mendel, J.M.: General type-2 fuzzy logic systems made simple: a tutorial. IEEE Trans. Fuzzy Syst. **22**(5), 1162–1182 (2014)
23. Mizumoto, M., Tanaka, K.: Some properties of fuzzy sets of type 2. Inf. Control **31**, 312–340 (1976)
24. Pomares, H., Rojas, I., Ortega, J., González, J., Prieto, A.: A systematic approach to a self-generating fuzzy rule-base for function approximation. IEEE Trans. Syst. Man Cybern. **30**(3), 431–447 (2000)
25. Pomares, H., Rojas, I., González, J., Damas, M., Pino, B., Prieto, A.: Online global learning in direct fuzzy controllers. IEEE Trans. Fuzzy Syst. **12**(2), 218–229 (2004)
26. Ruiz-Garcia, G., Hagras, H., Pomares, H., Rojas, I., Bustince, H.: Join and meet operations for type-2 fuzzy sets with non-convex secondary memberships. IEEE Trans. Fuzzy Syst. **24** (4), 1000–1008 (2016)
27. Tahayori, H., Tettamanzi, A.G.B., Antoni, G.D.: Approximated type-2 fuzzy sets operations. In: Proceedings of 2006 IEEE International Conference on Fuzzy Systems, Canada, pp. 1910–1917, July 2006
28. Trutschnig, W., González-Rodríguez, G., Colubi, A., Gil, M.A.: A new family of metrics for compact, convex (fuzzy) sets based on a generalized concept of mid and spread. Inf. Sci. **179**, 3964–3972 (2009)
29. Wagner, C., Hagras, H.: Evolving type-2 fuzzy logic controllers for autonomous mobile robots. In: Melin, P., Castillo, O., Gomez Ramírez, E., Kacprzyk, J., Pedrycz, W. (eds.) Analysis and Design of Intelligent Systems using Soft Computing Techniques, vol. 41, pp. 16–25. Springer, Heidelberg (2007)
30. Wagner, C., Hagras, H.: Towards general type-2 fuzzy logic systems based on zSlices. IEEE Trans. Fuzzy Syst. **18**(4), 637–660 (2010)

31. Walker, C., Walker, E.: The algebra of fuzzy truth value. Fuzzy Sets Syst. **149**, 309–347 (2005)
32. Zadeh, L.A.: The concept of linguistic variable and its application to approximate reasoning. Inf. Sci. **8**(3), 199–249 (1975)
33. Zhai, D., Hao, M., Mendel, J.M.: A non-singleton interval type-2 fuzzy logic system for universal image noise removal using quantum-behaved particle swarm optimization. In: Proceedings 2011 IEEE International Conference on Fuzzy Systems, pp. 957–964, June 2011

Algorithmic Developments of Information Granules of Higher Type and Higher Order and Their Applications

Witold Pedrycz[1,2,3(✉)]

[1] Department of Electrical and Computer Engineering, University of Alberta,
Edmonton, AB T6R 2V4, Canada
wpedrycz@ualberta.ca
[2] Department of Electrical and Computer Engineering, Faculty of Engineering,
King Abdulaziz University, Jeddah 21589, Saudi Arabia
[3] Systems Research Institute, Polish Academy of Sciences, Warsaw, Poland

Abstract. Information granules are conceptual entities using which experimental data are conveniently described and in the sequel their processing is realized at the higher level of abstraction. The central problem is concerned with the design of information granules. We advocate that a principle of justifiable granularity can be used as a sound vehicle to construct information granules so that they are (i) experimentally justifiable and (ii) semantically sound. We elaborate on the algorithmic details when forming information granules of type-1 and type-2. It is also stressed that the construction of information granule realized in this way follows a general paradigm of elevation of type of information granule, say numeric data (regarded as information granules of type-0) give rise to information granule of type-1 while experimental evidence coming as information granules of type-1 leads to the emergence of a single information granule of type-2. We discuss their direct applications to the area of system modeling, in particular showing how type-n information granules are used in the augmentation of numeric models.

Keywords: Granular computing · Information granules · Type and order of information granules · Principle of justifiable granularity · Coverage · Specificity

1 Introduction

Information granules are omnipresent. They are regarded as a synonym of abstraction. They support ways of problem solving through problem decomposition. Information granularity is central to perception and reasoning about complex systems. It also become essential to numerous pursuits in the realm of analysis and design of intelligent systems. Granular Computing forms a general conceptual umbrella, which embraces the well-known constructs including fuzzy sets, rough, sets, intervals and probabilities.

The ultimate objective of the study is to focus on the concepts, roles, and design of information granules of higher type and higher order. We offer a motivation behind the emergence of information granules of higher type. As of now, they become quite visible in the form of type-2 fuzzy sets – these constructs form the current direction of

© Springer International Publishing AG 2017
A. Petrosino et al. (Eds.): WILF 2016, LNAI 10147, pp. 27–41, 2017.
DOI: 10.1007/978-3-319-52962-2_2

intensive research in fuzzy sets, especially at its applied side. Several ways of forming (designing) information granules are outlined; it is demonstrated that clustering arises as a general way of transforming data into clusters (information granules). Another alternative comes in the form of the principle of justifiable granularity, which emphasizes a formation of information granules as a result of an aggregation of available experimental evidence and quantification of its diversity.

The structure of the paper reflects a top-down organization of the overall material. To make the study self-contained, we start with an exposure of the essential prerequisites (Sect. 2). In Sect. 3, we discuss main ways of building information granules; here the proposed taxonomy embraces a suite of key methods. Section 4 elaborates on the essence of information granules of higher type and higher order. We present them both in terms of their conceptual underpinnings and compelling motivating arguments as well as discuss ways of their construction. In Sect. 5, we focus on the direct usage of such information granules; it is advocated that the higher type of information granularity is associated with the realization of models, in particular fuzzy models, of increased experimental relevance.

2 Information Granules and Granular Computing: Essential Prerequisites

Information granules forming the Granular Computing are conceptual entities that support all processing realized in this environment. For the completeness of the study, we briefly recall some principles behind this paradigm.

2.1 Agenda of Granular Computing

Information granules are intuitively appealing and convincing constructs, which play a pivotal role in human cognitive and decision-making activities. We perceive complex phenomena by organizing existing knowledge along with available experimental evidence and structuring them in a form of some meaningful, semantically sound entities. In the sequel, such entities become central to all ensuing processes of describing the world, reasoning about the surrounding environment and supporting various decision-making activities. The term information granularity itself has emerged in different contexts and numerous areas of application. It carries various meanings. One can refer to Artificial Intelligence in which case information granularity is central to a way of problem solving through problem decomposition where various subtasks could be formed and solved individually. In general, as stressed by Zadeh [20], by information granule one regards a collection of elements drawn together by their closeness (resemblance, proximity, functionality, etc.) articulated in terms of some useful spatial, temporal, or functional relationships. In a nutshell as advocated in [9–13, 18], Granular Computing is about representing, constructing, processing, and communicating information granules.

We can refer here to some areas, which deliver compelling evidence as to the nature of underlying processing and interpretation in which information granules play a pivotal role. The applications include image processing, processing and interpretation of time series, granulation of time, design of software systems. Information granules are examples of abstractions. As such, they naturally give rise to hierarchical structures: the same problem or system can be perceived from different viewpoints and at different levels of specificity (detail) depending on the complexity of the problem, available computing resources, and particular needs and tasks to be addressed. A hierarchy of information granules is inherently visible in processing of information granules. The level of detail (which is represented in terms of the size of information granules) becomes an essential facet facilitating a way a hierarchical processing of information positioned at different levels of hierarchy and indexed by the size of information granules.

Such commonly encountered and simple examples presented above are convincing enough to highlight several essential features:

(a) information granules are the key components of knowledge representation and processing,
(b) the level of granularity of information granules (their size, to be more descriptive) becomes crucial to the problem description and an overall strategy of problem solving,
(c) hierarchies of information granules support an important aspect of perception of phenomena and deliver a tangible way of dealing with complexity by focusing on the most essential facets of the problem and,
(d) there is no universal level of granularity of information; the size of granules becomes problem-oriented and user dependent.

2.2   The Landscape of Information Granules

There are numerous well-known formal settings in which information granules can be expressed and processed. Here we identify several commonly encountered conceptual and algorithmic platform:

Sets (intervals) realize a concept of abstraction by introducing a notion of dichotomy: we admit element to belong to a given information granule or to be excluded from it. Along with set theory comes a well-developed discipline of interval analysis. Alternatively to an enumeration of elements belonging to a given set, sets are described by characteristic functions taking on values in $\{0,1\}$.

Fuzzy sets provide an important conceptual and algorithmic generalization of sets. By admitting partial membership of an element to a given information granule we bring an important feature which makes the concept to be in rapport with reality. It helps working with the notions where the principle of dichotomy is neither justified nor advantageous. The description of fuzzy sets is realized in terms of membership functions taking on values in the unit interval. Formally, a fuzzy set A is described by a membership function mapping the elements of a universe $\mathbf{X}$ to the unit interval $[0,1]$.

Shadowed sets [15] offer an interesting description of information granules by distinguishing among elements, which fully belong to the concept, are excluded from it and whose belongingness is completely *unknown*. Formally, these information granules are described as a mapping $X: \mathbf{X} \rightarrow \{1, 0, [0,1]\}$ where the elements with the membership quantified as the entire [0,ccc1] interval are used to describe a shadow of the construct. Given the nature of the mapping here, shadowed sets can be sought as a granular description of fuzzy sets where the shadow is used to localize unknown membership values, which in fuzzy sets are distributed over the entire universe of discourse. Note that the shadow produces non-numeric descriptors of membership grades.

Probability-oriented information granules are expressed in the form of some probability density functions or probability functions. They capture a collection of elements resulting from some experiment. In virtue of the concept of probability, the granularity of information becomes a manifestation of occurrence of some elements. For instance, each element of a set comes with a probability density function truncated to [0,1], which quantifies a degree of membership to the information granule.

Rough sets [7, 8] emphasize a roughness of description of a given concept X when being realized in terms of the indiscernibility relation provided in advance. The roughness of the description of X is manifested in terms of its lower and upper approximations of the resulting rough set.

2.3 Key Characterization of Information Granules

Information granules as being more general constructs as numeric entities, require a prudent characterization so that their nature can be fully captured. There are two main characteristics that are considered here.

Coverage
The concept of coverage of information granule, cov(.) is discussed with regard to some experimental data existing in R^n, that is $\{x_1, x_2,..., x_N\}$. As the name itself stipulates, coverage is concerned with an ability of information granule to represent (cover) these data. In general, the larger number of data is being "covered", the higher the coverage of the information granule. Formally, the coverage can be sought as a non-decreasing function of the number of data that are represented by the given information granule A. Depending upon the nature of information granule, the definition of cov(A) can be properly refined. For instance, when dealing with a multidimensional interval (hypercube) A, cov(A) in its normalized form is related with the normalized cardinality of the data belonging to A, $cov(A) = \frac{1}{N} card\{x_k | x_k \in A\}$. For fuzzy sets, the coverage is realized as a σ-count of A, where we combine the degrees of membership of x_k to A, $cov(A) = \frac{1}{N} \sum_{k=1}^{N} A(x_k)$.

Specificity
Intuitively, the specificity relates to a level of abstraction conveyed by the information granules. The higher the specificity, the lower the level of abstraction. The monotonicity property holds: if for the two information granules A and B one has $A \subset B$ (when the inclusion relationship itself is articulated according to the formalism in which A and B have been formalized) then specificity, sp(.), [16] satisfies the following inequality: $sp(A) \geq sp(B)$. Furthermore for a degenerated information granule comprising a single element x_0 we have a boundary condition $sp(\{x_0\}) = 1$. In case of a one-dimensional interval information granules, one can contemplate expressing

specificity on a basis of the length of the interval, say $sp(A) = exp(-length(A))$; obviously the boundary condition specified above holds here. If the range *range* of the data is available (it could be easily determined), say, then $sp(A) = 1 - |b-a|/length$ *(range)* where $A = [a, b]$, *range* $= [min_k x_k, max_k x_k]$.

The realizations of the above definitions can be augmented by some parameters to offer some additional flexibility. It is intuitively apparent that these two characteristics are associated: the increase in one of then implies a decrease in another: an information granule that "covers" a lot of data cannot be overly specific and vice versa. This is not surprising at all: higher coverage relates to the increasing level of abstraction whereas higher specificity is about more details being captured by the corresponding information granule.

3   Design of Information Granules

Before information granules can be used, they need to be constructed. There is an urgent need to build to come up with an efficient way of forming them to reflect the existing experimental evidence and some predefined requirement. Here we recall two categories of methods. Clustering is the one of them. Clustering techniques transform data into a finite number of information granules. The second class of methods involves the principle of justifiable granularity, which directly dwells on the characteristics of information granules (coverage and specificity) and builds an information granule, which offers an optimization of these characteristics.

3.1   Fuzzy C-Means – Some Brief Focused Insights

Objective function-based clustering is sought as one of the vehicles to develop information granules [1, 17]. In what follows, we briefly recall the essence of the method and elaborate on the format of the results. We consider a collection of n-dimensional numeric data $z_1, z_2, ..., z_N$. A formation of information granules is realized by minimizing an objective function expressing a spread of data around prototypes (centroids)

$$Q = \sum_{i=1}^{c} \sum_{k=1}^{N} u_{ik}^{m} \|x_k - v_i\|^2 \tag{1}$$

where c stands for the number of clusters (information granules). The description of the clusters is provided in the form of a family of prototypes $v_1, v_2, .., v_c$ defined in the data space and a partition matrix $U = [u_{ik}]$, $i = 1, 2..., c$; $k = 1, 2, ..., N$, $m > 1$. It is worth noting that the above-stated objective function is the same as being used in the Fuzzy C-Means (FCM) [1] however in the context of our discussion one could consider other forms of information granules. Note that in the FCM algorithm, the individual rows of the partition matrix are discrete membership functions of the information granules expressed by means of fuzzy sets. In this case, the parameter m standing in the above expression is referred to as a fuzzification coefficient. If one considers sets rather than fuzzy sets, one arrives at the Boolean partition matrix and the method comes as the K-Means algorithm. There are generalizations of the method engaging fuzzy sets of type-2 [2] or rough sets [5, 6].

There are two fundamental design issues that are inherently associated with fuzzy clustering (and clustering, in general) that is (a) a choice of the number of clusters and a selection of the value of the fuzzification coefficient (m), and (b) evaluation of the quality of the constructed clusters and interpretation of results. This task, which is highly relevant when dealing with the optimization of the parameters of the clustering algorithm, implies the usefulness of the clustering results used afterwards in fuzzy modeling and fuzzy classification. Various cluster validity indexes [17, 19] are used to assess the suitability of fuzzy clusters. Different cluster validity indexes can lead to quite distinct results. This is not surprising as each cluster validity index comes with some underlying rationale and in this way prefers a certain structure of clusters (and their ensuing number). On the other hand, a reconstruction criterion [14], emphasizes the quality of clusters being sought as information granules. The criterion is concerned with the evaluation of the quality of information granules (clusters) to describe the data. In essence, one described the available data in terms of information granules (clusters) and then using this characterization decodes (de-granulates) the original data. This transformation, referred to as a granulation-degranulation process leads to inevitable loses which are quantified in terms of a reconstruction error. The value of the error becomes minimized by optimizing the values of the key parameters of the clustering method (such as the fuzzification coefficient m and the number of clusters c). Crucial to the discovery of the structure is the data is a data space in which the clustering takes place.

3.2 The Principle of Justifiable Granularity

The principle of justifiable granularity [10, 12] delivers a comprehensive conceptual and algorithmic setting to develop an information granule. The principle is general as it shows a way of forming information granule without being restricted to certain formalism in which information granularity is expressed and a way experimental evidence using which this information granule comes from. For illustrative purposes, we consider a simple scenario. Let us denote one-dimensional numeric data of interest (for which an information granule is to be formed) by $Z = \{z_1, z_2, \ldots, z_N\}$. Denote the largest and the smallest element in Z by z_{min} and z_{max}, respectively. On a basis of Z we are form an information granule A so that it attempts to satisfy two intuitively requirements of coverage and specificity. The first one implies that the information granule is justifiable, viz. it embraces (covers) as many elements of Z as possible. The second one is to assure that the constructed information granule exhibits a well-defined semantics by being specific enough. For instance, when constructing a fuzzy set, say a one with a triangular membership function, we start with a numeric representative of Z, say a mean or a modal value (denoted here by m) and then separately determine the lower bound (a) and the upper bound (b). In case of an interval A, we start with a modal value and then determine the lower and upper bound, Fig. 1.

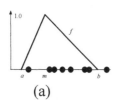
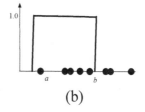

(a)   (b)

Fig. 1. Formation of information granules with the use of the principle of justifiable granularity: (a) triangular membership function, (b) interval (characteristic function). The design is realized by moving around the bounds a and b so that a certain optimization criterion is maximized

The construction of the bounds is realized in the same manner for the lower and upper bound so in what follows we describe only a way of optimizing the upper bound (b). The coverage criterion is expressed as follows

$$cov(A) = \sum_{z_k : z_k \in [m,b]} f(z_k) \tag{2}$$

where f is a decreasing linear portion of the membership function. For an interval (set) form of A, the coverage is expressed as a normalized count of the number of data included in the interval $[m, b]$,

$$cov(A) = card\{z_k | z_k \in [m, b]\} \tag{3}$$

The above coverage requirement states that we reward the inclusion of z_i in A. The specificity $sp(A)$ is realized as one of those specified in the previous section. As we intend to maximize coverage and specificity and these two criteria are in conflict, an optimal value of b is the one, which maximizes the product of the two requirements

$$Q(b) = cov(A) * sp(A)^{\gamma} \tag{4}$$

Furthermore the optimization performance index is augmented by an additional parameter γ used in the determination of the specificity criterion, $sp(A)^{\gamma}$ and assuming non-negative values. It helps control an impact of the specificity in the formation of the information granule. The higher the value of γ, the more essential the impact of specificity on A becomes. If γ is set to zero, the only criterion of interest is the coverage. Higher values of γ underline the importance of specificity as a resulting A gets more specific. The result of optimization comes in the form $b_{opt} = \arg \max_b Q(b)$. The optimization of the lower bound of the fuzzy set (a) is carried out in an analogous way as above yielding $a_{opt} = \arg \text{Max } Q(a)$.

Several observations are worth making here. First, the approach exhibits a general character and the principle is applicable to any formalism of information granules; here we just highlighted the case of sets and fuzzy sets. Second, it is visible that a single information granule represents a collection of many experimental data in a compact form.

4 Higher Type and Higher Order Information Granules

Information granules we discussed so far come with an inherent numeric description: intervals are described by two numeric bounds (a and b), fuzzy sets are described by *numeric* membership functions, probability functions (probability density functions) are *numeric* mappings. One may argue whether such a request is meaningful and does not create any restriction. In particular, with regard to fuzzy sets, this was a point of a visible criticism in the past: what is fuzzy about fuzzy sets? Obviously, the same issue could be formulated with respect to sets or probabilities. There are some interesting generalizations of information granules in which this type of requirement can be relaxed. This gives rise to the concept of information granules of type-2, type-3, and type-n, in general etc. The other direction of generalization deals with the nature of the space over which information granules are formed, which leads to information granules of higher order.

4.1 Higher Type Information Granules

By information granules of *higher type* (2nd type and nth type, in general) we mean granules in the description of whose we use information granules rather than numeric entities. For instance, in case of type-2 fuzzy sets we are concerned with information granules- fuzzy sets whose membership functions are granular. As a result, we can talk about interval-valued fuzzy sets, fuzzy fuzzy sets (or fuzzy^2 sets, for brief), probabilistic sets, uncertain probability, and alike. The grades of belongingness are then intervals in [0,1], fuzzy sets with support in [0,1], probability functions truncated to [0,1], etc. In case of type-2 intervals we have intervals whose bounds are not numbers but information granules and as such can be expressed in the form of intervals themselves, fuzzy sets, rough sets or probability density functions. Information granules have been encountered in numerous studies reported in the literature; in particular stemming from the area of fuzzy clustering in which fuzzy clusters of type-2 have been investigated [2] or they are used to better characterize a structure in the data and could be based upon the existing clusters. Fuzzy sets and interval-valued fuzzy sets form an intensive direction of research producing a number of approaches, algorithms, and application studies.

The development of information granules of higher type can be formed on a basis of information granules of lower type. The principle of justifiable granularity plays here a pivotal role as it realizes an elevation of type of information granularity. Refer to the discussion in the previous section. We started with experimental evidence formed by a *collection* of numeric data (viz. information granules of type-0) and form a *single* information granule of type-1. There is an apparent effect of elevation of the type of information granularity. If the available experimental evidence comes as information granules of type-1 then the result becomes an information granule of type-2. Likewise, if we start with a collection of type-2 information granules forming experimental evidence, the result becomes an information granule of type-3, etc. In particular, the principle of justifiable granularity can be regarded as a vehicle to construct type-2 fuzzy sets.

4.2   Higher Order Information Granules

Information granules, which are defined in the space (universe of discourse) whose elements are individual items, are called information granules of order-1. If the space itself is formed as a collection of information granules then any information granule defined over a space of information granules is referred to as information granules of order-2. The constructs could formed recursively thus forming information granules of order-3, 4, etc. It is worth noting that one can envision information granules of higher order and higher type.

The four alternatives that might arise here are displayed below, see Fig. 2. They capture the semantics of the resulting constructs.

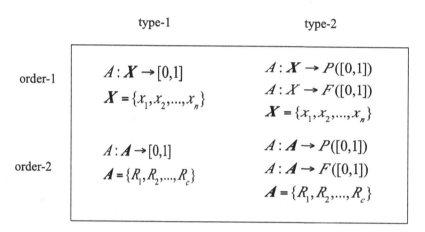

Fig. 2.   Examples of four categories of information granules of type-2 and order-2; P, F- families of intervals and fuzzy sets, respectively; $A = \{R_1, R_2, ..., R_c\}$- a collection of reference information granules

5   Selected Application Areas

In this section, we elaborate on several applications of information granules of higher type and higher order.

5.1   Fuzzy Modeling

The involvement of fuzzy sets of higher type, in particular type-2 fuzzy sets and interval–valued fuzzy sets have triggered a new direction in fuzzy modeling. A general motivation behind these models relates with the elevated generality of the concepts of fuzzy sets of higher type, which translates into a higher flexibility of type-2 fuzzy models. While this argument is valid, there are a number of ongoing challenges. This concerns an increase of complexity of the development schemes of such fuzzy models.

A significantly larger number of their parameters (in comparison with the previously considered fuzzy models) require more elaborate estimation mechanisms. This has immediately resulted in essential optimization challenges (which owing to the engagement of more advance population-based optimization tools have been overcome to some extent but at expense of intensive computing). At the end, type-2 fuzzy models are assessed as numeric constructs with the chain of transformations: type reduction (from type-2 to type-1) followed by defuzzification (reduction from type-1 to type-0 information granules, viz. numbers) thus resulting in a numeric construct.

Ironically, in spite of all significant progress being observed, fuzzy models seem to start losing identity, which was more articulated and visible at the very early days of fuzzy sets. While one may argue otherwise, there is a visible identity crisis: at the end of the day fuzzy models have been predominantly perceived and evaluated as numeric constructs with the quality expressed at numeric level (through accuracy measures).

5.2   Embedding Fuzzy Models: A Granular Parameter Space Approach

The concept of the granular models form a generalization of numeric models no matter what their architecture and a way of their construction are. In this sense, the conceptualization offered here are of general nature. They also hold for any formalism of information granules. A numeric model M_0 constructed on a basis of a collection of training data (x_k, target_k), $x_k \in R^n$ and $\text{target}_k \in R$ comes with a collection of its parameters a_{opt} where $a \in R^p$. Quite commonly, the estimation of the parameters is realized by minimizing a certain performance index Q (say, a sum of squared error between target_k and $M_0(x_k)$), namely $a_{\text{opt}} = \arg \text{Min}_a Q(a)$. To compensate for inevitable errors of the model (as the values of the index Q are never equal identically to zero), we make the parameters of the model information granules, resulting in a vector of information granules $A = [A_1\ A_2 \ldots\ A_p]$ built around original numeric values of the parameters a. In other words, the fuzzy model is embedded in the *granular* parameter space. The elements of the vector a are generalized, the model becomes granular and subsequently the results produced by them are information granules. Formally speaking, we have

- granulation of parameters of the model $A = G(a)$ where G stands for the mechanisms of forming information granules, viz. building an information granule around the numeric parameter
- result of the granular model for any x producing the corresponding information granule Y, $Y = M_1(x, A) = G(M_0(x)) = M_0(x, G(a))$.

Information granulation is regarded as an essential design asset [10]. By making the results of the model granular (and more abstract in this manner), we realize a better alignment of $G(M_0)$ with the data. Intuitively, we envision that the output of the granular model "covers" the corresponding target. Formally, let cov(*target*, Y) denote a certain coverage predicate (either Boolean or multivalued) quantifying an extent to which target is included (covered) in Y.

The design asset is supplied in the form of a certain allowable level of information granularity ε which is a certain non-negative parameter being provided in advance.

We allocate (distribute) the design asset across the parameters of the model so that the coverage measure is maximized while the overall level of information granularity serves as a constraint to be satisfied when allocating information granularity across the model, namely $\sum_{i-1}^{p} \varepsilon_i = e$ The constraint-based optimization problem reads as follows

$$\max_{\varepsilon_1,\varepsilon_2,\ldots,\varepsilon_p} \sum_{k=1}^{N} \mathrm{cov}(target_k \in Y_K)$$

subject to

$$\sum_{i=1}^{P} \varepsilon_i = e \text{ and } \varepsilon_i \geq 0 \tag{5}$$

The monotonicity property of the coverage measure is obvious: the higher the values of e, the higher the resulting coverage. Hence the coverage is a non-decreasing function of ε.

Along with the coverage criterion, one can also consider the specificity of the produced information granules. It is a non-increasing function of e. The more general form of the optimization problem can be established by engaging the two criteria leading to the two-objective optimization problem. The problem can be re-structured in the following form in which the objective function is a product of the coverage and specificity-determine optimal allocation of information granularity [ε₁ ε₂,..., εₚ] so that the coverage and specificity criteria become maximized.

Plotting these two characteristics in the coverage–specificity coordinates offers a useful visual display of the nature of the granular model and possible behavior of the behavior of the granular model as well as the original model. There are different patterns of the changes between coverage and specificity. The curve may exhibit a monotonic change with regard to the changes in e and could be approximated by some linear function. There might be some regions of some slow changes of the specificity with the increase of coverage with some points at which there is a substantial drop of the specificity values. A careful inspection of these characteristics helps determine a suitable value of ε – any further increase beyond this limit might not be beneficial as no significant gain of coverage is observed however the drop in the specificity compromises the quality of the granular model.

The global behavior of the granular model can be assessed in a global fashion by computing an area under curve (AUC) of the coverage-specificity curve. Obviously, the higher the AUC value, the better the granular model. The AUC value can be treated as an indicator of the global performance of the original numeric model produced when assessing granular constructs built on their basis. For instance, the quality of the original numeric models M_0 and M_0' could differ quite marginally but the corresponding values of their AUC could vary quite substantially by telling apart these two models. For instance, two neural networks of quite similar topology may exhibit similar performance however when forming their granular generalizations, those could differ quite substantially in terms of the resulting values of the AUC.

As to the allocation of information granularity, the maximized coverage can be realized with regard to various alternatives as far as the data are concerned: (a) the use of the same training data as originally used in the construction of the model, (b) use the testing data, and (c) usage of some auxiliary data.

5.3   Granular Input Spaces in Fuzzy Modeling

The underlying rationale behind emergence of granular input spaces deals with an ability to capture and formalize the problem at the higher level of abstraction by adopting a granular view of the input space in which supporting system modeling and model construction are located. Granulation of input spaces is well motivated and often implied by the computing economy or a flexibility and convenience they offer to they offer when capturing the. Here we would like to highlight some illustrative examples, especially those commonly visible in some temporal or spatial domains.

Granular input spaces deliver an important, unique, and efficient design setting for the construction and usage of fuzzy models: (i) information granulation of a large number of data (in case of streams of data) leads to a far smaller and semantically sound entities facilitating and accelerating the design of fuzzy models, and (ii) the results of fuzzy modeling are conveyed at a suitable level of specificity suitable for solving a given problem. In the sequel, information granules used to construct a model, viz. a mapping between input and output information granules.

5.4   Rule-Based Models and Their Augmentation with Schemes
of Allocation of Information Granularity

Functional rules (Takagi-Sugeno format of the conditional statements) link any input space with the corresponding local model whose relevance is confided to the region of the input space determined by the fuzzy set standing in the input space (A_i). The local character of the conclusion makes an overall development of the fuzzy model well justified: we fully adhere to the modular modeling of complex relationships. The local models (conclusions) could vary in their diversity; in particular local models in the form of constant functions (m_i) are of interest

$$- \text{ if } x \text{ is } A_i \text{ then } y \text{ is } m_i \tag{6}$$

These models are equivalent to those produced by the Mamdani-like rules with a weighted scheme of decoding (defuzzification). There has been a plethora of design approaches to the construction of rule-based models, cf. [3, 4].

Information granularity emerges in fuzzy models in several ways by being present in the condition parts of the rules, their conclusion parts and both. In a concise way, we can describe this in the following way (below the symbol $G(.)$ underlines the granular expansion of the fuzzy set construct abstracted from their detailed numeric realization or a granular expansion of the numeric mapping).

(i) *Information granularity associated with the conditions of the rules.* We consider the rules coming in the format

$$- \text{ if } G(A_i) \text{ then } f_i \tag{7}$$

where $G(A_i)$ is the information granule forming the condition part of the i-th rule. An example of the rule coming in this format is the one where the condition is described in terms of a certain interval-valued fuzzy set or type-2 fuzzy set, $G(A_i)$.

(ii) *Information granularity associated with the conclusion part of the rules.* Here the rules take on the following form

$$- \text{ if } x \text{ is } A_i \text{ then } G(f_i) \tag{8}$$

with $G(f_i)$ being the granular local function. The numeric mapping f_i is made more abstract by admitting their parameters being information granules. For instance, instead of the numeric linear function f_i, we consider $G(f_i)$ where $G(f_i)$ is endowed with parameters regarded as intervals or fuzzy numbers. In this way, we have $f_i(A_0, A_1, \ldots, A_n) = A_{i0} + A_{i1}x_1 + \ldots A_{in}x_n$ with the algebraic operations carried out on information granules (in particular adhering to the algebra of fuzzy numbers).

(iii) *Information granularity associated with the condition and conclusion parts of the rules.* This forms a general version of the granular model and subsumes the two situations listed above. The rules read now as follows

$$- \text{ if } G(A_i) \text{ then } G(f_i) \tag{9}$$

The augmented expression for the computations of the output of the model generalizes the expression used in the description of the fuzzy models (8). We have

$$Y = \sum_{\substack{i=1 \\ \oplus}}^{c} (G(A_i(x) \otimes G(f_i)) \tag{10}$$

where the algebraic operations shown in circles $\otimes$ and $\oplus$ reflect that the arguments are information granules instead of numbers (say, fuzzy numbers). The detailed calculations depend upon the formalism of information granules being considered. Let us stress that Y is an information granule. Obviously, the aggregation presented by (10) applies to (i) and (ii) as well; here we have some simplifications of the above stated formula.

There are no perfect models. Information granularity augmenting existing (numeric) models results in a granular model and makes it more in rapport with reality. In a general way, we can think of a certain general way of forming a granular model at successively higher levels of abstraction. Subsequently the representation (model) of a real system S can be symbolically described through the following relationship

$$S \approx M \oplus G(M) \oplus G^2(M) \oplus \ldots \oplus G^t(M) \tag{11}$$

where the symbols M, $G(M)$, $G^2(M)$, ... $G^t(M)$ stand for an original (numeric) model, granular model built with the use of information granules of type-1, $G(M)$, granular model realized with the use of information granules of type-2, $G^2(M)$,.., and information granules of type-t, etc. The symbol is used to denote the enhancements of the modeling construct aimed to model S. The models formed in this way are displayed in Fig. 3.

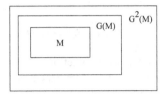

Fig. 3. A hierarchy of granular models: from numeric constructs to granular models with information granules of higher type

Noticeable is the fact that successive enhancements of the model emerge at the higher level of abstraction engaging information granules of the increasing type.

6 Conclusions

In the study, we have presented a general framework of Granular Computing and elaborated on their generalizations coming in the form of information granules of higher type and higher order. We offered a brief overview of fuzzy rule-based models and demonstrated that in light of new challenging modeling environments, there is a strongly motivated emergence of granular fuzzy models where the concept of information granularity and information granules of higher type/order play a pivotal role. The fundamentals of Granular Computing such as the principle of justifiable granularity and an optimal allocation of information granularity are instrumental in the construction of the granular models.

Acknowledgements. Support from the Canada Research Chair (CRC) and Natural Sciences and Engineering Research Council (NSERC) is gratefully acknowledged.

References

1. Bezdek, J.C.: Pattern Recognition with Fuzzy Objective Function Algorithms. Plenum Press, New York (1981)
2. Hwang, C., Rhee, F.C.H.: Uncertain fuzzy clustering: interval type-2 fuzzy approach to C-means. IEEE Trans. Fuzzy Syst. **15**(12), 107–120 (2007)

3. Jin, Y.: Fuzzy modeling of high-dimensional systems: complexity reduction and interpretability improvement. IEEE Trans. Fuzzy Syst. **8**, 212–221 (2000)
4. Johansen, T.A., Babuska, R.: Multiobjective identification of Takagi-Sugeno fuzzy models. IEEE Trans. Fuzzy Syst. **11**, 847–860 (2003)
5. Lai, J.Z.C., Juan, E.Y.T., Lai, F.J.C.: Rough clustering using generalized fuzzy clustering algorithm. Pattern Recogn. **46**(9), 2538–2547 (2013)
6. Li, F., Ye, M., Chen, X.: An extension to rough c-means clustering based on decision-theoretic rough sets model. Int. J. Approx. Reason. **55**(1), 116–129 (2014)
7. Pawlak, Z.: Rough sets. Int. J. Inf. Comput. Sci. **11**, 341–356 (1982)
8. Pawlak, Z.: Rough Sets. Theoretical Aspects of Reasoning About Data. Kluwer Academic Publishers, Dordecht (1991)
9. Pedrycz, W.: Granular computing - the emerging paradigm. J. Uncertain Syst. **1**(1), 38–61 (2007)
10. Pedrycz, W.: Granular Computing: Analysis and Design of Intelligent Systems. CRC Press/Francis Taylor, Boca Raton (2013)
11. Pedrycz, W., Bargiela, A.: An optimization of allocation of information granularity in the interpretation of data structures: toward granular fuzzy clustering. IEEE Trans. Syst. Man Cybern. Part B **42**, 582–590 (2012)
12. Pedrycz, W., Homenda, W.: Building the fundamentals of granular computing: a principle of justifiable granularity. Appl. Soft Comput. **13**, 4209–4218 (2013)
13. Pedrycz, W.: Knowledge-Based Fuzzy Clustering. Wiley, New York (2005)
14. Pedrycz, W., de Oliveira, J.V.: A development of fuzzy encoding and decoding through fuzzy clustering. IEEE Trans. Instrum. Meas. **57**(4), 829–837 (2008)
15. Pedrycz, W.: Shadowed sets: representing and processing fuzzy sets. IEEE Trans. Syst. Man Cybern. Part B **28**, 103–109 (1998)
16. Yager, R.R.: Ordinal measures of specificity. Int. J. Gen. Syst. **17**, 57–72 (1990)
17. Xu, R., Wunsch II, D.: Survey of clustering algorithms. IEEE Trans. Neural Netw. **16**(3), 645–678 (2005)
18. Yao, J.T., Vasilakos, A.V., Pedrycz, W.: Granular computing: perspectives and challenges. IEEE Trans. Cybern. **43**(6), 1977–1989 (2013)
19. Wang, W.N., Zhang, Y.J.: On fuzzy cluster validity indices. Fuzzy Sets Syst. **158**(19), 2095–2117 (2007)
20. Zadeh, L.A.: Towards a theory of fuzzy information granulation and its centrality in human reasoning and fuzzy logic. Fuzzy Sets Syst. **90**, 111–117 (1997)

Fuzzy Measures and Transforms

Cardiovascular Disease Risk Assessment Using the Choquet Integral

Luca Anzilli[1][(⊠)] and Silvio Giove[2]

[1] Department of Management, Economics, Mathematics and Statistics,
University of Salento, Lecce, Italy
luca.anzilli@unisalento.it
[2] Department of Economics, University Ca' Foscari of Venice, Venice, Italy

Abstract. In this paper we propose a cardiovascular risk diagnosis model based on non additive measures (*fuzzy* measures) and the Choquet integral. To this purpose, an ad hoc questionnaire was submitted to a set of doctors, from which a set of measures was elicited. The answers were then aggregated together in the spirit of consensus and an unique fuzzy measure was obtained. Again, the criteria used for the diagnosis were transformed using suitable membership functions. A cardiovascular disease risk index was then introduced as the Choquet integral of membership functions with respect to the fuzzy measure. A sensitivity analysis was performed too.

Keywords: Choquet integral · Fuzzy numbers · Consensus measures · Fuzzy measures · Cardiovascular disease risk

1 Introduction

In this contribution, we propose a fuzzy logic approach to measure the cardiovascular disease (CVD) risk for non smoker and non diabetic population. With respect to other approach based on statistics and data mining, see among other [1], which require the availability of a large set of data, in this paper we adopt an approach based on the elicitation of expert knowledge from a set of Experts (clinicians and doctors) to infer the basis of knowledge. Following previous proposal, [2] and limiting the attention to non smoker and non diabetic population, we consider as explicative variables for cardiovascular risk three main variables, namely the blood pressure, the cholesterol level and the body mass index. For each of the three variables, suitable membership functions were considered, w.t.a. to convert all the three to the common $[0, 1]$ numerical scale. The meaning of each membership can be interpreted as usually in fuzzy logic, as the degree of truth of the proposition "this value is critical". This approach avoids the application of *rigid* thresholds as sometimes done in medical application, permitting the introduction of ambiguity and uncertainty in the neighborhood of the critical threshold for the considered parameter. At the same time, a membership function can be intended in economic sense, the same way as an *utility* function; for the relationship among utility function and an uncertain threshold see [3]. The novelty of our proposal for cardiovascular risk relays into the way by which the *fuzzy* values of the criteria are aggregated.

© Springer International Publishing AG 2017
A. Petrosino et al. (Eds.): WILF 2016, LNAI 10147, pp. 45–53, 2017.
DOI: 10.1007/978-3-319-52962-2_3

Namely, instead of the application of different aggregation operators[1] as done in [2], we propose a method based on non additive measure and the Choquet integral [5], due to the wide generality of this approach, which includes as particular cases many aggregation operators. At the same time, the necessary parameters, *i.e.* the values of the measure, can be easily inferred by an ad hoc questionnaire which has been submitted to a set of doctors. The answers are then aggregated together in the spirit of consensus. This way, an unique fuzzy measure is obtained, which will be used to aggregate the fuzzy values of the three risk factors, and finally to evaluate the risk of cardiovascular disease for a certain risk population. A sensitivity analysis has been performed too.

2   Cardiovascular Disease and Decision Support Systems

Many clinical studies have identified possible factors that significantly increase the risk of hearth disease, the most common of which regards coronary arteries attack. Among them, we recall cholesterol and blood pressure high level, obesity, comorbidity, stress, and style of living, like inactivity, smoke, alcohol and drug. Moreover, the cardiovascular disease risk is higher in some age range, depending also on hereditary factors and is normally higher for males than for females. [6]. Anywise, the prevalence of many of them is not completely clear, see [7]. For this reason, in this contribution we propose a prototype of a Decision Support System (DSS) for hearth failure risk, based on clinical experience, instead that on statistical correlation among sampled data. The model is based on *fuzzy* measures, previously elicited by a set of expert doctors, using a suitable questionnaire developed *ad hoc*. Three of the major risk factors are considered here, in the line of what proposed in [2], even if the method could be extended to a higher number of risk factors[2]. The proposed DSS captures the expert knowledge by the interviewed doctors. As in [2], and following indications furnished by a pre-inquiry among the consulted specialist Experts, the considered major risk factors are: the total Cholesterol, the Body Mass Index and the systolic Blood Pressure, see the quoted references for a justification of this choice. Our approach is based on fuzzy logic, that is nowadays recognized an appropriate methodology to treat with vagueness and subjectivity uncertainty. For instance, the introduction of membership functions instead of *rigid* thresholds for risky versus non-risky parameter value, permits to capture the human uncertainty about the riskiness of the variable itself, see [8]. Thus a flexible risk score can be computed for a single individual to indicate his health condition, based on clinical symptoms and signs. At first, for each of these three factors, a suitable membership function was defined, representing the degree of membership "the value of the variable is satisfactory". Inside a precise range the membership will equal one, meaning that for value inside a complete satisfaction is reached. For values higher (lower)

---

[1] For a complete review of aggregation operators in the context of fuzzy logic see [4].

[2] It was underlined that the presence of too many risk factors makes the hearth risk evaluation a very complex task, also for the difficulty to find a cohort with a rigorous follow-up.

than an upper (lower) limit, the unacceptable value of the variable in set to zero. For values in between the two limits, a piecewise linear interpolation has been used. This way, we permit the introduction of subjective uncertainty, that is typical of human judgement, and easily modeled using fuzzy logic. To aggregate the fuzzy values of each variables, some aggregation operators were proposed by [2], but the obtained result is quite debatable given that no a priori justification were proposed about the choice of a particular operator instead than an other one. Thus we adopt a different strategy, aggregating the fuzzy values using the Choquet integral, using a set of non additive measures (NAM) elicited by the doctor's answers. To this purpose each doctor is interviewed and a questionnaire is submitted, formed by a set of questions which simulate suitable scenarios. Again, in the spirit of consensus reaching among the Experts (the interviewed doctors), the answers are aggregated building an average NAM obtained by a weighted linear combination where the weights depend on the distance between the single doctor's answer and the harmonic means of all the doctors, see below. In so doing, the doctor's answer weights more or less depending on the distance by the *common* opinion, representing the majority judgement about the considered item.

Definition 1. *Let $N = \{1, 2, \ldots, n\}$ be a finite set of criteria (factors). A monotonic measure (fuzzy measure) on N is a set function $m : 2^N \to \mathbb{R}_+$ such that*

(i) $m(\emptyset) = 0$;
(ii) $\forall T, S \subseteq N : \quad T \subseteq S \implies m(T) \leq m(S)$.

A monotonic measure m is said to be normalized if $m(N) = 1$.

The fuzzy measure m estimates the influence of every coalition of criteria on the cardiovascular disease risk. Let $V = \{v_1, v_2, \ldots, v_k\}$ be a set of k experts. For each $S \subseteq N$ we denote by $v_j(S)$ the quantified judgement of expert v_j on the influence of coalition S on the cardiovascular disease risk.

We may construct a consensus measure in the following way (see [9]). For $j = 1, \ldots, k$ we compute the total absolute distance, D_j, of j-th expert's evaluations to all other experts' evaluations, as follows:

$$D_j = \sum_{S \subseteq N} \sum_{\substack{\ell=1 \\ \ell \neq j}}^{k} |v_j(S) - v_\ell(S)|.$$

Then we compute the weight associated to the expert's valuation as

$$w_j = \frac{D_j^{-1}}{\sum_{\ell=1}^{k} D_\ell^{-1}}.$$

We observe that weights w_j can be expresses equivalently as

$$w_j = \frac{M(D_1, \ldots, D_k)}{k \, D_j}$$

being $M(D_1, \ldots, D_k)$ the harmonic mean of $D_1, \ldots, D_k$. Finally, the measure m is given by the weighted average of experts' valuations

$$m(S) = \sum_{j=1}^{k} w_j \, v_j(S), \qquad S \subseteq N. \qquad (1)$$

Definition 2. *The Möbius transform of m is the set function $\alpha_m : 2^N \to \mathbb{R}$ defined by*

$$\alpha_m(S) = \sum_{T \subseteq S} (-1)^{|S|-|T|} m(T), \qquad S \subseteq N$$

The overall importance of a criterion $i \in N$ with respect to m can be measured by means of its Shapley value (or Shapley index), which is defined by

$$\phi_m(i) = \sum_{T \subseteq N \backslash \{i\}} \frac{(n - |T| - 1)! \, |T|!}{n!} \left[m\left(T \cup \{i\}\right) - m\left(T\right) \right].$$

where $|T|$ is the cardinality of T, and $\sum_{i=1}^{n} \phi_m(i) = m(N)$. In terms of the Möbius representation of μ, the Shapley value of i can be rewritten as

$$\phi_m(i) = \sum_{T \subseteq N \backslash \{i\}} \frac{1}{|T| + 1} \, \alpha_m\left(T \cup \{i\}\right).$$

Murofushi and Soneda [10] suggested to measure the average interaction between two criteria i and j by means of the following interaction index:

$$I_m(i,j) = \sum_{T \subseteq N \backslash \{i,j\}} \frac{(n - |T| - 2)! \, |T|!}{(n-1)!} \left[m\left(T \cup \{i,j\}\right) \right.$$
$$\left. - m\left(T \cup \{i\}\right) - m\left(T \cup \{j\}\right) + m\left(T\right) \right].$$

2.1   Application to Cardiovascular Disease Risk

The three selected factors (denoted by $i = 1, 2, 3$) for the cardiovascular disease risk of a non smoker and non diabetic individual are: cholesterol level ($i = 1$), blood pressure ($i = 2$) and body mass index ($i = 3$). We prepared a questionnaire for $k = 8$ experts in order to find out the weights for criteria and interactions among them. In Table 1 we show the experts' valuation to each coalition of criteria. In Table 2 we have computed the weights w_j. In Table 3 we show the fuzzy measure m computed according to (1). In Table 4 we show the Möbius coefficients of the fuzzy measure m. Shapley values and Interaction indexes are computed in Table 5.

Table 1. Experts' answers to survey

S	$v_1(S)$	$v_2(S)$	$v_3(S)$	$v_4(S)$	$v_5(S)$	$v_6(S)$	$v_7(S)$	$v_8(S)$
{1}	0.30	0.30	0.20	0.10	0.20	0.10	0.20	0.20
{2}	0.30	0.30	0.20	0.10	0.20	0.10	0.20	0.20
{3}	0.30	0.40	0.30	0.20	0.30	0.30	0.40	0.40
{1,2}	0.50	0.40	0.40	0.20	0.40	0.30	0.30	0.30
{1,3}	0.60	0.40	0.50	0.40	0.50	0.40	0.50	0.40
{2,3}	0.60	0.40	0.60	0.40	0.50	0.40	0.50	0.40
{1,2,3}	0.80	0.80	0.80	0.80	0.90	0.70	0.90	0.70

Table 2. Weights w_j

w_1	w_2	w_3	w_4	w_5	w_6	w_7	w_8
0.09	0.12	0.14	0.10	0.14	0.11	0.14	0.14

Table 3. Fuzzy measure m

m({1})	m({2})	m({3})	m({1,2})	m({1,3})	m({2,3})	m({1,2,3})
0.20	0.20	0.33	0.35	0.46	0.48	0.80

Table 4. Möbius coefficients

$\alpha(\{1\})$	$\alpha(\{2\})$	$\alpha(\{3\})$	$\alpha(\{1,2\})$	$\alpha(\{1,3\})$	$\alpha(\{2,3\})$	$\alpha(\{1,2,3\})$
0.20	0.20	0.33	−0.05	−0.07	−0.05	0.24

Table 5. Shapley values and interaction indexes

$\phi_m(1)$	$\phi_m(2)$	$\phi_m(3)$	$I_m(1,2)$	$I_m(1,3)$	$I_m(2,3)$
0.22	0.23	0.35	0.07	0.05	0.06

3   The CVD Risk Model

In this section we present a cardiovascular disease risk (CVD) model for a non smoker and non diabetic individual. We consider as baseline factors for cardiovascular risk three main variables ("criteria"), namely the cholesterol level, the blood pressure and the body mass index. For each of the three variables a suitable membership function is considered. Each membership can be interpreted as the degree of truth of the proposition "this value is inside the optimal range". Individual's risk factor values are then computed by the complement to one of the membership degree; they can be interpreted as the degree of truth of the proposition "this value is critical". We define a CVD risk index combining the information on individual's risk factor values with the consensus measure representing the influence on CVD risk of every coalition of factors. The aggregation procedure is performed using the Choquet-integral.

3.1 Individual's Risk Factors

The three selected risk factors (denoted by $i = 1, 2, 3$) for the cardiovascular disease risk of a non smoker and non diabetic individual are: cholesterol level ($i = 1$), blood pressure ($i = 2$) and body mass index ($i = 3$). We denote by x_i the value of factor i. To each factor i we associate a fuzzy set A_i with membership function $\mu_i : [0, +\infty) \to [0, 1]$, where $\mu_i(x_i) = 1$ indicates that the value x_i of parameter i is normal and $\mu_i(x_i) = 0$ that the parameter is totally altered.

We model fuzzy sets A_i as *trapezoidal adaptive fuzzy numbers* (see [11]) $A_i = (a_i(1); a_i(2); a_i(3); a_i(4))_p$, with $a_i(1) < a_i(2) \le a_i(3) < a_i(4)$, defined by

$$
\mu_i(x) = \begin{cases}
\left(\frac{x - a_i(1)}{a_i(2) - a_i(1)}\right)^p & a_i(1) \le x \le a_i(2) \\
1 & a_i(2) \le x \le a_i(3) \\
\left(\frac{a_i(4) - x}{a_i(4) - a_i(3)}\right)^p & a_i(3) \le x \le a_i(4) \\
0 & x \le a_i(1) \ \text{ or } \ x \ge a_i(4)
\end{cases}
$$

where $p > 0$ is an optimistic ($0 < p < 1$) or pessimistic ($p > 1$) parameter. If $p = 1$, A is a trapezoidal fuzzy number; if $p > 1$ it is a concentration; if $0 < p < 1$ it is a dilation. Concentration by $p = 2$ may be interpreted as the linguistic hedge *very*, dilation by $p = 0.5$ as *more or less* [11]. The use of adaptive fuzzy numbers allows the decision maker to modify the shape of the membership functions without having to change the support or the core. Indeed, he can increase or decrease the fuzziness by adjusting the parameter p according to his subjective estimation. The core $[a_i(2), a_i(3)]$ is the optimal value range. We describe factors as fuzzy sets as follows:

- cholesterol level: $A_1 = (120, 140, 200, 240)_p$;
- blood pressure: $A_2 = (60, 90, 130, 170)_p$;
- body mass index: $A_3 = (17, 20, 25, 40)_p$.

In Fig. 1 we show membership functions of cholesterol level, blood pressure and body mass index for different values of parameter p.

3.2 Aggregation

We now define a CVD risk index combining membership functions by means of the Choquet integral, using the non additive measure elicited from question-naire's answers.

Definition 3. *The Choquet integral of a function $f : N \to \mathbb{R}_+$ with respect to a monotonic measure (fuzzy measure) m is defined by*

$$
C_m(f) = \sum_{i=1}^{n} [f(\sigma(i)) - f(\sigma(i-1))] \, m(E_{\sigma(i)})
$$

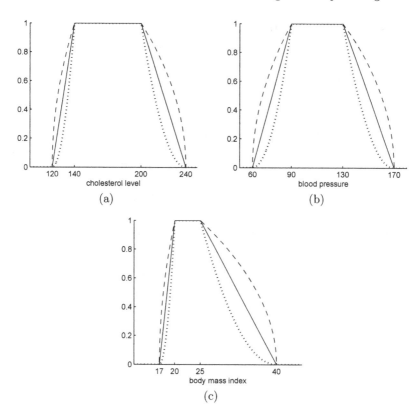

Fig. 1. Membership functions of cholesterol level (a), blood pressure (b) and body mass index (c) for $p = 1$ (continuous line), $p = 2$ (dotted line) and $p = 0.5$ (dashed line).

where σ is a suitable permutation of indices such that $f(\sigma(1)) \leq \cdots \leq f(\sigma(n))$, $E_{\sigma(i)} = \{\sigma(i), \ldots, \sigma(n)\}$ for $i = 1, \ldots, n$ and $f(\sigma(0)) = 0$.

In terms of Möbius transform of m the Choquet integral may be computed as

$$C_m(f) = \sum_{T \subseteq N} \alpha_m(T) \min_{i \in T} f(i).$$

In our CVD risk model the set of factors is $N = \{1, 2, 3\}$ and the function $f : N \rightarrow [0, 1]$ is defined by

$$f(i) = 1 - \mu_i(x_i), \qquad i = 1, 2, 3$$

where μ_i is the membership function associated to factor i. The meaning of $f(i)$ can be interpreted as the degree of truth of the proposition "this value is critical". The fuzzy measure m is the consensus measure elicited from doctors' answers. The CVD risk index is given by the Choquet integral

$$C_m(f) = \sum_{i=1}^{3} [f(\sigma(i)) - f(\sigma(i-1))] \, m(E_{\sigma(i)}) \tag{2}$$

where σ is a suitable permutation of indices such that $f(\sigma(1)) \le f(\sigma(2)) \le f(\sigma(3))$ and $E_{\sigma(i)} = \{\sigma(i), \ldots, \sigma(3)\}$ for $i = 1, 2, 3$ and $f(\sigma(0)) = 0$.

4   Numerical Results

We now perform a sensitivity analysis of the CVD risk index, defined as the Choquet integral (2), with respect to each factor for different values of parameter p. Numerical results are shown in Fig. 2. In Fig. 2(a) we have plotted CVD risk index as function of cholesterol level with $x_2 = 120$ (blood pressure) and $x_3 = 22$ (body mass index) fixed. In Fig. 2(b) we have plotted CVD risk index as function of blood pressure with $x_1 = 190$ (cholesterol level) and $x_3 = 22$ (body mass index) fixed. In Fig. 2(c) we have plotted CVD risk index as function of body mass index with $x_1 = 190$ (cholesterol level) and $x_2 = 120$ (blood pressure) fixed.

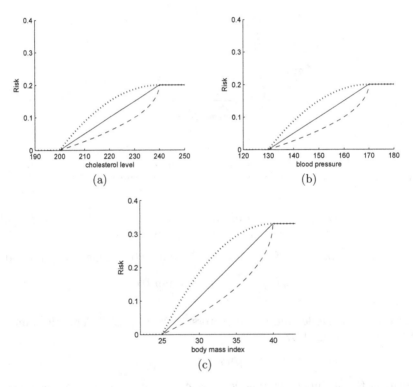

Fig. 2. CVD risk index as function of cholesterol level (a), blood pressure (b) and body mass index (c) for $p = 1$ (continuous line), $p = 2$ (dotted line) and $p = 0.5$ (dashed line).

5   Conclusion

In this paper we proposed a cardiovascular risk diagnosis model based on non additive measures (*fuzzy* measures) and the Choquet integral. A CVD risk index was introduced and some numerical results were presented. As a future development, we intend to investigate the properties of the proposed CVD risk index and to compare it with other typical and commonly used aggregation operators.

References

1. Soni, J., Ansari, U., Sharma, D., Soni, S.: Predictive data mining for medical diagnosis: an overview of heart disease prediction. Int. J. Comput. Appl. **17**(8), 43–48 (2011)
2. Lazzari, L.L., Moulia, P.I.: Fuzzy sets application to healthcare systems. Fuzzy Econ. Rev. **17**(2), 43 (2012)
3. Bordley, R., LiCalzi, M.: Decision analysis using targets instead of utility functions. Decisions Econ. Finan. **23**(1), 53–74 (2000)
4. Calvo, T., Mayor, G., Mesiar, R.: Aggregation Operators: New Trends and Applications, vol. 97, Physica (2012)
5. Choquet, G.: Theory of capacities. In: Annales de l'institut Fourier, vol. 5, pp. 131–295. Institut Fourier (1954)
6. Polat, K., Güneş, S., Tosun, S.: Diagnosis of heart disease using artificial immune recognition system and fuzzy weighted pre-processing. Pattern Recogn. **39**(11), 2186–2193 (2006)
7. Sanz, J.A., Galar, M., Jurio, A., Brugos, A., Pagola, M., Bustince, H.: Medical diagnosis of cardiovascular diseases using an interval-valued fuzzy rule-based classification system. Appl. Soft Comput. **20**, 103–111 (2014)
8. Tsipouras, M.G., Voglis, C., Fotiadis, D.I.: A framework for fuzzy expert system creation–application to cardiovascular diseases. IEEE Trans. Biomed. Eng. **54**(11), 2089–2105 (2007)
9. Pinar, M., Cruciani, C., Giove, S., Sostero, M.: Constructing the feem sustainability index: a choquet integral application. Ecol. Ind. **39**, 189–202 (2014)
10. Murofushi, T., Soneda, S.: Techniques for reading fuzzy measures (iii): interaction index. In: 9th Fuzzy System Symposium, pp. 693–696, Sapporo, Japan (1993)
11. Zadeh, L.A.: A fuzzy-set-theoretic interpretation of linguistic hedges. J. Cybern. **2**(2), 4–34 (1972)

Fuzzy Transforms and Seasonal Time Series

Ferdinando Di Martino and Salvatore Sessa[✉]

Dipartimento di Architettura, Università degli Studi di Napoli Federico II,
Via Toledo 402, 80134 Naples, Italy
{fdimarti,sessa}@unina.it

Abstract. Like in our previous papers, we show the trend of seasonal time series by means of polynomial interpolation and we use the inverse fuzzy transform for prediction of the value of an assigned output. As example, we use the daily weather dataset of the city of Naples (Italy) starting from data collected from 2003 till to 2015 making predictions on the the relative humidity parameter. We compare our method with the traditional F-transform based, the average seasonal variation and the famous ARIMA methods.

Keywords: ARIMA · Average seasonal variation · Forecasting · Fuzzy transform · Time series

1 Introduction

The seasonal time series are mainly studied for prediction of determined parameters with traditional statistical approaches like moving average method, Holt-Winters exponential smoothing, etc. [1, 2, 5], Autoregressive Integrated Moving Average (ARIMA) [1, 5, 9]. Also soft computing methods are able to deal such series like Support Vector Machine (SVM) [7, 8], Architecture Neural Networks (ANN) based like multi-layer Feed Forward Network (FNN) [12, 13], Time Lagged Neural Network (TLNN) [4]. We recall that the ANN based methods require an high and expensive computational effort. These shortcomings have been overcome from a forecasting method based on fuzzy transforms (F-tr) [3, 10, 11]. In [3] several forecasting indexes were proposed and we prefer to use the MADMEAN index which is more robust with respect to other ones [6]. Our Time Series Seasonal F-transform (TSSF) based method is given in Fig. 1. After the application of the best polynomial fitting for determining the trend in the training data, the time series is decomposed in S seasonal subsets on which F-tr method is applied as well. A coarse grained uniform fuzzy partition is fixed a priori. If the subset is not sufficiently dense with respect to this fuzzy partition (see Sect. 2), the F-tr process stops, otherwise the MADMEAN index is evaluated and it must be greater than an assigned threshold. If this holds, then the F-tr process is iterated by taking a finer uniform fuzzy partition and the inverse F-tr of every seasonal subset is considered as final output.

We recall the concept of F-tr in Sect. 2. Section 3 contains the F-tr based prediction method, in Sect. 4 we give our TSSF method, Sect. 5 contains the experiments.

© Springer International Publishing AG 2017
A. Petrosino et al. (Eds.): WILF 2016, LNAI 10147, pp. 54–62, 2017.
DOI: 10.1007/978-3-319-52962-2_4

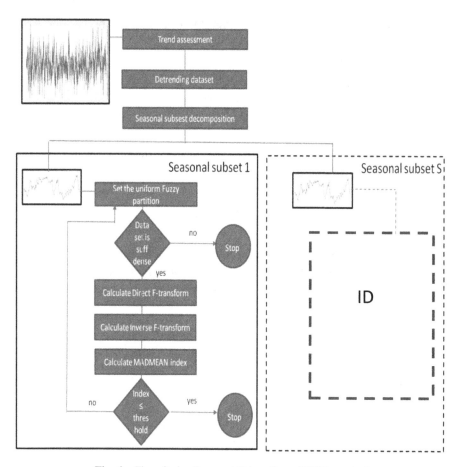

Fig. 1. Time Series Seasonal F-transform (TSSF) method

2 Direct and Inverse F-Transform

We recall definitions and results from [10]. Let $n \geq 2$ and x_1, x_2, ..., x_n be points of [a,b], called nodes, such that $x_1 = a < x_2 < ... < x_n = b$. Given fuzzy sets $A_1,...,A_n$: [a, b] $\rightarrow$ [0,1], called basic functions, we say that they constitute a fuzzy partition of [a,b] if $A_i(x_i) = 1$ for every $i = 1,2,...,n$; $A_i(x) = 0$ if $x \in]x_{i-1},x_{i+1}[$ for $i = 2,...,n-1$; $A_i(x)$ is a continuous function on [a,b]; $A_i(x)$ strictly increases on $[x_{i-1}, x_i]$ for $i = 2,...,$ n and strictly decreases on $[x_i,x_{i+1}]$ for $i = 1,...,n-1$; $A_1(x) + ... + A_n(x) = 1$ for every x $\in$ [a,b]. Moreover the partition is called uniform if $n \geq 3$ and $x_i = a + h \cdot (i-1)$, where $h = (b-a)/(n-1)$ and $i = 1, 2, ..., n$ (i.e. the nodes are equidistant); $A_i(x_i - x) = A_i(x_i + x)$ for every $x \in [0,h]$ and $i = 2,..., n-1$; $A_{i+1}(x) = A_i(x-h)$ for every $x \in [x_i, x_{i+1}]$ and $i = 1,2,...,n-1$. By treating only the discrete case, let's suppose that a function f assumes determined values in the set P of points $p_1,...,p_m$ of [a,b]. If P is sufficiently dense with respect to the fixed partition $\{A_1, A_2, ..., A_n\}$, that is for each $i \in \{1,...,n\}$ there exists an index $j \in \{1,...,m\}$ such that $A_i(p_j) > 0$, we can define the n-tuple

$\{F_1, F_2, ..., F_n\}$ as the (discrete) direct F-transform of f with respect to the basic functions $\{A_1, A_2, ..., A_n\}$, where F_i is given by

$$F_i = \frac{\sum_{j=1}^{m} f(p_j) A_i(p_j)}{\sum_{j=1}^{m} A_i(p_j)}$$

for i = 1,...,n. Then we define the discrete inverse F-transform of f with respect to $\{A_1, A_2, ..., A_n\}$ by setting for $j \in \{1,...,m\}$:

$$f_n^F(p_j) = \sum_{i=1}^{n} F_i A_i(p_j)$$

Extending the above definitions to functions in k (≥ 2) variables, let's suppose that $f(x_1, x_2, ..., x_k)$ assumes determined values in m points $p_j = (p_{j1}, p_{j2}, ..., p_{jk}) \in [a_1, b_1] \times [a_2, b_2] \times ... \times [a_k, b_k]$ for j = 1,...,m. We say that $P = \{(p_{11}, p_{12}, ..., p_{1k}), ..., (p_{m1}, p_{m2}, ..., p_{mk})\}$ is sufficiently dense with respect to the fuzzy partitions $\{A_{11}, A_{12}, ..., A_{1n_1}\}, ..., \{A_{k1}, A_{k2}, ..., A_{kn_k}\}$ of $[a_1, b_1], ..., [a_k, b_k]$, respectively, if for each k-tuple $\{h_1, ..., h_k\} \in \{1, ..., n_1\} \times ... \times \{1, ..., n_k\}$, there exists a point $p_j = (p_{j1}, p_{j2}, ..., p_{jk})$ in P, j = 1,...,m, such that $A_{1h_1}(p_{j1}) \cdot A_{2h_2}(p_{j2}) \cdot ... \cdot A_{kh_k}(p_{jk}) > 0$.

So we can define the $(h_1, h_2, ..., h_k)$-th component $F_{h_1 h_2 ... h_k}$ of the (discrete) direct F-transform of a

$$F_{h_1 h_2 ... h_k} = \frac{\sum_{j=1}^{m} f(p_{j1}, p_{j2}, ... p_{jk}) \cdot A_{1h_1}(p_{j1}) \cdot A_{2h_2}(p_{j2}) \cdot ... \cdot A_{kh_k}(p_{jk})}{\sum_{j=1}^{m} A_{1h_1}(p_{j1}) \cdot A_{2h_2}(p_{j2}) \cdot ... \cdot A_{kh_k}(p_{jk})}$$

Then we define the (discrete) inverse F-transform of f by setting for each point $p_j = (p_{j1}, p_{j2}, ..., p_{jk}) \in [a_1, b_1] \times ... \times [a_k, b_k]$, j = 1,...,m:

$$f_{n_1...n_k}^F(p_{j1}, ..., p_{jk}) = \sum_{h_1=1}^{n_1} \sum_{h_2=1}^{n_2} ... \sum_{h_k=1}^{n_k} F_{h_1 h_2 ... h_k} \cdot A_{1h1}(p_{j1}) ... A_{kh_k}(p_{jk}).$$

Theorem. Let $f(x_1, x_2, ..., x_k)$ be given on the set of points $P = \{(p_{11}, p_{12}, ..., p_{1k}), (p_{21}, p_{22}, ..., p_{2k}), ..., (p_{m1}, p_{m2}, ..., p_{mk})\} \in [a_1, b_1] \times [a_2, b_2] \times ... \times [a_k, b_k]$. Then for every $\varepsilon > 0$, there exist k integers $n_1 = n_1(\varepsilon), ..., n_k = n_k(\varepsilon)$ and k related fuzzy partitions (3) of $[a_1, b_1], ..., [a_k, b_k]$, respectively, such that the set P is sufficiently dense with respect to them and for every $p_j = (p_{j1}, p_{j2}, ..., p_{jk})$ in P, j = 1,...,m, the following inequality holds:

$$\left| f(p_{j1}, ..., p_{jk}) - f_{n_1...n_k}^F(p_{j1}, ..., p_{jk}) \right| < \varepsilon.$$

3   F-Transform Forecasting Method

For making this paper self-contained, we recall the F-tr based forecasting algorithm [3]. Let M be input-output pairs data $(x^{(j)}, y^{(j)})$, $x^{(j)}$ in R^n, $y^{(j)}$ in R (reals) for j = 1,2, ...,M. We must find a fuzzy rule-set, i.e. we must determine a mapping f from R^n to R. We assume that the *ith* component of any $x^{(j)}$, i.e. $x_i^{(1)}, \ldots, x_i^{(j)}, \ldots, x_i^{(M)}$ lie in $\left[x_i^-, x_i^+\right]$ for every i = 1,...,n, and $y^{(1)}, y^{(2)}, \ldots, y^{(m)}$ lie in $[y^-, y^+]$. We create a partition of n(i) fuzzy sets for each domain $\left[x_i^-, x_i^*\right]$ making the following steps:

(1)   Assign an uniform partition of n(i) fuzzy sets (n(i) $\geq$ 3) $\left\{A_{i1}, \ldots, A_{in(i)}\right\}$ of the domain $\left[x_i^-, x_i^*\right]$ of each variable x_i, i = 1,...,n. If $x_{i1}, \ldots, x_{is}, \ldots, x_{in(i)}$ are the nodes of $\left[x_i^-, x_i^*\right]$, each $A_{is(i)}$ is defined for s(i) = 1,...,n(i) as

$$A_{i1}(x) = \begin{cases} 0.5 \cdot \left(1 + \cos\frac{\pi \cdot (n(i)-1)}{(x_i^+ - x_i^-)}(x - x_{i1})\right) & \text{if } \quad x \in [x_{i1}, x_{i2}] \\ 0 & \text{otherwise} \end{cases}$$

$$A_{is(i)}(x) = \begin{cases} 0.5 \cdot \left(1 + \cos\frac{\pi \cdot (n(i)-1)}{(x_i^+ - x_i^-)}(x - x_{is(i)})\right) & \text{if } \quad x \in [x_{i(s-1)}, x_{i(s+1)}] \\ 0 & \text{otherwise} \end{cases}$$

$$A_{in(i)}(x) = \begin{cases} 0.5 \cdot \left(1 + \cos\frac{\pi \ (n(i)-1)}{(x_i^+ - x_i^-)}(x - x_{i(n(i)-1)})\right) & \text{if } \quad x \in [x_{i(n(i)-1)}, x_{in(i)}] \\ 0 & \text{otherwise} \end{cases}$$

where $x_{i1} = x_i^-$, $x_{in(i)} = x_i^+$;

(2)   If the dataset is not sufficiently dense with respect to the fuzzy partition, i.e. if there exists a variable x_i and a fuzzy set $A_{is(i)}$ of the corresponding fuzzy partition such that $A_{is(i)}(x_i^{(j)}) = 0$ for each j = 1,...,M, the process stops otherwise calculate the $n_1 \cdot n_2 \cdot \ldots \cdot n_k$ components of the direct F-tr of f via k = n, $p_{j1} = x_1^{(j)}, \ldots,$ $p_{jn} = x_n^{(j)}, y(j) = f(x_1^{(j)}, x_2^{(j)}, \ldots, x_n^{(j)})$:

$$F_{h_1 h_2 \ldots h_n} = \frac{\sum_{j=1}^{m} y^{(j)} \cdot A_{1h_1}(x_1^{(j)}) \cdot \ldots \cdot A_{nh_n}(x_n^{(j)})}{\sum_{j=1}^{m} A_{1h_1}(x_1^{(j)}) \cdot \ldots \cdot A_{nh_n}(x_n^{(j)})}$$

(3)   Calculate the inverse F-tr for approximating the function f:

$$f_{n_1 \ldots n_n}^F \left(x_1^{(j)}, \ldots, x_n^{(j)}\right) = \sum_{h_1=1}^{n_1} \sum_{h_2=1}^{n_2} \ldots \sum_{h_n=1}^{n_n} F_{h_1 h_2 \ldots h_n} \cdot A_{1h_1}(x_1^{(j)}) \cdot \ldots \cdot A_{nh_n}(x_n^{(j)})$$

(4)   calculate the forecasting index as

$$MADMEAN = \frac{\sum\limits_{j=1}^{M} \left| f^F_{n_1 n_2 \dots n_n}(x_1^{(j)}, \dots, x_n^{(j)}) - y^{(j)} \right|}{\sum\limits_{j=1}^{M} y^{(j)}}$$

If MADMEAN is less to an assigned threshold, then the process stops otherwise a finer fuzzy partition is taken restarting from the step 2.

4 TSSF Method

We consider time series of data formed from observations of a original parameter y_0 measured at different times. The dataset is formed by m pairs as $\left(t^{(0)}, y_0^{(0)}\right), \left(t^{(1)}, y_0^{(1)}\right), \dots, \left(t^{(M)}, y_0^{(M)}\right)$. Our aim is to evaluate seasonal fluctuations of a time series by using the F-transform method. First of all, we use a polynomial fitting for calculating the trend of the phenomenon with respect to the time. Afterwards we subtract the trend from the data obtaining a new dataset of the fluctuations $y^{(j)}$ being $y^{(j)} = y_0^{(j)} - trend(t^{(j)})$. After de-trending the dataset, it is partitioned in S subsets, being S seen as seasonal period. The seasonal data subset is composed by M_s pairs, expressing the fluctuation measures of the parameter y_0 at different times: $(t^{(1)}, y^{(1)})$, $(t^{(2)}, y^{(2)}) \dots (t^{(M)}_s, y^{(M)}_s)$, where $y^{(i)}$ is given by the original measure $y_0^{(j)}$ at the time $t^{(j)}$ minus the trend calculated at this time. The formulae of the corresponding one-dimensional direct and inverse F-transforms are as

$$F_h = \frac{\sum\limits_{j=1}^{M_S} y^{(j)} \cdot A_h(t^{(j)})}{\sum\limits_{j=1}^{M_S} A_h(t^{(j)})}, \quad f_n^F(t) = \sum\limits_{h=1}^{n} F_h \cdot A_h(t)$$

respectively. We start with three basic functions and control that the subset of data is sufficiently dense with respect to this fuzzy partition. For the sth fluctuation subset, we obtain the inverse F-tr by using the following n(s) basic functions:

$$f_{n(s)}^F(t) = \sum\limits_{h=1}^{n(s)} F_h \cdot A_h(t) \cdot$$

For evaluating the value of the parameter y_0 at the time t in the sth seasonal period, we add to $f_{n(s)}^F(t)$ the trend calculated at the time t obtaining $y_0(t) = f_{n(s)}^F(t) + trend(t)$. For evaluating the accuracy of the results, we can use another MADMEAN index [6] given as

$$MADMEAN = \frac{\sum_{j=1}^{M} \left| y_0(t^{(j)}) - y_0^{(j)} \right|}{\sum_{j=1}^{M} y_0^{(j)}} .$$

5   Experiments on Time Series Data

For reason of brevity, we avoid any comparison of our method with SVM and ANN based methods, however our results are compared with well known methods. The training dataset is composed from climate data of the city of Naples measured every thirty minutes (www.ilmeteo.it/portale/archivio-meteo/Napoli). For sake of brevity, we limit the results for the parameter Relative Humidity (RH). As training dataset, we consider the data recorded from 01/07/2003 to 31/12/2015. The number ID of the day is represented in the x-axis and the daily Relative humidity is represented on the y-axis in Fig. 2. We obtain the best fitting polynomial of ninth degree $y = a_0 + \sum_{i=1}^{9} a_i \cdot x^i$ (red colour in Fig. 2) whose coefficients are $a_9 = 6.55E\text{-}33$, $a_8 = -1.48E\text{-}27$, $a_7 = 1.34E\text{-}22$, $a_6 = -6.26E\text{-}18$, $a_5 = 1.5E\text{-}13$, $a_4 = -1.5E\text{-}09$, $a_3 = 0$, $a_2 = -0,067$, $a_1 = 0$, $a_0 = 3.00E\text{+}07$.

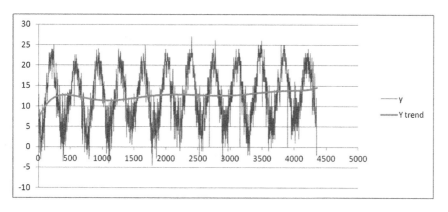

Fig. 2. Trend of RH in July and August (from 01/07/2003 till to 16/08/2015) with a 9° fitting polynomial (Color figure online)

We consider the month as seasonal period. We apply the TSSF method by setting the threshold MADMEAN index equal to 6%. The final results compared with three well known methods have been plotted in Figs. 3a, 3b, 3c and 3d, recalling that the Average seasonal variation method (avgSV) calculates the mean seasonal variation for each month and adds the mean seasonal variation to the trend value; moreover we used ForecastPro [5] for ARIMA and the traditional F-tr based prediction method [3].

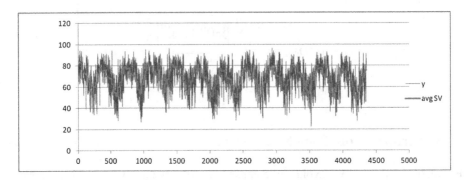

Fig. 3a. Results obtained for RH by using the avgSV method

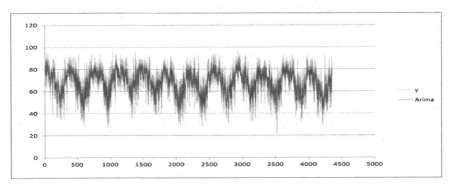

Fig. 3b. Results obtained for RH by using the ARIMA method

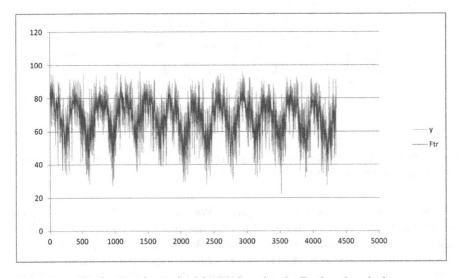

Fig. 3c. Results obtained for RH by using the F-tr based method

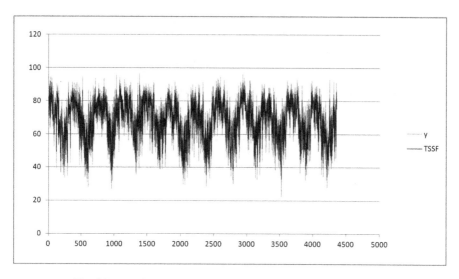

Fig. 3d. Results obtained for the RH by using the TSSF methods

Table 1 shows that the TFSS method improves the performances of ARIMA, avgSV and classical F-tr based forecasting methods.

Table 1. MADMEAN indices for RH in the 4 methods

Forecasting method	MADMEAN
avgSV	5.52%
ARIMA	4.81%
F-tr	5.06%
TSSF	4.23%

References

1. Box, G.E.P., Jenkins, G.M., Reinsel, G.C.: Time Series Analysis: Forecasting and Control. Prentice Hall, Englewood Cliffs (1994)
2. Chatfield, C.: Time Series Forecasting. Chapman & Hall/CRC, Boca Raton (2000)
3. Di Martino, F., Loia, V., Sessa, S.: Fuzzy transforms method in prediction data analysis. Fuzzy Sets Syst. **180**, 146–163 (2011)
4. Faraway, J., Chatfield, C.: Time series forecasting with neural networks: a comparative study using the airline data. Appl. Stat. **47**, 231–250 (1998)
5. Hymdam, R.J., Athanasopoulos, G.: Forecasting Principles and Practice. OText Publisher, Melbourne (2013)
6. Kolassa, W., Schutz, W.: Advantages of the MADMEAN ratio over the MAPE: foresight. Int. J. Appl. Forecasting **6**, 40–43 (2007)
7. Müller, K.-R., Smola, A.J., Rätsch, G., Schölkopf, B., Kohlmorgen, J., Vapnik, V.: Predicting time series with support vector machines. In: Gerstner, W., Germond, A., Hasler, M., Nicoud, J.-D. (eds.) ICANN 1997. LNCS, vol. 1327, pp. 999–1004. Springer, Heidelberg (1997). doi:10.1007/BFb0020283

8. Pai, P.F., Lin, K.P., Lin, C.S., Chang, P.T.: Time series forecasting by a seasonal support vector regression model. Expert Syst. Appl. **37**, 4261–4265 (2010)
9. Pankratz, A.: Forecasting with Dynamic Regression Models. Wiley, Hoboken (2012)
10. Perfilieva, I.: Fuzzy transforms: theory and applications. Fuzzy Sets Syst. **157**, 993–1023 (2006)
11. Štepnicka, M., Cortez, P., Peralta Donate, J., Štepnickova, L.: Forecasting seasonal time series with computational intelligence: on recent methods and the potential of their combinations. Expert Syst. Appl. **40**, 1981–1992 (2013)
12. Zhang, G., Patuwo, B.E., Hu, M.Y.: Forecasting with artificial neural networks: the state of the art. Int. J. Forecast. **14**, 35–62 (1998)
13. Zhang, G., Zhang, G.P.: Time series forecasting using a hybrid ARIMA and neural network model. Neurocomputing **50**, 159–175 (2003)

Improving Approximation Properties of Fuzzy Transform Through Non-uniform Partitions

Vincenzo Loia[1], Stefania Tomasiello[2], and Luigi Troiano[3(✉)]

[1] Dipartimento di Scienze Aziendali - Management & Innovation
Systems (DISA-MIS), Università degli Studi di Salerno,
via Giovanni Paolo II, 132, 84084 Fisciano, Italy
loia@unisa.it

[2] CORISA, Dipartimento di Ingegneria dell'Informazione ed Elettrica e Matematica
Applicata (DIEM), Università degli Studi di Salerno,
via Giovanni Paolo II, 132, 84084 Fisciano, Italy
stomasiello@unisa.it

[3] Dipartimento di Ingegneria (DING), Università degli Studi del Sannio,
Piazza Roma 21, 82100 Benevento, Italy
troiano@unisannio.it

Abstract. Function reconstruction is one of the valuable properties of the Fuzzy Transform and its inverse. The quality of reconstruction depends on how dense the fuzzy partition of the function domain is. However, the partition should be denser where the function exhibits faster variations, while the partition can be less dense where the function is moving slowly. In this paper, we investigate the possibility of having non-uniform fuzzy partitions, in order to better accommodate a different behavior of function across its domain.

1 Introduction

Fuzzy approximation deals with the approximation of a function to any degree of accuracy with a finite number of fuzzy rules. Fuzzy rules substantially define patches, which in an additive form cover the graph of the function, by averaging patches that overlap. There are many examples where such rules are fixed through neural networks or genetic algorithms (e.g. [1,2]).

Recently, a new fuzzy approximation technique was proposed by Perfilieva [3], namely the fuzzy transform (F–transform), which can be intended as an additive normal form [4]. Like other known transforms (e.g. Laplace, Fourier), it has two phases, that is direct and inverse, but unlike the other transforms, it uses a fuzzy partition of a universe. It uses a linear combination of basic functions in order to compute the approximate solution by means of its inverse.

Since F–transform was introduced, several papers devoted to it appeared. In particular, in [5] new types of F–transforms were presented, based on B–splines, Shepard kernels, Bernstein basis polynomials and Favard–Szasz–Mirakjan type operators for the univariate case. In [6] the relations between the least–squares (LS) approximation techniques and the F–transform for the univariate case were

© Springer International Publishing AG 2017
A. Petrosino et al. (Eds.): WILF 2016, LNAI 10147, pp. 63–72, 2017.
DOI: 10.1007/978-3-319-52962-2_5

investigated. In [7,8] some properties on the use of F-transform through the LS approximation for the bivariate case were discussed. In [9], it was proved that the accuracy of the inverse F-transform improves by making the partition tighter around a certain point. In [10] the F-transform was investigated from a neural network (NN) perspective, in order to find the best fuzzy partition for improving the accuracy.

In this paper we look at the problem from a more general perspective of optimization. Assuming a certain number and a certain type of basic functions, we minimize the error functional with respect to the position of the nodes of the partitions. In particular, this can be performed by non-linear programming techniques, such as sequential quadratic programming.

The numerical results are compared against the ones in literature and the ones obtained by means of the LS approach.

The paper is structured as follows: Sect. 2 provides theoretical foundations; in Sect. 3 the proposed methodology is presented; Sect. 4 is devoted to numerical experiments and finally Sect. 5 gives some conclusions.

2　Preliminaries

We briefly recall some definitions. Let $I = [a, b]$ be a closed interval and $x_1, x_2, \ldots, x_n$, with $n \geq 3$, be points of I, called nodes, such that $a = x_1 < x_2 < \ldots < x_n = b$.

A fuzzy partition of I is defined as a sequence $A_1, A_2, \ldots, A_n$ of fuzzy sets $A_i : I \to [0, 1]$, with $i = 1, \ldots, n$ such that

- $A_i(x) \neq 0$ if $x\epsilon(x_{i-1}, x_{i+1})$ and $A_i(x_i) = 1$;
- A_i is continuous and has its unique maximum at x_i;
- $\sum_{i=1}^{n} A_i(x) = 1, \quad \forall x\epsilon I$.

The fuzzy sets $A_1, A_2, \ldots, A_n$ are called basic functions and they form an uniform fuzzy partition if the nodes are equidistant.

In general, $h = \max_i |x_{i+1} - x_i|$ is the norm of the partition. For a uniform partition $h = (b - a)/(n - 1)$ and $x_j = a + (j - 1)h$, with $j = 1, \ldots, n$.

The fuzzy partition can be obtained by means of several basic functions. The most used are the hat functions

$$A_j(x) = \begin{cases} (x_{j+1} - x)/(x_{j+1} - x_j), & x\epsilon[x_j, x_{j+1}] \\ (x - x_{j-1})/(x_j - x_{j-1}), & x\epsilon[x_{j-1}, x_j] \\ 0, & otherwise \end{cases} \tag{1}$$

and the sinusoidal shaped basic functions

$$A_j(x) = \begin{cases} \frac{1}{2}\left(\cos(\pi\frac{x-x_j}{x_{j+1}-x_j}) + 1\right), & x\epsilon[x_j, x_{j+1}] \\ \frac{1}{2}\left(\cos(\pi\frac{x-x_j}{x_j-x_{j-1}}) + 1\right), & x\epsilon[x_{j-1}, x_j] \\ 0, & otherwise \end{cases} \tag{2}$$

Other basic functions are Bernstein basis polynomils and B–splines [5] and they are referred to fuzzy partitions with small support.

A fuzzy partition with small support has the additional property that there exists an integer $r \geq 1$ such that $suppA_i = \{x \in I : A_i(x) > 0\} \subseteq [x_i, x_{i+r}]$.

The fuzzy transform (F–transform) of a function $f(x)$ continuous on I with respect to $\{A_1, A_2, \ldots, A_n\}$ is the n–tuple $[F_1, F_2, \ldots, F_n]$ whose components are such that the following functional is minimum

$$\Phi = \int_a^b (f(x) - F_i)^2 A_i(x)dx, \tag{3}$$

that is

$$F_i = \frac{\int_a^b f(x)A_i(x)dx}{\int_a^b A_i(x)dx}. \tag{4}$$

The function

$$f_{F,n} = \sum_i^n F_i A_i(x), \qquad x \epsilon I \tag{5}$$

is called inverse F–transform of f with respect to $\{A_1, A_2, \ldots, A_n\}$ and it approximates a given continuous function f on I with arbitrary precision, as stated by Theorem 2 in [3].

In many real cases, where the function f is known only at a given set of points $\{p_1, p_2, \ldots, p_m\}$, the discrete F–transform can be used and Eq. 4 is replaced by

$$F_i = \frac{\sum_{j=1}^m f(p_j)A_i(p_j)}{\sum_{j=1}^m A_i(p_j)}, \qquad i = 1, \ldots, n \tag{6}$$

Similarly, Eq. 5 is replaced by

$$f_{F,n}(p_j) = \sum_i^n F_i A_i(p_j), \qquad j = 1, \ldots, m \tag{7}$$

giving the discrete inverse F–transform.

Let $\mathbf{v}$ and $\overline{\mathbf{v}}$ denote the m-sized vectors of known data and the approximate reconstruction through F-transform, then Eqs. 6 and 7 can be written in compact form as follows:

$$\mathbf{F} = \mathbf{SAv} \tag{8}$$

$$\overline{\mathbf{v}} = \mathbf{A}^T\mathbf{F} \tag{9}$$

where $\mathbf{A}$ is the $n \times m$ matrix of which ijth entry is $A_i(x_j)$, $\mathbf{S}$ is the diagonal matrix with entries $S_{ii} = 1/\sum_{j=1}^m A_i(p_j)$, $\mathbf{F}$ is the n–sized vector of F-transform components.

3   Methodology

In this section we outline the approach we aim to follow, first by stating the partition optimization problem, and then by refining the solution by means of the LS method.

3.1 Optimal Partition

Once, the cardinality n and the shape of the basic functions are fixed, the quality of reconstruction depends on where the partition nodes are placed. Indeed, ultimately the inverse F–transform $f_{F,n}(p_j)$ becomes parametric with respect to the partition nodes, as they represents the sole degree of freedom still available to control. In order to emphasize this, we will use the notation $f_{(\mathbf{x})}$ to indicate the inverse F-transform.

If we adopt MSE, as the error functional used to summarize the discrepancies between f and $f_{(\mathbf{x})}$ over the whole interval I, the optimization problem we aim to solve is

$$
\begin{aligned}
\underset{\mathbf{x}}{\text{minimize}} \quad & \sum_{j=1}^{m} \left(f(p_j) - f_{(\mathbf{x})}(p_j) \right)^2 \\
\text{subject to} \quad & x_i < x_{i+1}, \; i = 1..n-1 \\
& x_i \in [0,1], \; i = 1..n \\
& x_1 = a \\
& x_n = b
\end{aligned}
\tag{10}
$$

The solution to this problem can be found by means of well-known quadratic programming (QP) algorithms. Among them, *interior point methods* and sequential quadratic programming.

As result we obtain the location of nodes $x_1, \ldots, x_n$ within the interval I, that fully describes the partition made of basis functions $A_1, \ldots, A_n$. Once applied to Eq. (6), we get the vector $\mathbf{F} \equiv \langle F_1, \ldots, F_n \rangle$.

3.2 The LS Approach

The solution above can be further refined if we attempt to tune the $\mathbf{F}$ by means of the Least–Squares approximation.

Indeed, as argued in [6], the discrete F–transform of f with respect to $\{A_1, \ldots, A_n\}$ are unknowns λ_i to be obtained by means of the error vector $\mathbf{E}$

$$
\mathbf{E} = \mathbf{v} - \mathbf{A}^T \Lambda,
\tag{11}
$$

where Λ is the vector of which elements are λ_i. By minimizing $\mathbf{E}$ with respect to the λ_i, we get

$$
\Lambda = \mathbf{K}^{-1} \mathbf{A} \mathbf{v}
\tag{12}
$$

where

$$
\mathbf{K} = \mathbf{A} \mathbf{A}^T
\tag{13}
$$

The discrete inverse F–transform is given by:

$$
\bar{\mathbf{v}} = \mathbf{A}^T \Lambda
\tag{14}
$$

It should be pointed out that, since $\mathbf{K}$ is the Gram matrix associated to given sets of basis functions, it has full rank and $\mathbf{K}$ turns out to be positive definite.

4   Numerical Experiments

The aim of this section is to offer a quantitative and qualitative overview of how effective are optimal non-uniform partitions in approximating a function. In order to make our analysis comparable with other approaches, we considered two noticeable examples given by related literature.

We will use the following error measures in order to quantify the overall accuracy

– root mean square error (RMSE) $RMSE = \sqrt{\frac{\sum_{i=1}^{n}(\overline{v}_i - v_i)^2}{n}}$
– the maximum absolute error (MAE) $MAE = \max_i |\overline{v}_i - v_i|$

Instead, to outline accuracy in details, we will make use of two graphics: (i) Point-wise error, that measures the difference between the function and its approximation point-by-point; (ii) Cumulative Absolute Error, that shows how the error profile is going to be shaped by the increments given by the point-wise absolute error and sorted in decreasing order. In addition, we will provide the approximation and the partition plot as qualitative means to better understand benefits and limits of using non-uniform partitions.

4.1   First Example

The first example is taken from [10]. The function to be approximated is

$$f_1(x) = 2 \exp{-4(x - 0.5)} - 1, \qquad x \in [0, 1] \tag{15}$$

In [10], $m = 100$ was fixed, which for $n = 10$ and sinusoidal shaped basic functions provide an error (that is *simple normed least square criterion*) 0.462 for the original formula and 0.457 in the best case scenario through the NN approach.

For the same values of m and n, by means of the LS approach we find an error 0.173859. Instead, the solution obtained by means of optimal non-uniform partition return an error of 0.1474.

The error measured for different values of m and n is given in Table 1.

Table 1. Example 1: errors for different values of n and m

Approach	n	m	RMSE	MAE
LS	11	101	0.144454	0.441245
LS	11	1001	0.140057	0.509895
LS	21	1001	0.0744185	0.268912
LS	51	1001	0.0308599	0.122297

The benefits and issues of adopting a non-uniform optimal partition are depicted in Fig. 1, where a non-uniform partition is compared to a uniform partition when $n = 11$ and $m = 1$. Hat functions are used for the partition. We can

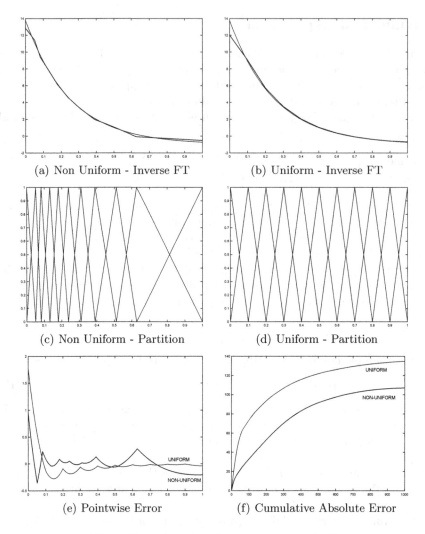

(a) Non Uniform - Inverse FT

(b) Uniform - Inverse FT

(c) Non Uniform - Partition

(d) Uniform - Partition

(e) Pointwise Error

(f) Cumulative Absolute Error

Fig. 1. Reconstruction (hat, $n = 11$, $m = 1001$)

notice how the partition is denser on the left side of charts, where the function f_1 moves faster. The cost to be paid is that the coverage becomes loose on the right side. This does not occur in the case of uniform partition, so that in this second case we have smaller errors in that side of the interval, as shown by the point-wise error. However, the number of benefits is larger than disadvantages.

What happens when we increase the cardinality of the partition is outlined in Fig. 2. The behavior is similar, although increasing the number of basis functions helps to improve the function reconstruction. We can notice, that when the number of basis functions is increased, advantages given by non-uniform partitions are reduced, as highlighted by the cumulative absolute error (shown on the last row).

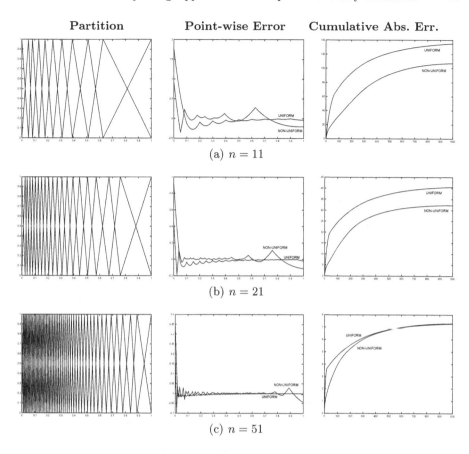

Fig. 2. Reconstruction at different partition cardinalities

Another advantage given by non-uniform partition is summarized by Fig. 3, where the profile given by the cumulative absolute error at different resolutions is compared to uniform partitions. The result is that a non-uniform partition makes a better use of points, so that increasing the number of points leads to have a smaller error. This is not the case of uniform partitions.

What happens when we use a different shape of basis functions is described by Fig. 4, where reconstruction is given by sinusoidal basis functions. Also in this case an optimal non-uniform partition provides a better result if compared to the uniform partition. In addition, the solution obtained so far, can be further optimized by means of LS. This leads to decrease the error. This is also confirmed when we compare the cumulative error to the one obtained with the same cardinality of hat functions. As depicted by Fig. 4-d, although the partition made of hat functions provides a smaller error, the application of LS might improve the situation for the sinusoidal partition.

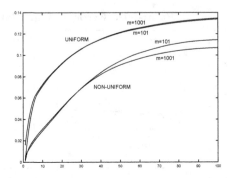

Fig. 3. Cumulative average absolute error (hat)

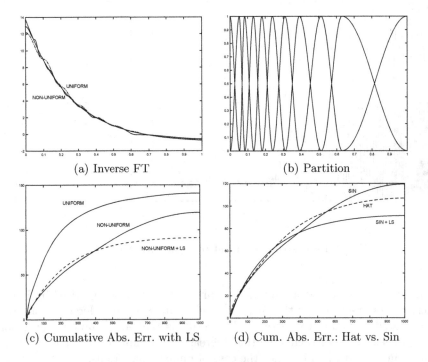

(a) Inverse FT   (b) Partition

(c) Cumulative Abs. Err. with LS   (d) Cum. Abs. Err.: Hat vs. Sin

Fig. 4. Reconstruction (sinusoidal, $n = 11$, $m = 1001$)

4.2 Second Example

This example was considered in [5]:

$$f_2(x) = 2 + \sin 1/(x + 0.15), \qquad x \in [0, 1] \tag{16}$$

In [5], the B-spline based F-transforms showed better approximation, even though the error was not quantified. Errors for different reconstruction scenarios are given in Table 2. Different reconstructions are summarized in Fig. 5. In this

Table 2. Example 2: errors for different values of n and m

Approach	n	m	RMSE	MAE
LS	11	101	0.111088	0.638991
LS	11	1001	0.0994629	0.728732
LS	21	1001	0.0278385	0.262915
LS	51	1001	0.0110888	0.0743088

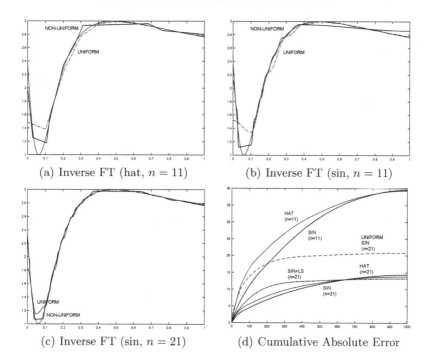

(a) Inverse FT (hat, $n = 11$)   (b) Inverse FT (sin, $n = 11$)

(c) Inverse FT (sin, $n = 21$)   (d) Cumulative Absolute Error

Fig. 5. Reconstruction of function f_2 ($m = 1001$)

case having a larger partition allows to better follow the shape of function f_2. According to the cumulative absolute error, non-uniform partitions performs better than the uniform partition. Sinusoidal functions are able to better follow the function on larger errors, but they accumulate more smaller errors, basically because of oscillations, so that on the overall interval they do not seem to go better than linear hat functions. LS optimization does not seem to be beneficial in this case.

5 Conclusions

In this paper, we investigated the effect of using non-uniform partitions on reconstructing the function as Inverse–FT. The problem of finding an optimal non-uniform partition is treated in terms of quadratic programming, using the nodes position within an interval $[a, b]$ as variables. As preliminary experimentation setting, we used two functions that have been used in literature. Results outline the following conclusions:

- non-uniform partitions seem to provide a better fitting to the function to be reconstructed;
- they become denser where the function is varying faster; this offers the possibility of better following the function, but at cost of becoming slower where the function moves slowly;
- a further optimization, such as that based on LS, seems to offer a further opportunity to better approximate the original function.

References

1. Dickerson, J., Kosko, B.: Fuzzy function approximation with ellipsoidal rules. IEEE Trans. Syst. Man Cybern. B: Cybern. **26**(4), 542–560 (1996)
2. Dankova, M., Stepnicka, M.: Genetic algorithms in fuzzy approximation. In: EUSFLAT - LFA 2005, pp. 651–656 (2005)
3. Perfilieva, I.: Fuzzy transforms: theory and applications. Fuzzy Sets Syst. **157**, 993–1023 (2006)
4. Dankova, M., Stepnicka, M.: Fuzzy transform as an additive normal form. Fuzzy Sets Syst. **157**, 1024–1035 (2006)
5. Bede, B., Rudas, I.J.: Approximation properties of fuzzy transforms. Fuzzy Sets Syst. **180**, 20–40 (2011)
6. Patane', G.: Fuzzy transform and least-squares approximation: analogies, differences, and generalizations. Fuzzy Sets Syst. **180**(1), 41–54 (2011)
7. Gaeta, M., Loia, V., Tomasiello, S.: Cubic B-spline fuzzy transforms for an efficient and secure compression in wireless sensor networks. Inf. Sci. **339**, 19–30 (2016)
8. Loia, V., Tomasiello, S., Vaccaro, A.: A fuzzy transform based compression of electric signal waveforms for smart grids. IEEE Trans. Syst. Man Cybern.: Syst. (2016). doi:10.1109/TSMC.2016.2578641
9. Troiano, L., Kriplani, P.: Supporting trading strategies by inverse fuzzy transform. Fuzzy Sets Syst. **180**, 121–145 (2011)
10. Stepnicka, M., Polakovic, O.: A neural network approach to the fuzzy transform. Fuzzy Sets Syst. **160**, 1037–1047 (2009)

Defining the Fuzzy Transform on Radial Basis

Elena Mejuto[1(✉)], Salvatore Rampone[2], and Alfredo Vaccaro[1]

[1] Dipartimento di Ingegneria, Università degli Studi del Sannio, Benevento, Italy
{mejuto,vaccaro}@unisannio.it
[2] Dipartimento di Scienze E Tecnologie, Università degli Studi del Sannio,
Benevento, Italy
ramnpone@unisannio.it

Abstract. Fuzzy transform (F-transform) is a functional operator that proved to be effective in multiple processing tasks concerning time series and images, including summarization, compression, filtering and information fusion. Its operational definition is originally based on Ruspini partitions that, if on one side provides an easier interpretation of results, on the other it might limit the application of F-transform for higher dimensional data spaces. In this paper we propose a radial definition of F-transform that is based on the distance between points in a p-normed space.

1 Introduction

Since its inception by Perfilieva [4], the *fuzzy transform* (F-transform, FT) has been widely investigated and applied to different fields [1–3,5–7,9–12]. The idea is to map a function $f : D \rightarrow \mathbb{R}$ to a finite vector space by partitioning the domain D and averaging the function value over the domain fuzzy subsets. The result is a vector of reals that offers a compact representation of the initial function. From this vector it is possible to obtain an approximation of the original function by reconstructing the value assumed at each point over the domain D. Approximation is due to the loss of information produced by mapping the function to a lower dimensional space.

The partitioning of D is generally performed by computing the Cartesian product of fuzzy partitions defined over each dimension of D. Although in principle this approach might be followed for domains of any dimension, on practice it is feasible on low-dimensional spaces such as time-series and images.

In this paper we propose to compute the fuzzy partition on a radial basis according to the distance between the points and the partition nodes. The remainder of this paper is organized as follows: Sect. 2 provides preliminary definitions; Sect. 3 introduces the definition of *radial fuzzy transform*; Sect. 4 characterizes the radial F-transform in terms of approximation; Sect. 5 outlines conclusions and future work.

2 Preliminaries

Given the set of *basic functions* $A_1, \ldots, A_n$, we can define *F-transform* as follows.

© Springer International Publishing AG 2017
A. Petrosino et al. (Eds.): WILF 2016, LNAI 10147, pp. 73–81, 2017.
DOI: 10.1007/978-3-319-52962-2_6

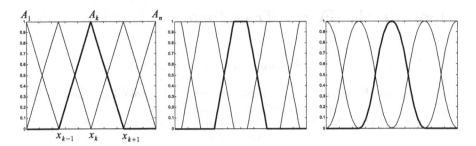

Fig. 1. Examples of uniform partitions (from left: triangular, trapezoidal and Z-shape difference).

Definition 1 (F-transform). *Let $f : D \subseteq \mathbb{R}^q \to \mathbb{R}$ be continuous, that is $f \in C(D)$. Given a partition $A_1, \ldots, A_n$ on D, we define the (integral) F-transform of f the set of values $[F_1, \ldots, F_n]$ given by*

$$F_i = \frac{\int_D f(\boldsymbol{x}) A_i(\boldsymbol{x}) d\boldsymbol{x}}{\int_D A_i(\boldsymbol{x}) d\boldsymbol{x}} \qquad i = 1..n \tag{1}$$

Basic functions are a fuzzy partition of D. Let $\boldsymbol{x}_1, \ldots, \boldsymbol{x}_n in \mathbb{R}^q$ be a collection of nodes, and $A_1, \ldots, A_n$ a collection fuzzy sets centered on them, i.e., $A_i(\boldsymbol{x}_i) = 1$ and $A_i(\boldsymbol{x}_j) = 0$ if $i \neq j$. Following the definition given by Ruspini [8], $A_1, \ldots, A_n$ is a fuzzy partition of D if

1. $D \subseteq \bigcup\limits_{i=1}^{n} supp(A_i)$
2. $\forall \boldsymbol{x} \in D, \sum\limits_{i=1}^{n} A_i(\boldsymbol{x}) = 1$

If nodes are uniformly spaced, the partition is said to be *uniform*. Some examples of (uniform) Ruspini partitions over $\mathbb{R}$ are given in Fig. 1.

When processing time series, images and other sources of data, it is useful to introduce the concept of *discrete F-transform*.

Definition 2 (Discrete F-transform). *Let f be a function defined at points $\boldsymbol{\xi}_1, \boldsymbol{\xi}_2, \ldots, \boldsymbol{\xi}_s \in D$. Given the basic functions $A_1, \ldots, A_n$ covering D, with $n < s$, we define the discrete F-transform of f, the set of values $[F_1, \ldots, F_n]$ computed as*

$$F_i = \frac{\sum\limits_{j=1}^{s} f(\boldsymbol{\xi}_j) A_i(\boldsymbol{\xi}_j)}{\sum\limits_{j=1}^{s} A_i(\boldsymbol{\xi}_j)} \qquad i = 1..n \tag{2}$$

We assume that points $\boldsymbol{\xi}_1, \ldots, \boldsymbol{\xi}_s$ are enough dense with respect to the given partition. We denote the F-transform of a function f with respect to $A_1, \ldots, A_n$ by

$$\mathbf{F}_n[f] = [F_1, \ldots, F_n] \tag{3}$$

As argued in [4], the components of $\boldsymbol{F}_n[f]$ are the weighted mean values of the function f, where weights are given by the basic functions.

The function f can be reconstructed by its *inverse F-transform* (IFT) defined as

Definition 3 (Inverse F-transform). *Let $\boldsymbol{F}_n[f] = [F_1, \ldots, F_n]$ be the integral (or discrete) F-transform of function $f \in C(D)$ with respect to basic functions $A_1, \ldots, A_n$, forming a fuzzy partition of D. Then the function*

$$f_{F,n}(\boldsymbol{x}) = \sum_{i=1}^{n} F_i A_i(\boldsymbol{x}) \tag{4}$$

is called the inverse F-transform *of f.*

In the case of univariate functions, $D \subseteq \mathbb{R}$ and a Ruspini partition can be obtained by assuming on the left and right side of A_i functions that are symmetric with respect to the middle points between nodes, as those presented in Fig. 1. For multivariate functions, $D \subseteq \mathbb{R}^q$, with $q > 1$. In this case, a common procedure consists in:

1. building a grid of nodes, so that $\boldsymbol{X} = \{\boldsymbol{x}_1, \ldots, \boldsymbol{x}_n\}$ is obtained as the Cartesian product $X_{[1]} \times \cdots \times X_{[q]}$, where $X_{[h]}$ is a set of nodes defined over the h-th dimension
2. defining a Ruspini partition over each dimension h
3. assuming as basic functions $A_i(\boldsymbol{x}) = \prod_{h=1}^{q} A_{i[h]}(x_{[h]})$

It is easy to prove that for any $\boldsymbol{x} \in D$, we have $\sum_{i=1}^{n} A_i(x) = 1$. An example of basic function obtained by combining two triangular basic functions is given in Fig. 2.

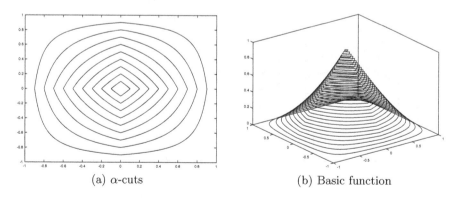

(a) α-cuts   (b) Basic function

Fig. 2. An example of basic function obtained by combining two triangular basic functions

3   Radial Fuzzy Transform

We can re-define the Fuzzy Transform in terms of distance from the nodes. This leads to the definition of *radial F-transform*.

Definition 4 (Radial F-transform). *Given a function* $f : D \subseteq \mathbb{R}^q \to \mathbb{R}$, *continuous, we define* radial F-transform *the vector* $F = [\{F_1, \ldots, F_n\}]$, *with*

$$F_i = \frac{\int_D f(\boldsymbol{x})\Lambda(|\boldsymbol{x} - \boldsymbol{x}_i|)d\boldsymbol{x}}{\int_D \Lambda(|\boldsymbol{x} - \boldsymbol{x}_i|)d\boldsymbol{x}} \qquad i = 1..n \tag{5}$$

where $\Lambda : \mathbb{R} \to [0,1]$ *is a monotonically decreasing function, such that* $\Lambda(0) = 1$, *it is almost everywhere continuous, except is some points where it is left-continuous.*

Examples of Λ are given in Fig. 3. The function Λ describes the "closeness" of $\boldsymbol{x}$ to the some node $\boldsymbol{x}_i$, therefore it takes as argument a measure of distance denoted by $|\boldsymbol{x} - \boldsymbol{x}_i|$. In the remainder of this paper we will refer to distances based on p-norm, i.e., the Manhattan distance and the Euclidean distance, both special cases of the Minkowski distance defined as:

$$||\boldsymbol{x} - \boldsymbol{x}_i||_p = \left(\sum_{h=1}^{q} |x_h - x_{i,h}|^p \right)^{1/p} \tag{6}$$

It is easy to prove that the support of Λ, i.e., the subset of points $d \in \mathbb{R}_0^+$ such that $\Lambda(d) > 0$, is a semi-open interval $[0, d_m)$[1]. In this paper we assume d_m limited although d_m can be $+\infty$.

Fuzzy sets that are forming the partition on which the fuzzy transform is computed are determined by (i) Λ and (ii) by the definition of distance $|\boldsymbol{x} - \boldsymbol{x}_i|$. In the case $q = 1$, i.e., D is one-dimensional, the latter does not play any role, and the shape of fuzzy sets is only determined by λ. When $q > 1$, which norm we use (or more generally which definition of distance we assume) has clearly an

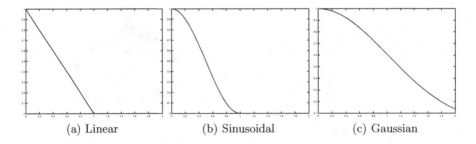

(a) Linear   (b) Sinusoidal   (c) Gaussian

Fig. 3. Examples of function Λ

---

[1] Except when d is a point of discontinuity for Λ. In that case the interval is closed because of right-continuity.

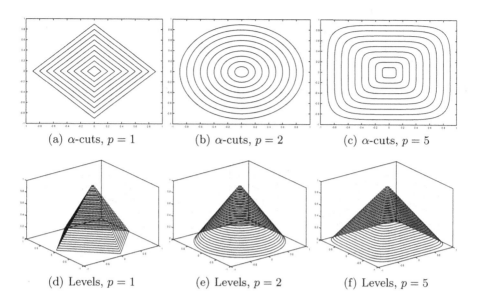

(a) α-cuts, $p = 1$ (b) α-cuts, $p = 2$ (c) α-cuts, $p = 5$

(d) Levels, $p = 1$ (e) Levels, $p = 2$ (f) Levels, $p = 5$

Fig. 4. Fuzzy sets obtained by different p-norms (linear Λ, $q = 2$)

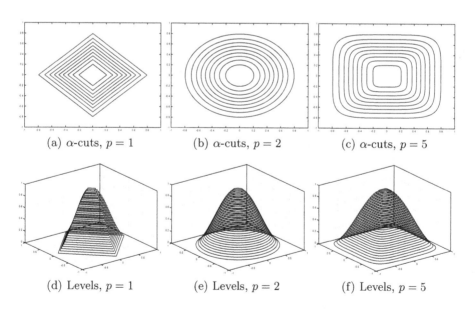

(a) α-cuts, $p = 1$ (b) α-cuts, $p = 2$ (c) α-cuts, $p = 5$

(d) Levels, $p = 1$ (e) Levels, $p = 2$ (f) Levels, $p = 5$

Fig. 5. Fuzzy sets obtained by different p-norms (linear Λ, $q = 2$)

impact on distances as outlined by Fig. 4 in the case of $q = 2$ and linear Λ. If we assume a sinusoidal Λ we obtain the basic functions depicted in Fig. 5.

Basic functions obtained by this method does not meet requirements to form a Ruspini partition, as it is not guaranteed that in each point $x \in D$ they sum up to 1. We need to redefine the inverse F-transform.

Definition 5 (Inverse Radial F-transform). *Let $F_n[f] = [F_1, \ldots, F_n]$ be the integral (or discrete) radial F-transform of function $f \in C(D)$ with respect to function Λ and a given p-norm. Then the function*

$$f_{F,n}(x) = \frac{\sum\limits_{i=1}^{n} F_i \Lambda(|x - x_i|)}{\sum\limits_{i=1}^{n} \Lambda(|x - x_i|)} \qquad x \in D \tag{7}$$

is called the inverse radial F-transform *of f.*

Comparing Eq. (7) to Eq. (4), we can rewrite the first as

$$f_{F,n}(x) = \sum_{i=1}^{n} F_i A_i^{\Lambda}(x) \qquad x \in D \tag{8}$$

where

$$A_i^{\Lambda}(x) = \frac{\Lambda(|x - x_i|)}{\sum\limits_{i=1}^{n} \Lambda(|x - x_i|)} \qquad x \in D \tag{9}$$

This offers the possibility of a direct comparison with the basic function forming a Ruspini partition in $\mathbb{R}^q$, and illustrated in Fig. 2. The basic function is drastically different from that used in the ordinary F-transform, as outlined by Fig. 6.

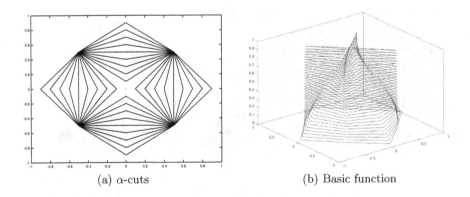

(a) α-cuts   (b) Basic function

Fig. 6. Adjusted basic function with linear Λ and $p = 1$

4   Approximation

As ordinary F-transform, also RFT is characterized by loss of information, that is its inverse can only provide an approximation of the original function. However, the following proposition proves shows that radial inverse F-transform $f_{F,n}$ can approximate the original function with any arbitrary precision.

Proposition 1 (Universal Approximation). *Let f be a continuous function on D. Then for any $\epsilon > 0$ there exist n_ϵ and a collection of nodes $\boldsymbol{x}_1, \ldots, \boldsymbol{x}_{n_\epsilon}$ in D such that for all $\boldsymbol{x} \in D$*

$$|f(\boldsymbol{x}) - f_{F,n_\epsilon}(\boldsymbol{x})| \leq \epsilon \tag{10}$$

for any choice of Λ and p.

Proof. It is obtained straightforward by extending to multivariate functions the proof given in [4]. □

Similarly to the ordinary F-transform, RFT inverse is a weighted mean of the original points, that acts for example as a blurring filter on images.

Proposition 2. *For any $\boldsymbol{x} \in D$, assuming $\boldsymbol{x} \in [x_k, x_{k+1}]$ we have*

$$\min_{\boldsymbol{x} \in \cup R_k \cap D} f(\boldsymbol{x}) \leq f_{F,n}(\boldsymbol{x}) \leq \max_{\boldsymbol{x} \in \cup R_k \cap D} f(x) \tag{11}$$

where R_k is the support of node $\boldsymbol{x}_k$.

Proof. From definition of RFT, we have

$$F_i = \frac{\int_{\cup R_k \cap D} f(\boldsymbol{x}) \Lambda(|\boldsymbol{x} - \boldsymbol{x}_i|) d\boldsymbol{x}}{\int_{\cup R_k \cap D} \Lambda(|\boldsymbol{x} - \boldsymbol{x}_i|) d\boldsymbol{x}} \quad i = 1..n$$

Therefore F_i is given by the weighted mean of $f(\boldsymbol{x})$ over D restricted to $\cup R_k$, as the other nodes do not affect the computation of F_i being out of range. As any weighted mean, because of compensation property. This is become more evident if we consider the discrete RFT, that is

$$F_i = \frac{\sum\limits_{\cup R_k \cap D} f(\boldsymbol{x}) \Lambda(|\boldsymbol{x} - \boldsymbol{x}_i|)}{\sum\limits_{\cup R_k \cap D} \Lambda(|\boldsymbol{x} - \boldsymbol{x}_i|)} \quad i = 1..n$$

We have

$$\min_{\boldsymbol{x} \in \cup R_k \cap D} f(\boldsymbol{x}) \leq F_i \leq \max_{\boldsymbol{x} \in \cup R_k \cap D} f(\boldsymbol{x})$$

When we apply the inverse we get

$$f_{F,n}(\boldsymbol{x}) = \frac{\sum\limits_{i=1}^{n} F_i \Lambda(|\boldsymbol{x} - \boldsymbol{x}_i|)}{\sum\limits_{i=1}^{n} \Lambda(|\boldsymbol{x} - \boldsymbol{x}_i|)} \quad \boldsymbol{x} \in D$$

that is again a weighted average of F_i, therefore

$$\min_{x \in \cup R_k \cap D} f(x) \leq \min_{x \in \cup R_k \cap D} F_i \leq f_{F,n}(x) \leq \max_{x \in \cup R_k \cap D} F_i \leq \max_{x \in \cup R_k \cap D} f(x) \qquad \square$$

Corollary 1. *Approximation of $f_{F,n}(x)$ improves by making denser the partition around x. In particular,*

$$\lim_{n \to \infty} f_{F,n}(x) = f(x) \qquad (12)$$

Proof. Proof is consequence of Proposition 2. If we assume $x \in R$, where $R \equiv \cup R_k \cap D$, we have

$$|f(x) - f_{F,n}(x)| \leq \max_{x \in R} f(x) - \min_{x \in R} f(x)$$

By making the partition denser around x, we have $R' \subseteq R$, therefore

$$|f(x) - f_{F,n}(x)| \leq \max_{x \in R'} f(x) - \min_{x \in R'} f(x) \leq \max_{x \in R} f(x) - \min_{x \in R} f(x)$$

If $n \to \infty$, the partition becomes densest around any $x \in D$ and $R \to \{x\}$, thus entailing the limit expressed by Eq. 12. $\qquad \square$

5   Conclusions

In this paper we introduced the definition of radial F-transform, which assume the distance from node as means to generate the basic functions. In particular we considered the Minkowski distance as generalization of the Euclidean distance. The basic functions obtained by this method do not meet the requirements to form a Ruspini partition. However, they better represent fuzzy sets in higher dimensions, preserving the original shape given on one-dimensional domains. This requires to change how the inverse F-transform is computed. This does not affect the property of providing a universal approximation. Future work is aimed at testing the application of radial F-transform to practical examples in high-dimensional spaces, comparing results with the conventional F-transform. Since RFT is based on distance, the curse of dimensionality might affect the output over a given high number of dimension. This aspect should be taken into consideration when RFT is applied to typical problems of data analysis. Besides time series and image processing.

References

1. Martino, F.D., Loia, V., Perfilieva, I., Sessa, S.: An image coding/decoding method based on direct and inverse fuzzy transforms. Int. J. Approx. Reason. **48**(1), 110–131 (2008)
2. Perfilieva, I.: Approximating models based on fuzzy transforms. In: Proceedings of the 4th EUSFLAT Conference, pp. 645–650 (2005)

3. Perfilieva, I.: Fuzzy transforms and their applications to image compression. In: Bloch, I., Petrosino, A., Tettamanzi, A.G.B. (eds.) WILF 2005. LNCS (LNAI), vol. 3849, pp. 19–31. Springer, Heidelberg (2006). doi:10.1007/11676935_3

4. Perfilieva, I.: Fuzzy transforms: theory and applications. Fuzzy Sets Syst. **157**(8), 993–1023 (2006)

5. Perfilieva, I., Novák, V., Dvořák, A.: Fuzzy transform in the analysis of data. Int. J. Approx. Reason. **48**(1), 36–46 (2008)

6. Perfilieva, I., Novák, V., Pavliska, V., Dvorák, A., Stepnicka, M.: Analysis and prediction of time series using fuzzy transform. In: IJCNN, pp. 3876–3880. IEEE (2008)

7. Perfilieva, I., Valásek, R.: Data compression on the basis of fuzzy transforms. In: Proceedings of the 4th EUSFLAT Conference, pp. 663–668 (2005)

8. Ruspini, E.H.: A new approach to clustering. Inf. Control **15**(1), 22–32 (1969)

9. Stefanini, L.: Fuzzy transform and smooth functions. In: Carvalho, J.P., Dubois, D., Kaymak, U., da Costa Sousa, J.M. (eds.) IFSA/EUSFLAT Conference, pp. 579–584 (2009)

10. Troiano, L.: Fuzzy co-transform and its application to time series. In: 2010 International Conference of Soft Computing and Pattern Recognition, SoCPaR 2010, 7–10 December 2010, pp. 379–384. Cergy-Pontoise (2010). Conference Code: 83752

11. Troiano, L., Kriplani, P.: Supporting trading strategies by inverse fuzzy transform. Fuzzy Sets Syst. **180**(1), 121–145 (2011)

12. Troiano, L., Kriplani, P.: A mean-reverting strategy based on fuzzy transform residuals. In: 2012 IEEE Conference on Computational Intelligence for Financial Engineering and Economics, CIFEr 2012, New York City, NY, 29–30 March 2012, pp. 11–17 (2012). Conference Code: 93992

Granularity and Multi-logics

Cronholm and Multi-logia

Reasoning with Information Granules to Support Situation Classification and Projection in SA

Angelo Gaeta[1]([✉]), Vincenzo Loia[2], and Francesco Orciuoli[2]

[1] Dipartimento di Ingegneria dell'Informazione ed Elettrica e Matematica Applicata,
University of Salerno, 84084 Fisciano, SA, Italy
agaeta@unisa.it
[2] Dipartimento di Scienze Aziendali, Management and Innovation Systems,
University of Salerno, 84084 Fisciano, SA, Italy
{loia,forciuoli}@unisa.it

Abstract. We present our results on the adoption of a set theoretic framework for Granular Computing in the domain of Situation Awareness. Specifically, we present two cases of reasoning with granules and granular structures devoted, respectively, to support human operators in *(i)* classifying situations on the basis of a set of incomplete observations using abductive reasoning, and *(ii)* obtaining early warning information on situation projections reasoning on granular structures.

Keywords: Situation Awareness · Computational intelligence · Granular computing

1 Introduction

Situation Awareness (SA) is defined by Endsley as the perception of the elements in the environment within a volume of time and space, the comprehension of their meaning, and the projection of their status in the near future [5]. A situation is an abstract state of affairs interesting to specific applications, and provide a human understandable representation of data and information to support rapid decision making [2]. This paper leverages and homogenizes authors previous works on SA and GrC [3,4,7], showing how these results can be can be systematized into a common framework aligned to the three perspectives of GrC.

The paper is organized as follows. We report in Sect. 2 an high level view of how GrC, in its different perspectives, can enforce SA. This high level view is then deepened in Sect. 3 with respect to the application of GrC for structured problem solving, and in Sect. 4 with regards to the information processing perspective, presenting also the set theoretic framework we adopt. Two cases of applications are described in Sect. 5 and related subsections. Lastly, conclusion and future works are presented in Sect. 6.

2 GrC for SA: The Three Perspectives

Yao [10] in his triarchic theory of granular computing discusses GrC according to three complementary perspectives: (i) a general method of structured

© Springer International Publishing AG 2017
A. Petrosino et al. (Eds.): WILF 2016, LNAI 10147, pp. 85–94, 2017.
DOI: 10.1007/978-3-319-52962-2_7

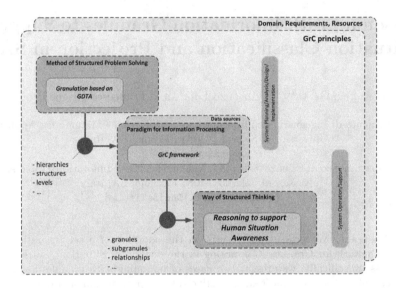

Fig. 1. Three perspectives of GrC and situation awareness (adapted from [7])

problem solving, (ii) a paradigm of information processing, and (iii) a way of structured thinking.

Figure 1 integrates these three perspectives of GrC on the phases of the life cycle of SA. In particular, the first perspective is used when SA domain experts execute the design tasks that provide a first granulation of the environment according to the goals and requirements. The results are exploited in the second phase that employs another perspective of GrC, i.e. paradigm for information processing. In this case, specific techniques and frameworks of GrC can be employed to design and develop granules and granular structures suitable for SA applications. The third phase considers the deploy of the constructed hierarchy of granules to deal with SA issues, via tools and techniques to support human reasoning with granules.

The following Sects. 3 and 4 provide additional information on the fist two phase. Illustrative examples on reasoning with granules and granular structures to support human SA are reported in Sect. 5.

3   A Method of Structured Problem Solving

GrC promotes a general approach for describing, understanding, and solving real world problems based on hierarchies and multilevel structures. An example of such kind of process is the Goal-Directed Task Analysis (GDTA), a cognitive task proposed by Endsley in [5] and widely adopted in SA. GDTA focuses on the goals and sub-goals the human operator must achieve and the requirements that are needed in order to make appropriate decisions. The result of GDTA is an abstract structure, depicted in Fig. 2, establishing requirements for all the

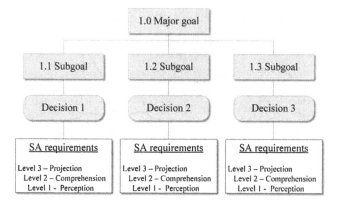

Fig. 2. Goal, decisions and SA requirements resulting from GDTA

levels of an SA application (that are SA L1 *perception*, SA L2 *comprehension*, and SA L3 *projection*). The value of a GDTA for GrC is that a well formalized GDTA gives correct requirements for a granulation process. Usually, in fact, experts perform a top-down analysis of the domain decomposing a whole into its parts. This kind of goal-driven approach is balanced by a data-driven one that develop a bottom-up analysis of the domain via integration of low level elements into a whole. These two approaches (goal-driven and data-driven) essentially resemble the human cognitive capabilities of analysis (i.e. from whole to parts) and synthesis (i.e. from parts to whole).

3.1   Granulation Process Following a GDTA

Figure 3 shows how we can leverage on a GDTA for granulation and creation of granular structures in the context of SA. Starting from the requirements at the perception level (SA L1) of the GDTA, giving information on the elements to perceive for a specific SA objective, we can create type 1 granules. Information on the correct number and level of abstraction of granules to be created depends on different factors such as the number and kind of elements to be perceived, and their relative importance for the objective. In [7] we presented an overview of GrC techniques that can be used to solve common issues at this level such as object recognition, feature reduction and outlier detection. In order to accommodate SA L2 requirements for comprehension, type 1 granules can be abstracted and fused to create type 2 granules. Granular relationships of coarsening and partial coarsening can be used depending on the specific SA L2 requirements. The result is a granular structure representing the elements of the environment created following SA L1 and L2 requirements. This structure is built with a degree of imprecision and uncertainty that can be measured with the concept of information granulation, *IG*, giving a measure of how much information is granulated in a structure.

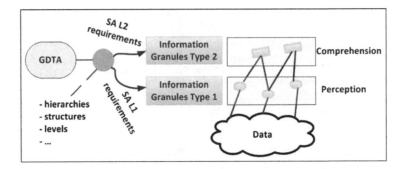

Fig. 3. The process of granulation from a GDTA (from [4])

4   A Paradigm for Information Processing

GrC can be also considered as a paradigm for representing and processing information in the form of granules. There are different formal settings for GrC (e.g. set theory, interval calculus, fuzzy sets, rough sets, shadowed sets, probabilistic granules) and, in general, all these settings support and promote the creation of granular structures using a wide set of relationships [10] to organize granules in hierarchies, trees, networks, and so on.

The following subsection presents an approach, based on a set-theoretic framework for GrC [11], to build granules and granular structure according to the GDTA requirements.

4.1   A GrC Framework for SA

We use the concepts of neighbour and neighbourhood systems described by Yao in [11]. For each element x of an universe U, we can define a subset $n(x) \subseteq U$ that we call neighbourhood of x. More formally, given a distance function $D :$ $U \times U \to R^+$, for each $d \in R^+$ we can define the neighbourhood of x:

$$n_d(x) = \{y | D(x, y) \leq d\} \tag{1}$$

We define (1) as a (type 1) granule, and this definition is flexible enough to support several types of granulation required by SA L1 requirements: spatial proximity, similarity, indistinguishability. From (1) high order granules and granular structures may be constructed. Let us consider a neighbourhood system of x as a non empty family of neighbourhoods:

$$NS(x) = \{n_d(x) | d \in R^+\} \tag{2}$$

Neighbourhood systems like (2) can be used to create multi-layered granulations. Specifically, a nested system $NS(x) = \{n_1(x), n_2(x), ..., n_j(x)\}$ with $n_1(x) \subset n_2(x) \subset ... \subset n_j(x)$ can induce a hierarchy such that we can define refinement and coarsening relationships on granules $n_1(x) \prec n_2(x) \prec ... \prec n_j(x)$. The union of neighbourhood systems for all the elements of an universe defines a granular structure:

$$GS = \cup_{i=1}^{|U|} NS(x_i) \tag{3}$$

If $NS(x_i)$ is a hierarchy, GS is a hierarchical granular structure.

For a granular structure GS we define the information granularity as:

$$IG = \frac{1}{|U|} \sum_{i=1}^{|U|} \frac{|NS(x_i)|}{|U|} \tag{4}$$

For the information granularity the two extremes (finest and coarser granularity) are $\frac{1}{|U|} \leq IG \leq 1$.

Lastly, we formalize the distance between two granular structures using a measure proposed for both rough sets [6] and fuzzy sets [8]. Given two granular structures GS_1 and GS_2 we define their distance as follows:

$$D(GS_1, GS_2) = \frac{1}{|U|} \sum_{i=1}^{|U|} \frac{|NS_1(x_i) \bigtriangleup NS_2(x_i)|}{|U|} \tag{5}$$

where $|NS_1(x_i) \bigtriangleup NS_2(x_i)|$ is the cardinality of a symmetric difference between the neighbourhood systems: $|NS_1(x_i) \cup NS_2(x_i)| - |NS_1(x_i) \cap NS_2(x_i)|$.

5   Reasoning on Situation with Granules and Granular Structures

In this section, we report two illustrative examples on how the formal framework presented in Sect. 4.1 can support reasoning on situations for SA applications. These examples are based on our previous works [3,4].

5.1   Abductive Reasoning for Situation Classification

Abduction is a form of reasoning that allows to infer hypotheses or theories that can explain some observations and, for our purposes, can support the classification of the current situation. This form of reasoning can be enabled by granules and granular structures designed with the framework presented in Sect. 4.1. In this example, we use rough sets that can be understood in the context of the framework presented in Sect. 4.1. Specifically, if the distance D defined for (1) is an equivalence relation, a neighbourhood specializes into a rough set. The idea is to create granules of hypothesis that are equivalent on the basis of a subset of observations. This allows SA operators to infer what are the groups of hypotheses that are equivalent with respect to the available (and usually incomplete) set of observations. Abduction can be supported by reasoning on granular structures in incremental way. When, for instance, a new observation is recognized by an operator a coarse granule, such as h_1, can be refined in finer granules, such as h_{11} and h_{12}, allowing to discern hypotheses. If SA operators have also information

Table 1. Decision table for hypotheses

	o1	o2	o3	d
h1	0	0	0	D
h2	0	0	1	S
h3	0	1	0	D
h4	0	1	1	D
h5	1	0	0	S
h6	1	0	1	D
h7	1	1	0	D
h8	1	1	1	S

on hypotheses supporting specific situations, we can apply three-way decisions [9] to understand how well a specific class, e.g. the class of hypotheses supporting dangerous situations, can be approximated by the created granules.

Let us consider a case with 8 hypotheses and 3 observations with binary values, i.e. $0 =$ inconsistent with the hypothesis, $1 =$ consistent. We can define the Hypotheses-Observations HO matrix reported in Table 1.

Using rough set terminology, if we exclude for the moment the decisional attribute d, HO is an information table. We can define a granule of equivalent hypotheses as:

$$[h]_O = \{h^I | h^I \in U, f(h) = f(h^I)\} \tag{6}$$

where f is an information function for each attribute. (6) can be considered as a specialization of (1), and means that given a set O of observations the hypotheses h and h^I are equivalent because have the same value on every observation $o \in O$. In analogy with the definition of (2) and (3), we can consider the following subsets of attributes for HO: $A_1 = \{o_1\}$, $A_2 = \{o_1, o_2\}$, and $A_3 = \{o_1, o_2, o_3\}$ that satisfy a nested sequence, $A_3 \supset A_2 \supset A_1$. We define the following equivalence relations on this sequence of subsets $I = O_{A_3} \subset O_{A_2} \subset O_{A_1} \subset O_0 = U \times U$, and build the hierarchical granular structure presented in Table 2 where granules at the top level (L4) correspond to the equivalence relation O_0. L3 corresponds to the equivalence relation O_{A_1} that are the subsets of hypotheses that are equivalent considering only one observation o_1, and so on. L1 considers all the observations of HO and provides the finest granulation (allowing to discern all the hypotheses) while L4 provides the coarsest one.

If we consider all the subsets of observations of HO the resulting granular structure is a lattice of granules such as the one of Fig. 4, where we do not depict the extreme cases L4 and L1, and for each granule of hypotheses we report the values of the observations.

By inspecting this structure a SA operator has a view of the hypotheses that can be equivalent on the basis of the partial set of observations derived from the environment. If we add the decisional attribute d to Table 1 we can classify hypothesis with respect to supporting situations, e.g. S for Safe situations

Table 2. Granular structure

L4	$\{h1, h2, h3, h4, h5, h6, h7, h8\}$
L3	$\{\{h1, h2, h3, h4\}, \{h5, h6, h7, h8\}\}$
L2	$\{\{h1, h2\}, \{h3, h4\}, \{h5, h6\}, \{h7, h8\}\}$
L1	$\{\{h1\}, \{h2\}, \{h3\}, \{h4\}, \{h5\}, \{h6\}, \{h7\}, \{h8\}\}$

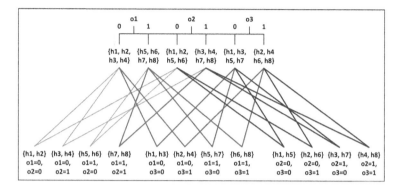

Fig. 4. Lattice of partitions for HO (adapted from [3])

and D for dangerous ones, and use three-way decisions to classify granules of hypotheses.

Let be, for instance, $D = \{h_1, h_3, h_4, h_6, h_7\}$ be the set of all hypotheses related to dangerous situations. We can evaluate if a granule of equivalent hypotheses belongs to $POS(D)$, $NEG(D)$ or $BND(D)$, that are the regions that approximate positively or negatively the set of hypotheses related to dangerous situations, or advise for a deferring the decision. Let us consider the equivalence relation O_{A_2} corresponding to the granular level L2 of Table 2.

The universe of 8 hypotheses is partitioned in 4 granules, each of them with 2 hypotheses and, following [9], we evaluate the conditional probability $P(D|[h]_O)$, and use $\alpha = 0.63$ and $\beta = 0.25$, that are the values used in [9]. We have the results reported in Table 3. The positive region of D is $POS(D) = \{h_3, h_4\}$, while all the other granules belong to boundary region $BND(D) = \{h_1, h_2\} \cup \{h_5, h_6\} \cup \{h_7, h_8\}$. This means that at L2 the granule $\{h_3, h_4\}$ is a good approximation of hypotheses supporting dangerous situations. We can evaluate the three regions for all the levels of a multi level structure, and enhance the Fig. 4 with information on how each granule approximates the class of safe or dangerous situations.

5.2 Situation Projections with *IG* and Distance

In some cases SA operators have good mental models for projections. They have expertise and knowledge to foresee possible evolutions of a situation but,

Table 3. Evaluation result

	$D \cap [h]_O$	$P(D|[h]_O)$
$[h]_O = \{h1, h2\}$	1	0.5
$[h]_O = \{h3, h4\}$	2	1
$[h]_O = \{h5, h6\}$	1	0.5
$[h]_O = \{h7, h8\}$	1	0.5

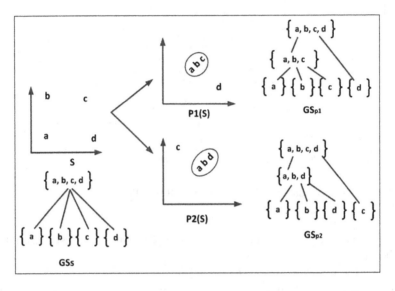

Fig. 5. Situations and Granular structures (adapted from [4])

in most cases, they are not able to reason on differences among possible evolutions and/or between a recognized situation and a possible projection. Measuring and understanding if and how evolutions of situations differ is and important and challenging task. In this example we show how the concepts of IG and distance between granular structures can give early warning information on the differences between two possible projections of a recognized situation.

The example is shown in Fig. 5. Let us suppose $U = \{a, b, c, d\}$ is a set of objects to monitor, and the current situation at $t = t_0$ is S. Let us suppose that from S the SA operator expects two projections, $P1(S)$ and $P2(S)$.

Using the framework described in Sect. 4.1, we create granular structures that can support reasoning on this scenario.

For the situation S we have:

$$NS(a) = \{\{a\}\}$$
$$NS(b) = \{\{b\}\}$$
$$NS(c) = \{\{c\}\}$$
$$NS(d) = \{\{d\}\}$$

The granular structure GS_S is, in this case, the union of four singletons corresponding the specific objects.

For $P1(S)$ we have the following neighbourhood systems:

$$NS(a) = \{\{a\}, \{a, b, c\}\}$$
$$NS(b) = \{\{b\}, \{a, b, c\}\}$$
$$NS(c) = \{\{c\}, \{a, b, c\}\}$$
$$NS(d) = \{d\}$$

with $NS(a)$, $NS(b)$ and $NS(c)$ that are nested systems and induce the hierarchy we can see in $GS_{P1(S)}$ with the creation of a coarse granule $\{a, b, c\}$ reporting information on groups of objects.

For $P2(S)$ we have the following neighbourhood systems:

$$NS(a) = \{\{a\}, \{a, b, d\}\}$$
$$NS(b) = \{\{b\}, \{a, b, d\}\}$$
$$NS(c) = \{c\}$$
$$NS(d) = \{\{d\}, \{a, b, d\}\}$$

$GS_{P2(S)}$ seems to be similar to $GS_{P1(S)}$. However a closer look to the three situations in terms of IG and distance shows some differences.

The IG of these structures is:

$$IG(GS_S) = \tfrac{1}{4}[\tfrac{1}{4} + \tfrac{1}{4} + \tfrac{1}{4} + \tfrac{1}{4}] = \tfrac{1}{4}$$
$$IG(GS_{P1(S)}) = \tfrac{1}{4}[\tfrac{2}{4} + \tfrac{2}{4} + \tfrac{2}{4} + \tfrac{1}{4}] = \tfrac{7}{16}$$
$$IG(GS_{P2(S)}) = \tfrac{1}{4}[\tfrac{2}{4} + \tfrac{2}{4} + \tfrac{1}{4} + \tfrac{2}{4}] = \tfrac{7}{16}$$

and the distances between these structures are:

$$D(GS_S, GS_{P1(S)}) = \tfrac{1}{4}[\tfrac{2-1}{4} + \tfrac{2-1}{4} + \tfrac{2-1}{4} + 0] = \tfrac{3}{16}$$
$$D(GS_{P1(S)}, GS_{P2(S)}) = \tfrac{1}{4}[\tfrac{3-1}{4} + \tfrac{3-1}{4} + \tfrac{2-1}{4} + \tfrac{2-1}{4}] = \tfrac{6}{16}$$
$$D(GS_S, GS_{P2(S)}) = \tfrac{1}{4}[\tfrac{2-1}{4} + \tfrac{2-1}{4} + 0 + \tfrac{2-1}{4}] = \tfrac{3}{16}$$

The situation S, represented with GS_S, has a value of $IG(GS_S) = \tfrac{1}{|U|}$ that is the finest granulation. So we can understand that in this case the SA operator has precise information on the position of the four objects. There is a difference between the situation S and its two projections $P1(S)$ and $P2(S)$ because there are different values of IG, and this is explained by the fact that the projected situations bring a different information, i.e. the creation of an higher order granule. However, in this case, since both the projections have an higher order granule, only with IG we are not able to differentiate between these two projections, i.e. $IG(GS_{P1(S)}) = IG(GS_{P2(S)})$. With the evaluation of IG we can inform the SA operator that the projected situations bring a new kind of information but we can not provide early warnings on the fact that the two projections represent two different situations. Also the values of the distances between S and the projections is the same, i.e. $D(GS_S, GS_{P1(S)}) = D(GS_S, GS_{P2(S)})$. On these bases one could be tempted to consider the two projections as similar, but they are different and the distance between the two projections $D(GS_{P1(S)}, GS_{P2(S)}) \neq 0$ evidences this fact.

6   Conclusion and Future Works

Starting from our previous results, we presented a framework that allows creation of granules and granular structures following GDTA requirements in order to reason on situations. We presented two illustrative examples on how the constructs of the framework can support abduction for situation classification and provision of early warning information on situation projections. Future works concern the adoption of the GrC framework into a complete SA system, such as [1], and an evaluation using real SA operators and assessment methods for SA.

References

1. D'Aniello, G., Gaeta, A., Loia, V., Orciuoli, F.: Integrating GSO and saw ontologies to enable situation awareness in green fleet management. In: 2016 IEEE International Multi-Disciplinary Conference on Cognitive Methods in Situation Awareness and Decision Support (CogSIMA), pp. 138–144 (2016)
2. D'Aniello, G., Gaeta, A., Gaeta, M., Lepore, M., Orciuoli, F., Troisi, O., et al.: A new dss based on situation awareness for smart commerce environments. J. Ambient Intell. Humanized Comput. **7**(1), 47–61 (2016)
3. D'Aniello, G., Gaeta, A., Gaeta, M., Loia, V., Reformat, M.Z.: Application of granular computing and three-way decisions to analysis of competing hypotheses. In: 2016 IEEE International Conference on Systems, Man, and Cybernetics (SMC 2016), Budapest, Hungary, pp. 1650–1655. IEEE (2016)
4. D'Aniello, G., Gaeta, A., Loia, V., Orciuoli, F.: A granular computing framework for approximate reasoning in situation awareness. Granular Comput., 1–18 (2016). doi:10.1007/s41066-016-0035-0
5. Endsley, M.R.: Designing for Situation Awareness: An Approach to User-Centered Design. CRC Press, Boca Raton (2011)
6. Liang, J.: Uncertainty and feature selection in rough set theory. In: Yao, J.T., Ramanna, S., Wang, G., Suraj, Z. (eds.) RSKT 2011. LNCS (LNAI), vol. 6954, pp. 8–15. Springer, Heidelberg (2011). doi:10.1007/978-3-642-24425-4_2
7. Loia, V., D'Aniello, G., Gaeta, A., Orciuoli, F.: Enforcing situation awareness with granular computing: a systematic overview and new perspectives. Granular Comput. **1**(2), 127–143 (2016)
8. Qian, Y., Li, Y., Liang, J., Lin, G., Dang, C.: Fuzzy granular structure distance. IEEE Trans. Fuzzy Syst. **23**(6), 2245–2259 (2015)
9. Yao, Y.: The superiority of three-way decisions in probabilistic rough set models. Inf. Sci. **181**(6), 1080–1096 (2011)
10. Yao, Y.: A triarchic theory of granular computing. Granular Comput. **1**(2), 145–157 (2016)
11. Yao, Y.: Granular computing using neighborhood systems. In: Roy, R., Furuhashi, T., Chawdhry, P.K. (eds.) Advances in Soft Computing, pp. 539–553. Springer, Heidelberg (1999)

Sequences of Orthopairs Given by Refinements of Coverings

Stefania Boffa$^{(\boxtimes)}$ and Brunella Gerla

Università dell' Insubria, Varese, Italy
sboffa@uninsubria.it

Abstract. In this paper we consider sequences of orthopairs given by refinements of coverings and partitions of a finite universe. While operations among orthopairs can be fruitfully interpreted by connectives of three-valued logics, we investigate the algebraic structures that are the counterpart of operations among sequences of orthopairs.

Keywords: Orthopairs · Forests · Many-valued algebraic structures · Gödel algebras

1 Introduction

Coverings and partitions of a universe allows to approximate subsets by means of rough sets and orthopairs: by refining such coverings we get better approximations, hence refinements naturally arise in knowledge representation and in the rough set framework.

An *orthopair* is a pair of disjoint subsets of a universe U. Despite their simplicity, orthopairs are used in several situations of knowledge representation [8,9]. They are commonly used to model uncertainty and to deal with approximation of sets. In particular, any rough approximation of a set determines an orthopair. Indeed, given a covering C of a universe U, every subset X of U determines the orthopair $(\mathcal{L}_C(X), \mathcal{E}_C(X))$, where $\mathcal{L}_C(X)$ is the *lower approximation* of X, i.e., the union of the blocks of C included in X, and $\mathcal{E}_C(X)$ is the *impossibility domain* or *exterior region*, namely the union of blocks of C with no elements in common with X [8]. Several kinds of operations have been considered among rough sets [9], corresponding to connectives in three-valued logics, such as Łukasiewicz, Nilpotent Minimum, Nelson and Gödel connectives [4,6].

In [1] the authors started the investigation of the algebraic structures related to sequences of orthopairs corresponding to refinements of partitions, focusing on a generalization of the three-valued Sobociński operation and on IUML-algebras.

In this paper we extend such approach considering coverings instead of partitions. In this case, the Sobociński operation is no more useful and other operations must be considered. We establish here a first partial result regarding the minimum conjunction and algebraic structures arising as rotation of Gödel algebras: such result is achieved by imposing rather strong conditions on the considered sequence of coverings, and it is a first step for a more general result.

© Springer International Publishing AG 2017
A. Petrosino et al. (Eds.): WILF 2016, LNAI 10147, pp. 95–105, 2017.
DOI: 10.1007/978-3-319-52962-2_8

By *refinement sequence* we mean a sequence $\mathcal{C} = C_0, \ldots C_n$ of coverings of a universe U such that every block of C_i is contained in an unique block of C_{i-1}, for each i from 1 to n. We notice that in our approach we will deal with refinements built on partial coverings, that is coverings that do not cover all the universe [11]. This is a more general case of that defined in [1], in which the elements of $\mathcal{C}$ are partitions of the given universe. We can assign a refinement sequence $\mathcal{C}_K$ to a formal context $K = (X, (Y, \leq), I)$ with hierarchically ordered attributes (see [5]), with the additional information that $(Y, \leq)$ is a forest: for each attribute y of Y, the set $\{x \in X | (x, y) \in I\}$ is a block of a covering in $\mathcal{C}_K$.

Example 1. The table below records some students $s_1, \ldots, s_7$ of engineer's degree that has passed some exams: Algebra, Physics and Geometry. Exactly, it represents the relation of the formal context with hierarchically ordered attributes, where $\{s_1, \ldots, s_7\}$ is the set of the objects and *Algebra, Physics* and *Geometry* are the attributes.

	Algebra	Physics	Geometry
s_1	×	×	
s_2	×	×	
s_3	×	×	×
s_4	×		
s_5	×		
s_6	×		×
s_7	×	×	×

Assume that each student have to pass Algebra before to Physics or Geometry, then the attributes are ordered in the following way:

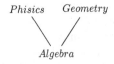

Phisics   *Geometry*

Algebra

Therefore, we consider the refinement sequence (C_0, C_1) of partial coverings of $\{s_1, \ldots, s_7\}$, where $C_0 = \{\{s_1, \ldots, s_7\}\}$ and $C_1 = \{\{s_1, s_2, s_3, s_7\}, \{s_3, s_6, s_7\}\}$.

As shown in [1], we can use refinement sequences for representing classifications in which we want to better specify some classes while ignoring others, or also for representing temporal evolution, when new attributes are added to an information table.

Given a refinement sequence $\mathcal{C}$ of U, we denote by $O_{\mathcal{C}}$ the set of the sequences of orthopairs of U determined by the coverings of $\mathcal{C}$.

The *Sobociński* conjunction $*_S$ is defined on $\{0, 1/2, 1\}$ by $\min(x, y)$ for $x \leq 1 - y$ and $\max(x, y)$ otherwise. Such operation (like all three-valued operations) can be defined also on orthopairs on a universe U [9]. In this way we get an operation between orthopairs defined by $(A, B) *_S (C, D) = (((A \setminus D) \cup (C \setminus B)), B \cup D)$. On the other hand, *Kleene* three-valued conjunction is defined by setting $1 * 1/2 = 1/2 * 1/2 = 1/2 * 1 = 1/2$, $1 * 1 = 1$ and $0 * x = 0$ for every x.

An *idempotent uninorm mingle logic algebra (IUML-algebra)* [12] is an idempotent commutative bounded residuated lattice $\mathcal{A} = (A, \wedge, \vee, *, \rightarrow, \bot, \top, e)$, satisfying the following proprieties: $(x \rightarrow y) \vee (y \rightarrow x) \geq e$, and $(x \rightarrow e) \rightarrow e = x$, for every $x, y \in A$. In any IUML-algebra, if we define the unary operation $\neg$ as $\neg x = x \rightarrow e$, then $\neg \neg x = x$ ($\neg$ is involutive) and $x \rightarrow y = \neg(x * \neg y)$. The set $\{0, 1/2, 1\}$ with the Sobociński conjunction is an example of three-valued IUML-algebra.

In [2] a dual categorical equivalence is described between finite forests F with order preserving open maps and finite IUML-algebras with homomorphisms. For any finite forest F, denote by $SP(F)$ the set of pairs of disjoint upsets of F and define the following operations: if $X = (X^1, X^2)$ and $Y = (Y^1, Y^2) \in SP(F)$, we set (see also [1]):

1. $(X^1, X^2) \sqcap (Y^1, Y^2) = (X^1 \cap Y^1, X^2 \cup Y^2)$,
2. $(X^1, X^2) \sqcup (Y^1, Y^2) = (X^1 \cup Y^1, X^2 \cap Y^2)$,
3. $(X^1, X^2) * (Y^1, Y^2) = ((X^1 \cap Y^1) \cup (X \diamond Y), (X^2 \cup Y^2) \setminus (X \diamond Y))$ where, for each $U = (U^1, U^2), V = (V^1, V^2) \in SP(F)$, letting $U^0 = F \setminus (U^1 \cup U^2)$, we set $U \diamond V =\uparrow ((U^0 \cap V^1) \cup (V^0 \cap U^1))$, and
4. $(X^1, X^2) \rightarrow (Y^1, Y^2) = \neg((X^1, X^2) * (Y^2, Y^1))$, where $\neg(X^1, X^2) = (X^2, X^1)$.

Theorem 1. [2] *For every finite forest F, $(SP(F), \sqcap, \sqcup, *, \rightarrow, (\emptyset, F), (F, \emptyset), (\emptyset, \emptyset))$ is an IUML-algebra. Vice-versa, each finite IUML-algebra is isomorphic with $SP(F)$ for some finite forest F.*

Also recall that a *De Morgan algebra* $(A, \wedge, \vee, \neg, \bot, \top)$ (see [7]) is a bounded distributive lattice such that $\neg(x \wedge y) = \neg x \vee \neg y$ and $\neg \neg x = x$, for each $x, y \in A$. A *Kleene algebra* is a De Morgan algebra further satisfying the Kleene equation: $x \wedge \neg x \leq y \vee \neg y$ for every $x, y \in A$. A *Gödel algebra* is an integral commutative bounded residuated lattice $(G, \wedge, \vee, *, \rightarrow, \bot, \top)$ that is idempotent ($x * x = x$) and prelinear ($(x \rightarrow y) \vee (y \rightarrow x) = \top$). A dual categorical equivalence for finite Gödel algebras and finite forests with open maps holds, and for each finite forests the set of its downsets can be equipped with a structure of Gödel algebra. If $\mathcal{G} = (G, \wedge, \vee, \rightarrow, \bot, \top)$ is a Gödel algebra, by *rotation of $\mathcal{G}$* we mean the structure $\mathcal{G}^* = (G^*, \otimes, \oplus, \sim, (\bot, \top), (\top, \bot))$, where $G^* = \{(x, \top) | x \in G\} \cup \{(\top, x) | x \in G\}$, and for every $x, y \in G$:

- $(x, \top) \otimes (y, \top) = (x \wedge y, \top)$, $(\top, x) \otimes (\top, y) = (\top, x \vee y)$, $(x, \top) \otimes (\top, y) = (x, \top)$,
- $\sim (x, \top) = (\top, x)$ and $\sim (\top, x) = (x, \top)$
- $(x, y) \oplus (x_1, y_1) = \sim (\sim (x, y) \otimes \sim (x_1, y_1))$.

Note that with such definition, every element of the form $(x, \top)$ is smaller than any element of the form $(\top, y)$.

In the Sects. 1 and 2, using Theorem 1, we survey the relationship between sequences of successive refinements of orthopairs (over a finite universe) and (not necessarily three-valued) IUML-algebras and we adapt it to the case of coverings.

In the Sect. 3, we impose the constraint on $\mathcal{C}$ that all blocks of every covering has at least one element in common. In Example 1 this condition holds: the students s_3 and s_7 are in each block of C_0 and C_1. Then, we equip $O_{\mathcal{C}}$ with the operations $\wedge_{SO}$, $\vee_{SO}$ and $\neg_{SO}$ obtained by applying for each covering of $\mathcal{C}$ the Kleene operations and, at the end, we characterize $O_{\mathcal{C}}$ as a rotation of a Gödel algebra. In particular, we show that the set of sequences of orthopairs is the rotation of the Gödel algebra made by the sequences of lower approximations and we show that such case can be equipped with a structure of Kleene algebra.

2   Sequences of Coverings and Orthopairs

Definition 1. *A* partial covering *of U is a covering of a subset of U. A sequence $\mathcal{C} = C_0, \ldots, C_n$ of partial coverings of U is a* refinement sequence *if each element of C_i is contained in only one element of C_{i-1}, for $i = 1, \ldots, n$.*

Example 2. If $U = \{a, b, c, d, e, f, g, h, i, j\}$ then $C_0 = \{\{a, b, c, d, e\}, \{f, g, h, i\}\}$, $C_1 = \{\{a, b, c\}\{c, d\}, \{f, g\}, \{h, i\}\}$ is a refinement sequence of partial coverings of U.

A refinement sequence of partial partitions of a universe (defined in [1]) is a special case of refinement sequence of partial coverings in which each partial covering is a partial partition. Without danger of confusion, by refinement sequence of U we shall mean a refinement sequence $\mathcal{C}$ of partial coverings of U and we shall specify when $\mathcal{C}$ has partitions as elements. In order to prove our results (see Proposition 2) we do not consider coverings containing singletons, that is blocks with only one element. Let us fix a refinement sequence $\mathcal{C} = C_0, \ldots, C_n$ of partial coverings of U. For any $X \subseteq U$ and for every $i = 0, \ldots, n$ we consider the *orthopair* $(\mathcal{L}_i(X), \mathcal{E}_i(X))$ determined by C_i.

Definition 2. *Let $\mathcal{C} = C_0, \ldots, C_n$ be a refinement sequence of U and let $X \subseteq U$. Then we let $O_{\mathcal{C}}(X) = ((\mathcal{L}_0(X), \mathcal{E}_0(X)), \ldots, (\mathcal{L}_n(X), \mathcal{E}_n(X)))$.*

Example 3. Given $U = \{a, b, c, d, e, f, g, h, i, j\}$, $X = \{a, b, c, d, e\}$ and the refinement sequence of partial coverings of U given by $C_0 = \{\, \{a, b, c, d, e, f, g, h, i, j\} \,\}$, $C_1 = \{\, \{a, b, c, d, e\}, \{e, f, g, h, i\} \,\}$, $C_2 = \{\, \{a, b, c\}, \{c, d\}, \{e, f, g\}, \{g, h\} \,\}$, then $O_{\mathcal{C}}(X) = ((\emptyset, \emptyset), (\{a, b, c, d, e\}, \emptyset), (\{a, b, c, d\}, \{g, h\}))$.

Definition 3. *Let $\mathcal{C}$ be a refinement sequence of U. We associate with $\mathcal{C}$ a forest $(F_{\mathcal{C}}, \leq_{F_{\mathcal{C}}})$, where:*

1. *$F_{\mathcal{C}} = \bigcup_{i=0}^{n} C_i$ (the set of nodes is the set of all subsets of U belonging to the coverings $C_0, \ldots, C_n$), and*
2. *for $N, M \in F_{\mathcal{C}}$, $N \leq_{F_{\mathcal{C}}} M$ if and only if there exists $i \in \{0, \ldots, n-1\}$ such that $N \in C_i$, $M \in C_{i+1}$ and $M \subseteq N$ (the partial order relation is the reverse inclusion).*

Example 4. The forest associated with the refinement sequence (C_0, C_1) of partitions of $\{a, b, c, d, e, f, g, h, i, j\}$, where $C_0 = \{\{a, b, c, d, e\}, \{e, f, g, h, i\}\}$ and $C_1 = \{\{a, b\}, \{c, d\}, \{e, f, g\}, \{g, h\}\}$, is shown in the following figure:

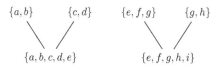

For any $X \subseteq U$ the sequence $\mathcal{O}_C(X)$ of orthopairs with respect to C determines two subsets of the forest F_C, obtained by considering the blocks contained in $\mathcal{L}_i(X)$ and the blocks contained in $\mathcal{E}_i(X)$. This observation leads to the following definition.

Definition 4. *For every refinement sequence* $C = C_0, \ldots, C_n$ *of* U *and any* $X \subseteq U$ *we let* (X_C^1, X_C^2) *be such that* $X_C^1 = \{N \in F_C : N \subseteq X\}$ *and* $X_C^2 = \{N \in F_C : N \cap X = \emptyset\}$. *Let* $SO(F_C)$ *be the set* $\{(X_C^1, X_C^2) \mid X \subseteq U\}$.

Example 5. Given U and C of Example 4, if $X = \{a, b, e, g\}$ then $X_C^1 = \{\{a, b\}\}$ and $X_C^2 = \{\{c, d\}\}$.

We write (X^1, X^2) instead of (X_C^1, X_C^2), when C is clear from the context.

Proposition 1. *Let* $C = C_0, \ldots, C_n$ *be a refinement sequence of* U *and* $X \subseteq U$. *Then* $(\mathcal{L}_i(X), \mathcal{E}_i(X)) = (\cup\{N \in C_i | N \in X^1\}, \cup\{N \in C_i | N \in X^2\})$, *for each* $i \in \{0, \ldots, n\}$.

Proof. An element x of U belongs to $\mathcal{L}_i(X)$ if and only if there exists $N \in C_i$ such that $x \in N$ and $N \subseteq X$. This is equivalent to say that $x \in \cup\{N \in C_i | N \in X^1\}$. Similarly, $x \in \mathcal{E}_i(X)$ if and only if there exists $N \in C_i$ such that $x \in N$ and $N \cap X = \emptyset$, namely $N \in \cup\{N \in C_i | N \in X^2\}$.

The proof of the following theorem can be found in [1]:

Theorem 2. *Given a set* U *and a refinement sequence* C *of* U, *the map*

$$h : \mathcal{O}_C(X) \in \{\mathcal{O}_C(X) \mid X \subseteq U\} \mapsto (X^1, X^2) \in SO(F_C)$$

is a bijection. Further, $SO(F_C) \subseteq SP(F_C)$.

In particular, the second part of Theorem 2 states that pairs of $SO(F_C)$ are pairs of upsets of the forests F_C.

Proposition 2. *Let* C *be a refinement sequence of* U, *then the pair* (A, B) *of* $SP(F_C)$ *belongs to* $SO(F_C)$ *if and only if* $\{\bigcup_{N \in A} N\} \cap \{\bigcup_{N \in B} N\} = \emptyset$ *and if a node* N *of* F_C *is contained in the union of some nodes of* A *(resp.* B*), then* $N \in A$ *(resp.* B*).*

Proof. ($\Rightarrow$). Let $(A, B) \in SO(F_C)$, then $A = X^1$ and $B = X^2$, for some $X \subseteq U$. Trivially, the nodes of X^1 are included in X, and the nodes of X^2 are disjoint with X, hence $X^1 \cap X^2 = \emptyset$. Furthermore, let $N, N_1, \ldots, N_k \in F_C$ such that $N \subseteq N_1 \cup \ldots \cup N_k$ and $N_i \subseteq X$ (or $N_i \cap X = \emptyset$), for i from 1 to n. Trivially, $N \subseteq X$ ($N \cap X = \emptyset$).

($\Leftarrow$). Let $(A, B) \in SP(F_C)$. We set $\mathcal{A} = \{\bigcup_{N \in A} N\}$, $\mathcal{B} = \{\bigcup_{N \in B} N\}$, $\mathcal{D} = \{N \in F_C \setminus (A \cup B) \mid N \cap \mathcal{A} = \emptyset$ and $M \in B$ for each $M \subset N\}$ and $\mathcal{E} = \{N \in F_C \setminus (A \cup B) \mid N \cap \mathcal{A} = \emptyset\} \cap F_L$, where F_L is the set of the leaves of F_C. Let D be the set obtained by picking an element $x_{D_N} \in N \setminus \mathcal{B}$, for each node N of $\mathcal{D}$. Also, we call $\mathcal{E}$ the set obtained by taking an element in every node of E disjointed with $\mathcal{D}$ and such that if $N \cap N' \neq \emptyset$, then $x_{E_N} = x_{E_{N'}}$, for each $N, N' \in E$. Setting $X = \mathcal{A} \cup \mathcal{D} \cup \mathcal{E}$, we can prove that $A = X^1$ and $B = X^2$, using the hypothesis and the condition of C that the blocks are not singletons.

Definition 5. *A refinement sequence C is* safe *if for each $N \in \bigcup_{i=0}^{n} C_i$, if $N \subseteq N_1 \cup \ldots \cup N_k$ with $N_1, \ldots, N_k \in \bigcup_{i=0}^{n} C_i$, then there exists $i \in \{1, \ldots k\}$ such that $N \subseteq N_i$. A refinement sequence C is* pairwise overlapping *if there are no disjoint blocks in coverings of C.*

Corollary 1. *Let C be a refinement sequence of U. Then, $SO(F_C) = SP(F_C)$ if and only if C is a safe refinement sequence of partial partitions of U.*

Proof. Condition 1 of Proposition 2 holds for each pair of $SO(F_C)$ when each element of C is a partial partition of U, and condition 2 holds for each pairs of $SO(F_C)$ when C is safe.

Example 6. We consider the following forest F_C of the universe U:

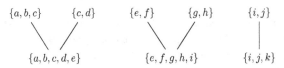

Then, the pair $(\{\{a, b, c\}, \{\{c, d\}\})$, not satisfying condition 1 of Proposition 2, does not belong to $SO(F_C)$, since if a set X includes $\{a, b, c\}$, then it is not disjoint with $\{c, d\}$. Also the pair $(\{\{e, f\}, \{g, h\}, \{i, j\}\}, \emptyset)$ is not an element of $SO(F_C)$, since a subset of U that includes the nodes $\{e, f\}, \{g, h\}$ and $\{i, j\}$, must include the node $\{e, f, g, h, i\}$ too (we note that $(\{\{e, f\}, \{g, h\}, \{i, j\}\}, \emptyset)$ does not satisfy condition 2 of Proposition 2).

3   Refinement Sequences and IUML-Algebras

When every covering of the refinement sequence C is a partial partition, $SO(F_C)$ can be equipped with a structure of IUML-algebra (see [1]): $SO(F_C)$ is isomorphic to the IUML-algebra $SP(F)$, where F is obtained from F_C. This result does not hold in the general case of partial coverings. Now, we consider a particular kind of refinement sequence, in which there are not pairs of disjoint blocks.

Proposition 3. *Let* C *be a pairwise overlapping refinement sequence of* U. *Then* $SO(F_C) \subseteq \{(A, B) \in SP(F_C) | A = \emptyset \text{ or } B = \emptyset\}$.

Proof. Let $(A, B) \in SO(F_C)$. By the definition of $SO(F_C)$, there exists a subset X of U that includes each node of A and it is disjointed with each node of B. Then, A or B is equal to the empty set, since any element of u belongs to X or doesn't belong to X.

Proposition 4. *Let* C *be a pairwise overlapping refinement sequence of* U. *Then,* $SO(F_C) = \{(A, B) \in SP(F_C) | A = \emptyset \text{ or } B = \emptyset\}$ *if and only if* C *is safe.*

Proof. ($\Rightarrow$). Let $N \in F_C$. If $N = N_1 \cup \ldots \cup N_k$ where $N \not\subseteq N_i$ for each $i = 1, \ldots, k$, then the pairs $(\uparrow \{N_1, \ldots, N_k\}, \emptyset)$ and $(\emptyset, \uparrow \{N_1, \ldots, N_k\})$ are excluded from $SO(F_C)$. ($\Leftarrow$). Let $(A, B) \in SP(F_C)$ such that $A = \emptyset$ or $B = \emptyset$. If $A = \emptyset$, then $(A, B) = (\emptyset, X^2)$, where $X = \bigcup_{N \in B} \{N\}$. Viceversa, if $B = \emptyset$, then $(A, B) = (X^1, \emptyset)$, where $X = \bigcup_{N \in A} \{N\}$.

Theorem 3. *Let* C *be a pairwise overlapping refinement sequence of* U. *Then, there exists a finite forest* F *such that* $(SO(F_C), \subseteq) \simeq (SP(F), \subseteq)$ *if and only if* F_C *is a chain.*

Proof. ($\Rightarrow$). We note that every element of $SO(F_C)$ is comparable with $(\emptyset, \emptyset)$. If L^1 and L^2 are incomparable leaves of F, then $(\{L^1\}, \{L^2\})$ is incomparable with $(\emptyset, \emptyset)$. Consequently, if $SO(F_C) \simeq SP(F_C)$ then F_C must be a chain.
  ($\Leftarrow$). If F_C is a chain, then $SO(F_C) = SP(F_C)$, since C is a refinement sequence of partial partitions of U such that every node is not equal to the union of its successors (see [1]).

As shown in [2], each finite IUML-algebra is the set of the pairs of disjoint upsets of a finite forest. Then, the following corollary holds:

Corollary 2. *Let* C *be a pairwise overlapping refinement sequence of* U. *Then, there exist* $*$ *and* $\rightarrow$ *such that* $(SO(F_C), \sqcap, \sqcup, *, \rightarrow, (\emptyset, F_C), (F_C), (\emptyset, \emptyset))$ *is a finite IUML-algebra if and only if* F_C *is a chain.*

Example 7. Let F_C be the following forest:

As showed in this figure, the IUML-algebras $SP(F_1)$ and $SP(F_2)$, where F_1 is the chain of 4 nodes and F_2 is made by two disjoint nodes, are 9 elements as $SO(F_C)$, but their Hasse's diagrams are different from that of $SO(F_C)$.

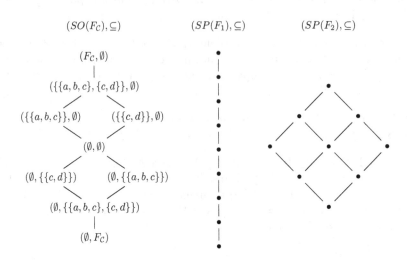

$$(SO(F_{\mathcal{C}}), \subseteq) \qquad (SP(F_1), \subseteq) \qquad (SP(F_2), \subseteq)$$

4 Gödel Algebras e Sequences of Orthopairs

In this section, we focus on the structure $(SO(F_{\mathcal{C}}), \sqcap, \sqcup, \neg, (\emptyset, F_{\mathcal{C}}), (F_{\mathcal{C}}, \emptyset))$ with the restriction that $\mathcal{C}$ is a safe and pairwise overlapping sequence, i.e., there are no disjoint blocks in $\bigcup_{i=0}^{n} C_i$, and where $\sqcap$, $\sqcup$ and $\neg$ are the operations defined in Sect. 1.

For any finite forest F, we denote by $Up(F)$ the structure $(U(F), \underline{\wedge}, \underline{\vee}, \rightarrow, F, \emptyset)$, where $U(F)$ is the set of the upsets of F, $\underline{\wedge}$ and $\underline{\vee}$ are respectively the set union and intersection, and for any $X, Y \in U(F)$, $X \rightarrow Y = \emptyset$ if $Y \subseteq X$ and $X \rightarrow Y = Y$ otherwise. We note that $Up(F)$ is dually isomorphic to the Gödel algebra $Sub(F)$ (of downsets) defined in [2], hence it is a Gödel algebra.

Theorem 4. *Let $\mathcal{C}$ be a safe and pairwise overlapping refinement sequence of U. Then, $(SO(F_{\mathcal{C}}), \sqcap, \sqcup, \neg, (\emptyset, F_{\mathcal{C}}), (F_{\mathcal{C}}, \emptyset))$ is isomorphic to the rotation of $Up(F_{\mathcal{C}})$ and it is a Kleene algebra.*

Proof. By Theorem 2, elements of $SO(F_{\mathcal{C}})$ are pair of disjoint upsets of $F_{\mathcal{C}}$, and by Proposition 4 such pairs always have one of the components equal to $\emptyset$. Hence $G = \{(\emptyset, X) \mid X \in U(F_{\mathcal{C}})\}$ can be equipped with a structure of Gödel algebra isomorphic to $U(F_{\mathcal{C}})$. Considering that the order in $U(F_{\mathcal{C}})$ is the reverse inclusion and $\emptyset$ is the top element, the rotation of $U(F_{\mathcal{C}})$ has as domain the set $\{(X, \emptyset) \mid X \in U(F_{\mathcal{C}})\} \cup \{(\emptyset, X) \mid X \in U(F_{\mathcal{C}})\}$ and the operations are given by considering the reverse inclusion with respect to the first component and inclusion with respect to the second. Such structure is clearly isomorphic to $(SO(F_{\mathcal{C}}), \sqcap, \sqcup, \neg, (\emptyset, F_{\mathcal{C}}), (F_{\mathcal{C}}, \emptyset))$. In order to check that it is a Kleene algebra, we focus to check that it satisfies the Kleene equation, since the other properties of De Morgan algebras are consequences of the lattice properties. So, if for example $(X, \emptyset), (\emptyset, Y) \in SO(F_{\mathcal{C}})$, then:

$$(X, \emptyset) \sqcap \neg(X, \emptyset) = (X, \emptyset) \sqcap (\emptyset, X) = (\emptyset, X)$$

$$(\emptyset, Y) \sqcup \neg(\emptyset, Y) = (\emptyset, Y) \sqcup (Y, \emptyset) = (Y, \emptyset)$$

and $(\emptyset, X) \le (Y, \emptyset)$. The other cases are similar.

Now, to enunciate the following corollary, we consider the maps π_1 and π_2 from $SO(F_C)$ to $Up(F_C)$ such that $\pi_1((X^1, X^2)) = X^1$ and $\pi_2((X^1, X^2)) = X^2$, for each $X \subseteq U$.

Corollary 3. *Let C be a safe and pairwise overlapping refinement sequence of U. Then, $\pi_1(SO(F_C)) = \pi_2(SO(F_C)) = U(F_C)$.*

Given a refinement sequence C of pairwise overlapping coverings of U and $X \subseteq U$, we denote by $X_{\mathcal{L}}$ the sequence $(\mathcal{L}_0(X), \ldots, \mathcal{L}_n(X))$ and by $X_{\mathcal{E}}$ the sequence $(\mathcal{E}_0(X), \ldots, \mathcal{E}_n(X))$. Moreover, we set $\mathcal{L} = \{X_{\mathcal{L}} | X \subseteq U\}$ and $\mathcal{E} = \{X_{\mathcal{E}} | X \subseteq U\}$. Then, as a consequence of Theorem 2, the following proposition holds:

Proposition 5. *Let C be a safe and pairwise overlapping refinement sequence of U. Then both $\mathcal{L}$ and $\mathcal{E}$ can be equipped with a structure of Gödel algebras and with such structure they are both isomorphic to $U(F_C)$.*

In our last Theorem we describe the operation between sequences of orthopairs by means of Kleene operations over orthopairs:

Theorem 5. *Let $C = C_0, \ldots, C_n$ be a refinement sequence of U. Then, the structure $(SO(F_C), \sqcap, \sqcup, \neg, (\emptyset, F_C), (F_C, \emptyset))$ induces on sequences of orthopairs the operations*

$$\mathcal{O}_C(X) \wedge_{SO} \mathcal{O}_C(Y) = h^{-1}(h(\mathcal{O}_C(X)) \sqcap h(\mathcal{O}_C(Y))),$$
$$\mathcal{O}_C(X) \vee_{SO} \mathcal{O}_C(Y) = h^{-1}(h(\mathcal{O}_C(X)) \sqcup h(\mathcal{O}_C(Y))),$$
$$\neg_{SO}\mathcal{O}_C(X) = h^{-1}(h(\neg\mathcal{O}_C(X)))$$

for every $X, Y \subseteq U$. Furthermore, $\mathcal{O}_C(X) \wedge_{SO} \mathcal{O}_C(Y)$ and $\neg_{SO}\mathcal{O}_C(X)$ are equal respectively to the sequences of orthopairs $((A_0, B_0), \ldots, (A_n, B_n))$ and $((C_0, D_0), \ldots, (C_n, D_n))$ such that $(A_i, B_i) = (\mathcal{L}_i(X), \mathcal{E}_i(X)) * (\mathcal{L}_i(Y), \mathcal{E}_i(Y))$, and $(D_i, E_i) = (\mathcal{E}_i(X), \mathcal{L}_i(X))$ for each $i = 1, \ldots, n$ and where $*$ is the Kleene conjunction.

Proof. Let $X, Y, Z \subseteq U$ such that $\mathcal{O}_C(Z) = \mathcal{O}_C(X) \wedge_{SO} \mathcal{O}_C(Y) = h^{-1}((X^1 \cup Y^1, X^2 \cap Y^2))$. Therefore, $Z^1 = X^1 \cup Y^1$ and $Z^2 = X^2 \cap Y^2$. Now, let us fix $i \in \{0, \ldots, n\}$, then $(\mathcal{L}_i(Z), \mathcal{E}_i(Z)) = (\cup\{N \in F_C \mid N \in C_i \text{ and } N \in X^1 \cup Y^1\}, \cup\{N \in F_C \mid N \in C_i \text{ and } N \in X^2 \cap Y^2\})$ (by Proposition 1) $= (\{\cup\{N \in F_C \mid N \in C_i \text{ and } N \in X^1\}\} \cup \{\cup\{N \in F_C \mid N \in C_i \text{ and } N \in Y^1\}\}, \{\cup\{N \in F_C \mid N \in C_i \text{ and } N \in X^2\}\} \cap \{\cup\{N \in F_C \mid N \in C_i \text{ and } N \in Y^2\}\})$ (by Proposition 1) $= (\mathcal{L}_i(X) \cup \mathcal{L}_i(Y), \mathcal{E}_i(X) \cap \mathcal{E}_i(Y))$. For the operations $\vee_{SO}$ and $\neg_{SO}$ the proof is analogous.

Example 8. Let $\mathcal{C}$ be as in Example 7 and let $X = \{a, b\}$ and $Y = \{a, b, c\}$. Then, $O(X) = ((\emptyset, \emptyset), (\emptyset, \{c, d\}))$ and $O(Y) = ((\emptyset, \emptyset), (\{a, b, c\}, \emptyset))$. We have $O(X) \wedge_{SO} O(Y) = ((\emptyset, \emptyset), (\emptyset, \{c, d\}))$, $O(X) \vee_{SO} O(Y) = ((\emptyset, \emptyset), (\{a, b, c\}, \emptyset))$ and $\neg O(X) = ((\emptyset, \emptyset), (\{c, d\}, \emptyset))$.

Corollary 1. *Let $\mathcal{C}$ be a safe and pairwise overlapping refinement sequence of U and $M_{\mathcal{C}} = \{x \in N | N \in \bigcup_{i=0}^{n} C_i\}$. Then, $(O_{\mathcal{C}}, \wedge_{SO}, \vee_{SO}, \neg_{SO}, (\emptyset, M_{\mathcal{C}}), (M_{\mathcal{C}}, \emptyset))$ is the rotation of the Gödel algebra $(\mathcal{L}, \wedge_{\mathcal{L}}, \vee_{\mathcal{L}}, \rightarrow_{\mathcal{L}}, \emptyset, M_{\mathcal{C}})$ and it is a Kleene algebra.*

5   Conclusions

In this work we continue the investigation started in [1] about the algebraic structures and the operations between sequences of orthopairs, given by considering different refinements of coverings of the universe. We stated here that under some conditions of the sequence of coverings, the resulting structure is a Kleene algebra. The defined operations can be indeed defined by considering Kleene conjunction for each level of covering. By using a duality result for finite Kleene algebras (see [3]) we plan in the future to associate to each Kleene algebra an algebra of sequences of orthopairs and to weak the conditions on coverings. Anyway, it appears that while operations between orthopairs are suitably descripted by three-valued logics, in order to considered refinements of orthopairs we need to look at algebraic structures (and related logical systems) that arise as rotations of lattices.

References

1. Aguzzoli, S., Boffa, S., Ciucci, D., Gerla, B.: Refinements of Orthopairs and IUML-algebras. In: Flores, V., et al. (eds.) IJCRS 2016. LNCS (LNAI), vol. 9920, pp. 87–96. Springer, Heidelberg (2016). doi:10.1007/978-3-319-47160-0_8
2. Aguzzoli, S., Flaminio, T., Marchioni, E.: Finite Forests. Their Algebras and Logics (submitted)
3. Aguzzoli, S., Cabrer, L., Marra, V.: MV-algebras freely generated by finite Kleene algebras. Algebra Univers. **70**, 245–270 (2013)
4. Banerjee, M., Chakraborty, K.: Algebras from rough sets. In: Pal, S., Skowron, A., Polkowski, L. (eds.) Rough-Neural Computing, pp. 157–188. Springer, Heidelberg (2004)
5. Belolahvek, R., Sklenar, V., Zacpal, J.: Formal concept analysis with hierarchically ordered attributes. Int. J. Gen. Syst. 1–12 (2004)
6. Bianchi, M.: A temporal semantics for nilpotent minimum logic. Int. J. Approx. Reason. **55**(1, part 4), 391–401 (2014)
7. Cignoli, R.: Injective de Morgan and Kleene algebras. Proc. Am. Math. Soc. **47**(2), 269–278 (1975). doi:10.1090/S0002-9939-1975-0357259-4
8. Ciucci, D.: Orthopairs: a simple and widely used way to model uncertainty. Fundam. Inform. **108**(3–4), 287–304 (2011)

9. Ciucci, D., Dubois, D.: Three-valued logics, uncertainty management and rough sets. In: Peters, J.F., Skowron, A. (eds.) Transactions on Rough Sets XVII. LNCS, vol. 8375, pp. 1–32. Springer, Heidelberg (2014). doi:10.1007/978-3-642-54756-0_1

10. Comer, S.D.: On connections between information systems, rough sets, and algebraic logic. In: Algebraic Methods in Logic and Computer Science, pp. 117–124. Banach Center Publications, no. 28 (1993)

11. Csajbók, Z.E.: Approximation of sets based on partial covering. In: Peters, J.F., Skowron, A., Ramanna, S., Suraj, Z., Wang, X. (eds.) Transactions on Rough Sets XVI. LNCS, vol. 7736, pp. 144–220. Springer, Heidelberg (2013). doi:10.1007/978-3-642-36505-8_9

12. Metcalfe, G., Montagna, F.: Substructural fuzzy logics. J. Symb. Log. **72**(3), 834–864 (2007)

Minimally Many-Valued Extensions of the Monoidal t-Norm Based Logic MTL

Stefano Aguzzoli and Matteo Bianchi[✉]

Department of Computer Science, via Comelico 39/41, 20135 Milano, Italy
aguzzoli@di.unimi.it, matteo.bianchi@unimi.it

Abstract. In this paper we shall deal with those axiomatic extensions of the monoidal *t*-norm based logic MTL which are *minimally many-valued*, that is, those logics extending MTL such that any further extension collapses them to Boolean Logic. We shall prove some characterisation results concerning the algebraic semantics of these logic, and completely classify the minimally many-valued logics extending Hájek's Basic Logic and Weak Nilpotent Minimum Logic. For the latter logics, we shall use our results to evaluate the complexity of deciding whether a formula is *Booleanising* for some extensions of them, or whether it provides a non-classical extension.

Keywords: Minimally many-valued logics · Almost minimal varieties · MTL-algebras · Single chain generated varieties · Booleanising formulas

1 Introduction

Esteva and Godo's Monoidal t-norm based logic, MTL for short, is widely considered as fundamental in mathematical fuzzy logic, as it is the logic of all left-continuous *t*-norms and their residua. Classical Boolean logic can be seen as an axiomatic extension of MTL, and of every one of its non-classical extensions, by means, for instance, of the excluded middle axiom. In this paper we shall deal with those axiomatic extensions of MTL which are the closest to Boolean logic, that is, those logics extending MTL such that any further extension collapses them to Boolean logic. We call them *minimally many-valued extensions*. In particular we shall completely classify minimally many-valued extensions of Hajek's Basic logic BL, which is the logic of all continuous t-norms, and of the logic WNM of weak nilpotent minimum, which is a major extension of MTL containing major left-continuous *t*-norm based logics, such as Gödel, Nilpotent Minimum, Revised Drastic Product. We shall provide necessary conditions that an extension of MTL must satisfy to be minimally many-valued. Moreover, for logics admitting finite-valued semantics we shall identify necessary and sufficient conditions. Further we shall exploit our classification results to introduce and prove some complexity results about some class of formulas related to classical and non-classical extensions of MTL. Namely, a formula φ is called *Booleanising for some extension* L' of a logic L iff L' extendend with φ collapses to Boolean

© Springer International Publishing AG 2017
A. Petrosino et al. (Eds.): WILF 2016, LNAI 10147, pp. 106–115, 2017.
DOI: 10.1007/978-3-319-52962-2_9

logic. We shall prove that deciding whether φ is Booleanising for some extension of either BL or WNM is in the complexity class DP. A formula φ is called a *non-classical extension formula* for a logic L iff every non-classical extension of L can be further extended with φ remaining non-classical. We shall prove that deciding whether φ is a non-classical extension formula for BL is a Σ_2^p problem in the polynomial hierarchy, while for WNM it is in $coNP$.

2 Preliminaries

MTL, introduced in [16] is the logic of all left-continuous t-norms and their residua, as proved in [21]. Its formulas are defined in the usual inductive way from a denumerable set of variable and the connectives $\{*, \rightarrow, \wedge, \vee, \bot, \neg\}$. MTL and its axiomatic extensions are all algebraizable in the sense of [8], and the equivalent algebraic semantics of MTL is given by the variety $\mathbb{MTL}$ of MTL-algebras. For this reason we will use the same symbols for the algebraic operations and the logical connectives. We refer the reader to [15,16] for background on MTL and $\mathbb{MTL}$. As the lattices of subvarieties of $\mathbb{MTL}$ (ordered by inclusion) is isomorphic with the order dual of the lattice of axiomatic extensions of MTL (ordered by strength), we shall write L and $\mathbb{L}$ respectively for an axiomatic extension of MTL and for the corresponding subvariety of $\mathbb{MTL}$. A totally ordered MTL-algebra is called MTL-chain. If $\mathcal{A}$ is an MTL-algebra, with $\mathbf{V}(\mathcal{A})$ we will denote the variety generated by $\mathcal{A}$.

BL, introduced in [20] is the logic of all continuous t-norms and their residua, as proved in [12]. It corresponds with the axiomatic extension of MTL via *divisibility*: $(\varphi \wedge \psi) \rightarrow (\varphi * (\varphi \rightarrow \psi))$.

WNM, the logic of weak nilpotent minimum t-norms [16], is obtained extending MTL via $\neg(\varphi * \psi) \vee ((\varphi \wedge \psi) \rightarrow (\varphi * \psi))$.

Classical Boolean logic B is obtained extending MTL via the *excluded middle law*: $\varphi \vee \neg\varphi$.

Other major axiomatic extensions of MTL are product logic Π, Łukasiewicz logic Ł, and Gödel logic G. With $\mathbb{P}, \mathbb{MV}, \mathbb{G}$ we will, respectively, denote their corresponding varieties [20]. Moreover, with $\mathbb{C}$ we will denote the variety generated by Chang's MV-algebra [11]. Moving to the finitely-valued logics and their corresponding varieties, with $\mathbb{MV}_k, \mathbb{G}_k, \mathbb{NM}_k$ we will indicate the varieties generated, respectively, by the k-element MV-chain $\mathbf{Ł}_k$, the k-element Gödel-chain $\mathbf{G}_k$, and the k element nilpotent minimum chain $\mathbf{NM}_k$. The corresponding logics are denoted, respectively, by $\mathrm{Ł}_k$, G_k and NM_k: we refer the reader to [11,15,19,20] for more details. Finally, with $\mathbf{2}$ we will denote the two-element Boolean algebra.

2.1 Bipartite, Simple and Semihoop-Simple MTL-Algebras

We recall that an algebra is *simple* when its only congruences are the identity and the total relation. We shall use x^n to denote $x * \cdots * x$, where x occurs n-times. A variety of MTL-algebras is such that all its chains are simple iff it belongs to a variety $\mathbb{S}_n\text{MTL}$, for some $n \geq 3$, where $\mathbb{S}_n\text{MTL}$ is the class of MTL-algebras satisfying the n-weakening of the excluded middle law (see [23, 26]):

$$x \vee \neg(x^{n-1}) = 1. \tag{EM_n}$$

Clearly, the case $n = 2$ gives $\mathbb{B}$.

For each $n \geq 3$, $\mathbb{S}_n^-\text{MTL}$ is the subvariety of $\mathbb{S}_n\text{MTL}$ whose class of chains is given by the chains in $\mathbb{S}_n\text{MTL}$ without the negation fixpoint (*i.e.* an element x such that $\neg x = x$), and it is axiomatised by $\neg(((x \rightarrow \neg x) * (\neg x \rightarrow x))^{n-1}) = 1$ (see [26]).

The *radical $Rad(\mathcal{A})$* of an MTL-algebra $\mathcal{A}$ is the intersection of all maximal filters of $\mathcal{A}$. Clearly, when $\mathcal{A}$ is a chain, then $Rad(\mathcal{A})$ coincides with its unique maximal filter. The *coradical $\overline{Rad}(\mathcal{A})$* of $\mathcal{A}$ is the set $\{a \in A \mid \neg a \in Rad(\mathcal{A})\}$.

$\mathcal{A}$ is said *bipartite* when $A = Rad(\mathcal{A}) \cup \overline{Rad}(\mathcal{A})$. The variety generated by all bipartite MTL-algebras is called $\mathbb{BP}_0$ in [27], and it is axiomatised by extending MTL via the identity

$$(\neg((\neg x)^2))^2 = \neg((\neg(x^2))^2). \tag{BP_0}$$

The radical of an MTL-algebra is the most general universe of a *prelinear semi-hoop* [17][1], in the sense that $Rad(\mathcal{A})$ is a *prelinear subsemihoop* of $\mathcal{A}$, and that each prelinear semihoop is a radical of some MTL-algebra (or even, of some bipartite MTL-algebra). When $Rad(\mathcal{A})$ is a simple prelinear semihoop, we say that $\mathcal{A}$ is *semihoop-simple*.

Recall further from [1] that each BL-chain $\mathcal{A}$ is the *ordinal sum* $\bigoplus_{i \in I} \mathcal{W}_i$ of a family of *Wajsberg hoops* (that is, $\perp$-free subreducts of MV-algebras), where I is a chain with minimum i_0, and $\mathcal{W}_{i_0}$ is bounded.

2.2 Minimally Many-Valued Logics and Almost Minimal Varieties

In this paper we shall deal with those extensions of MTL which are distinct from B but closest to it.

Definition 1. *A consistent extension L of MTL is called* minimally many-valued *(m.m.v., for short) if it is distinct from B, but it is such that any of its proper consistent extensions coincides with B.*

Using the antiisomorphism between the lattices of axiomatic extensions of MTL and of subvarieties of MTL we can formulate the algebraic equivalent of Definition 1. The following notion was introduced and firstly studied in [18, 22].

Definition 2. *A variety of MTL-algebras is said* almost minimal *(a.m.) whenever the variety of Boolean algebras is its only proper non-trivial subvariety.*

Clearly, L is m.m.v. iff $\mathbb{L}$ is almost minimal.

Naively enough, one can think that being minimally many-valued amounts to have the smallest number of truth-values greater than two. Actually, the only two three-valued extensions of MTL, namely G_3 and $Ł_3$, are m.m.v. On the other hand we shall see that logics with arbitrarily large finite sets of truth-values, and even some infinitely valued logics, are m.m.v., too.

---

[1] In [17] such structures are called *basic semihoops*, but we use the terminology of [15].

3   MTL Algebras: The Search for Almost Minimal Varieties

In this section we present some result concerning almost minimal varieties of MTL-algebras.

Theorem 1. *Let* $\mathbb{L}$ *be an almost minimal variety of MTL-algebras. Then* $\mathbf{V}(\mathcal{C}) = \mathbb{L}$ *for each non-Boolean chain* $\mathcal{C} \in \mathbb{L}$. *Whence, every m.m.v. logic is single chain complete, in the sense of [3, 24].*

Proof. Let $\mathbb{L}$ be an almost minimal variety of MTL-algebras. Clearly, $\mathbb{B} \subsetneq \mathbf{V}(\mathcal{C}) \subseteq \mathbb{L}$ for any non-Boolean chain $\mathcal{C} \in \mathbb{L}$. If $\mathbf{V}(\mathcal{C}) \neq \mathbb{L}$ then $\mathbb{L}$ would not be almost minimal. Whence $\mathbf{V}(\mathcal{C}) = \mathbb{L}$. □

The following result provides some information about the structure of the chains of almost minimal varieties of MTL-algebras.

Theorem 2. *Let* $\mathbb{L}$ *be an almost minimal variety of MTL-algebras. Then either every L-chain is simple or every L-chain is bipartite.*

Proof. Let $\mathbb{L}$ be an almost minimal variety of MTL-algebras. By Theorem 1 $\mathbf{V}(\mathcal{A}) = \mathbb{L}$, for some L-chain $\mathcal{A}$. We first prove that each L-chain is either simple or bipartite. Assume by contradiction that $\mathbb{L}$ contains an L-chain $\mathcal{B}$ being neither simple nor bipartite. By [27, Proposition 3.17] the set $Rad(\mathcal{B}) \cup \overline{Rad}(\mathcal{B})$ is the carrier of a subalgebra of $\mathcal{B}$, say $\mathcal{S}$, which is clearly bipartite. Whence $\mathbf{V}(\mathcal{S}) \subsetneq \mathbf{V}(\mathcal{B})$, as they are distinguished by the identity (BP$_0$). Since $\mathcal{B}$ is not simple then $\{1\} \subsetneq Rad(\mathcal{B})$, whence $\mathcal{S} \not\simeq \mathbf{2}$. Then $\mathbb{B} \subsetneq \mathbf{V}(\mathcal{S}) \subsetneq \mathbf{V}(\mathcal{B}) \subseteq \mathbb{L}$, contradicting the almost minimality of $\mathbb{L}$.

By Theorem 1 if $\mathcal{A}$ simple, then every other L-chain is simple, for otherwise, $\mathbb{L}$ would contain a bipartite chain generating a non-Boolean proper subvariety. If $\mathcal{A}$ is bipartite, then every other L-chain is bipartite. □

Theorem 3. *Let* $\mathbb{L}$ *be an almost minimal variety of MTL-algebras. Then either* $\mathbb{L} \subset \mathbb{BP}_0$ *or* $\mathbb{L} = \mathrm{MV}_3$ *or* $\mathbb{L} \subset \mathbb{S}_n^- \mathrm{MTL}$, *for some* $n \geq 4$.

Proof. By Theorem 2, [27, Theorem 3.20], [26, Theorem 8.18] either $\mathbb{L} \subset \mathbb{BP}_0$ or $\mathbb{L} \subset \mathbb{S}_n \mathrm{MTL}$ for some $n \geq 3$. Then, assume the latter case. If $\mathbb{L} \neq \mathrm{MV}_3$, then it cannot contain the chain $\mathbf{L}_3$, otherwise $\mathbb{L}$ would not be almost minimal, as $\mathrm{MV}_3 = \mathbf{V}(\mathbf{L}_3)$. Whence, each chain $\mathcal{A} \in \mathbb{L}$ must lack the negation fixpoint, otherwise $\mathbf{L}_3$ would be a subalgebra of $\mathcal{A}$. We conclude $\mathbb{L} \subset \mathbb{S}_n^- \mathrm{MTL}$ for some $n \geq 4$. □

Proposition 1. *Let* $\mathbb{L}$ *be an almost minimal variety of MTL-algebras generated by a finite chain* $\mathcal{A}$. *Then each non-Boolean chain in* $\mathbb{L}$ *is isomorphic with* $\mathcal{A}$.

Proof. It is known that every variety of MTL-algebras is congruence distributive, and then since $\mathcal{A}$ is finite, by [9, Ch. IV, Corollary 6.10] we have that all the subdirectly irreducible algebras in $\mathbf{V}(\mathcal{A})$ belong to $\mathbf{HS}(\mathcal{A})$. By Theorem 1, every non-Boolean chain $\mathcal{B}$ in $\mathbb{L}$ generates $\mathbb{L}$. Using [14, Proposition 4.18] we derive that $|\mathcal{A}| = |\mathcal{B}|$, and hence $\mathcal{B}$ is finite and then subdirectly irreducible. Whence $\mathcal{B} \in \mathbf{HS}(\mathcal{A})$ and we conclude that $\mathcal{A} \simeq \mathcal{B}$. □

Proposition 2. *Let $\mathcal{A}$ be an MTL-chain with more than two elements. If every element $a \in A \setminus \{0, 1\}$ generates $\mathcal{A}$, then $\mathcal{A}$ is simple or bipartite. Moreover, if $\mathcal{A}$ is bipartite then it is semihoop simple.*

Proof. Let $\mathcal{A}$ be an MTL-chain with more than two elements, singly generated by every element $a \in A \setminus \{0, 1\}$. Assume that $\mathcal{A}$ is not simple. Then $Rad(\mathcal{A}) \neq \{1\}$, that is, there exists $a \in Rad(\mathcal{A}) \setminus \{1\}$. Whence $Rad(\mathcal{A})$ generates $\mathcal{A}$, but this means $Rad(\mathcal{A}) \cup \overline{Rad}(\mathcal{A}) = A$, that is, $\mathcal{A}$ is bipartite. Assume now $\mathcal{A}$ is bipartite. By way of contradiction assume further that $\mathcal{A}$ is not semihoop simple, then there is a filter $\mathfrak{f}$, with $\{1\} \subsetneq \mathfrak{f} \subsetneq Rad(\mathcal{A})$. Whence, by [27, Proposition 3.17], $\mathfrak{f}$ generates the subalgebra $\mathfrak{f} \cup \bar{\mathfrak{f}}$, where $\bar{\mathfrak{f}} = \{a \in A \mid \neg a \in \mathfrak{f}\}$. Observe that $a < b$ for each $a \in \bar{\mathfrak{f}}$ and each $b \in Rad(\mathcal{A})$, as $\neg a * b \in Rad(\mathcal{A})$, $\neg a * b > 0 = \neg a * a$. We conclude that $\mathfrak{f} \cup \bar{\mathfrak{f}} \subsetneq A$, contradicting the assumption that every element $a \in A \setminus \{0, 1\}$ generates $\mathcal{A}$.   □

We can now state the characterisation theorem for the almost minimal varieties generated by a finite MTL-chain.

Theorem 4. *Let $\mathcal{A}$ be a finite MTL-chain. Then the variety $\mathbb{L} = \mathbf{V}(\mathcal{A})$ is almost minimal if and only if $|\mathcal{A}| > 2$, and every element $0 < a < 1$ singly generates $\mathcal{A}$.*

Proof. Let $\mathcal{A}$ be a finite MTL-chain. Assume first that $|\mathcal{A}| > 2$, and every element $0 < a < 1$ generates $\mathcal{A}$. Then, by Proposition 2, $\mathcal{A}$ is either simple or bipartite and semihoop simple. Since every element $0 < a < 1$ generates $\mathcal{A}$, the only subalgebras of $\mathcal{A}$ are $\mathbf{2}$ and $\mathcal{A}$ itself. If $\mathcal{A}$ is simple then $\mathbf{H}(\mathcal{A})$ contains only the trivial one-element algebra and $\mathcal{A}$ itself. If $\mathcal{A}$ is bipartite and semihoop simple, then by [27, Theorem 3.20] $\mathbf{H}(\mathcal{A})$ contains only the trivial one-element algebra, the algebra $\mathbf{2}$ and $\mathcal{A}$ itself. Since $\mathbb{L}$ has the congruence extension property, and every L-chain $\mathcal{B}$ is finite, and hence subdirectly irreducible, by [9, Ch. IV, Corollary 6.10] $\mathcal{B} \in \mathbf{HS}(\mathcal{A})$. As a consequence, the subvarieties of $\mathbb{L}$ are the trivial one, $\mathbb{B}$ and $\mathbb{L}$ itself. That is, $\mathbb{L}$ is almost minimal.

Finally, suppose that $\mathbb{L}$ is almost minimal. Then the only elements (up to isomorphism) in $\mathbf{S}(\mathcal{A})$ must be $\mathbf{2}$ and $\mathcal{A}$, as any subchain of $\mathcal{A}$ having cardinality c such that $2 < c < |A|$, would generate a proper non-Boolean subvariety of $\mathbb{L}$. Hence, every element $0 < a < 1$ generates $\mathcal{A}$.   □

As a corollary, we get a strengthening of Theorem 2, for the almost minimal varieties generated by a finite chain.

Theorem 5. *Let $\mathbb{L}$ be an almost minimal variety of MTL-algebras generated by a finite chain $\mathcal{A}$. Then $\mathcal{A}$ is singly generated by each one of its non-Boolean elements. Further, $\mathcal{A}$ is simple or bipartite and semihoop simple, and the only other non-trivial chain in $\mathbb{L}$ is (up to isomorphisms) $\mathbf{2}$.*

Proof. Immediate by Theorem 4, Proposition 2 and Proposition 1.   □

4   The Cases of WNM-Algebras and of BL-Algebras

In this section we classify all the almost minimal varieties of WNM-algebras and BL-algebras.

Lemma 1. *Let $\mathcal{A}$ be an MTL-chain containing $0 < a < 1$ such that $a * a = a$ and $\neg\neg a = a$. Then the subalgebra of $\mathcal{A}$ generated by a is isomorphic to $\mathbf{NM}_4$.*

Proof. Let $\mathcal{A}$ be an MTL-chain containing $0 < a < 1$ such that $a * a = a$ and $\neg\neg a = a$. Clearly the set $B = \{0, a, \neg a, 1\}$ is closed under $\neg$ and $*$: we now show that $a \to \neg a = \neg a$, and this proves that B is closed also under $\to$, and that $a \neq \neg a$. We have that $a \to \neg a \geq \neg a$: assume by contradiction that $a \to \neg a > \neg a$, and set $c = a \to \neg a$. Since $\neg a < c$ we have $(a * a) * c = a * c > 0$: however $a * (a * c) = 0$, since $a * c \leq \neg a$, and hence $*$ is not associative, a contradiction. Hence the subalgebra of $\mathcal{A}$ generated by a has B as carrier, and clearly it is isomorphic to $\mathbf{NM}_4$.   □

Theorem 6. *The almost minimal varieties of WNM-algebras are $\mathbb{MV}_3$, $\mathbb{G}_3$ and $\mathbb{NM}_4$.*

Proof. By Theorem 1, we have that every almost minimal variety of WNM-algebras is generated by a single WNM-chain. Since the variety of WNM-algebras is locally finite (see [19]), every infinite WNM-chain $\mathcal{C}$ contains finite subchains (with more than two elements), singly generating proper subvarieties of $\mathbf{V}(\mathcal{C})$. Whence, no variety generated by an infinite WNM-chain could be almost minimal.

Let then $\mathcal{A}$ be a finite WNM-chain such that $\mathbf{V}(\mathcal{A})$ is almost minimal. By Theorem 5, $\mathcal{A}$ is simple or bipartite and semihoop simple, and by Theorem 4, $|A| > 2$.

If $\mathcal{A}$ is simple, then by [26, Proposition 8.13] it is a DP-chain. From [4] and [7, Theorem 1] we have that every DP-chain with at least three elements contains $\mathbf{L}_3$ as a subalgebra. Then, $\mathcal{A} \simeq \mathbf{L}_3$, and hence $\mathbf{V}(\mathcal{A}) = \mathbb{MV}_3$.

Assume now that $\mathcal{A}$ is bipartite and semihoop simple. Since $\mathcal{A}$ is not Boolean, $Rad(\mathcal{A})$ contains at least two elements. Let us call m its minimum. As $\mathcal{A}$ is bipartite, then by [27, Theorem 3.20] $Rad(\mathcal{A}) = A^+ = \{a \in A \mid a > \neg a\}$. Notice that A^+, being $\mathcal{A}$ a WNM-chain, contains only idempotent elements. Then, as $\mathcal{A}$ is finite and semihoop simple, we must have that $Rad(\mathcal{A}) = \{m, 1\}$. We analyse three exhaustive cases, depending on the negation of m.

If $\neg m = 0$, then it is easy to check that $\mathcal{A}$ contains $\mathbf{G}_3$ as subalgebra, and since $\mathbf{V}(\mathcal{A})$ is almost minimal by hypothesis, we must conclude that $\mathcal{A} \simeq \mathbf{G}_3$, and $\mathbf{V}(\mathcal{A}) = \mathbb{G}_3$.

If $\neg\neg m = m$, then, by Lemma 1, $\mathcal{A}$ contains $\mathbf{NM}_4$ as subalgebra, and since $\mathbf{V}(\mathcal{A})$ is almost minimal by hypothesis, we must conclude that $\mathcal{A} \simeq \mathbf{NM}_4$, and $\mathbf{V}(\mathcal{A}) = \mathbb{NM}_4$.

The last case is when $\neg m > 0$ and $\neg\neg m > m$. As $m \in A^+$ we have that $\neg m < m$, and hence $\neg m < m < \neg\neg m$. But then $\neg\neg m = 1$, as $A^+ = Rad(\mathcal{A}) = \{m, 1\}$, and this is a contradiction, since $\neg m > 0$ by hypothesis.

The proof is settled.   □

The almost minimal varieties of BL-algebras have already been classified in [22]. To render the paper self-contained we provide here an alternative proof.

Theorem 7. *The almost minimal varieties of MV-algebras are* $\mathbb{C}$ *and the family* $\{\mathbb{MV}_k : k - 1$ *is prime*$\}$.

Proof. By [11, Theorems 8.3.5, 8.4.4] we have that every non-trivial variety of MV-algebras different from $\mathbb{C}$ contains $\mathbb{MV}_k$ as subvariety, for some $k \geq 3$. Since $\mathbb{MV}_m$ is a subvariety of $\mathbb{MV}_n$ iff $m - 1$ divides $n - 1$, the claim is settled.   □

Theorem 8 [22, Theorem 6]. *The almost minimal varieties of BL-algebras are* $\mathbb{C}$, $\mathbb{P}$, $\mathbb{G}_3$ *and the family* $\{\mathbb{MV}_k \mid k - 1$ *is prime*$\}$.

Proof. Let $\mathcal{A}$ be a BL-chain such that $\mathbf{V}(\mathcal{A})$ is almost minimal. By Theorem 7, if $\mathcal{A}$ is an MV-chain, then $\mathbf{V}(\mathcal{A}) = \mathbb{C}$ or $\mathbf{V}(\mathcal{A}) \in \{\mathbb{MV}_k : k - 1$ is prime$\}$. Assume now that $\mathcal{A}$ is not an MV-chain. Then, $\mathcal{A}$ is isomorphic with an ordinal sum of Wajsberg hoops $\mathcal{W}_0 \oplus \bigoplus_{i \in I} \mathcal{W}_i$, for a totally ordered index set $I \neq \emptyset$, and $\mathcal{W}_0$ bounded. Fix $i \in I$. If $\mathcal{W}_i$ is bounded, then $\mathbf{G}_3 \cong \mathbf{2} \oplus \mathbf{2}$ is a subalgebra of $\mathcal{A}$, whence $\mathcal{A} \cong \mathbf{G}_3$, for otherwise $\mathbf{V}(\mathcal{A})$ would not be almost minimal. If $\mathcal{W}_i$ is unbounded, then $\mathcal{A}$ contains as a subalgebra of infinite cardinality $\mathbf{2} \oplus \mathcal{W}_i$. By [13, Corollaries 2.9, 2.10], $\mathbf{V}(\mathbf{2} \oplus \mathcal{W}_i) = \mathbb{P}$, which is almost minimal. Whence, $\mathbf{V}(\mathcal{A}) = \mathbb{P}$. The proof is settled.   □

In [6], the authors characterise Łukasiewicz and Gödel logic among a class of $[0, 1]$-valued continuous t-norm based logics, by means of a pair of *metalogical* principles. Being deliberately rather vague on what counts as metalogical principle, we offer here the following characterisation for Product logic.

Theorem 9. *The only minimally many-valued logic which is complete with respect to a continuous t-norm is product logic.*

Proof. Immediate, from Theorem 8.   □

In the forthcoming paper [2] we shall offer a characterisation analogous to Theorem 9 for the logic of Chang's MV-algebra.

One may wonder how much the knowledge of all minimally many-valued extensions of a logic L provides information on the very logic L. In this regards, Theorems 6 and 8 yield the following observation.

Theorem 10. *There are infinitely many distinct extensions of BL which have the same set of minimally many-valued extensions as BL does. The same holds for WNM.*

Proof. First recall that a formula is a tautology of every $Ł_k$, for $k - 1$ a prime number iff it is a tautology of Ł [11, Proposition 8.1.2]. Whence, it is immediate to note from Theorem 8 that $\mathbf{G}_3 \wedge \varPi \wedge$ Ł (where $\wedge$ denotes the meet in the lattice of axiomatic extensions of MTL, ordered by strength) has the same set of m.m.v. extensions as BL does. The proof is settled by further noticing that every member of the infinite family of distinct logics $\{\mathbf{G}_i \wedge \varPi \wedge$ Ł $\mid i \geq 3\}$, together with G $\wedge$ $\Pi\wedge$ L, share the same set of minimally many-valued extensions as BL does. Indeed, if $3 \leq i < j$, then $\mathbb{G}_i \subsetneqq \mathbb{G}_j \subsetneqq \mathbb{G}$, and hence $G_i \wedge \Pi \wedge$ L, $G_j \wedge \Pi\wedge$L and G $\wedge$ $\Pi\wedge$ L are pairwise distinct. For WNM the proof is the same, once we replace $\{G_i \wedge \Pi\wedge \text{ L} \mid i \geq 3\}$ with $\{G_i \wedge \text{NM}_4\wedge \text{ L}_3 \mid i \geq 3\}$. $\square$

5 Booleanising and Non-classical Extension Formulas

We recall the reader that the complexity class DP is defined as the class whose members are of the form $L_1 \cap L_2$, for $L_1 \in NP$ and $L_2 \in coNP$. DP coincides with BH_2, that is, the second step in the so-called Boolean Hierarchy BH, which in turns is entirely contained in Δ_2^p in the polynomial hierarchy [10, 28].

A formula φ is a *non-classical extension formula* for L if the extension of L with φ *as an axiom scheme*, which we denote $L + \varphi$, is a consistent logic distinct from Boolean logic. Clearly, a Boolean tautology φ is not a non-classical extension formula for L iff $L + \varphi$ collapses to B. We can then generalise in a natural way the problem of determining whether a formula is a non-classical extension for L as follows. Given an extension L of MTL, a formula φ is *Booleanising* for some non-classical extension L' of L, if $L' + \varphi$ coincides with Boolean logic.

In the sequel, if a formula φ is not a tautology of a logic L, we shall say that φ *fails* in L. In the following theorems we shall use the fact that the lattice of extensions of a logic L (extending MTL) is algebraic, this fact implying that each extension L' of L is weaker than at least one m.m.v. extension.

The problem $L-$Booleanising is the set of formulas φ such that φ is Booleanising for some non-classical extension of L. The problem $L - $N.C.Extension is the set of formulas φ such that φ is a non-classical extension for L.

Theorem 11. *BL$-$Booleanising and WNM$-$Booleanising are in DP.*

Proof. We start observing that for any L we have that $\varphi \in L - $ Booleanising iff φ is a Boolean tautology and φ fails in some extension L' of L, such that $L' + \varphi$ is Boolean logic. The latter fact implies that φ fails in some minimally many-valued logic L'' extending L', otherwise $L' + \varphi$ would not be a proper extension of L''.

We shall deal first with the BL case. Recall that φ fails in some finitely-valued Łukasiewicz logic $Ł_k$, with $k - 1$ prime iff it fails in the infinitely-valued Łukasiewicz logic Ł [11, Proposition 8.1.2]. Whence, by Theorem 8, φ fails in some m.m.v. extension of BL iff it fails in $G_3 \wedge \Pi\wedge$Ł. Deciding that φ is a Boolean tautology has complexity $coNP$, while deciding that φ fails in $G_3 \wedge \Pi\wedge$ Ł has complexity NP, since it is well-known that the tautology problem for G_3, Π and Ł are all $coNP$-problems (see [5]). Whence, BL$-$Booleanising is in DP.

For what regards the WNM case, we just observe that, by Theorem 6, φ fails in some m.m.v. extension of WNM iff it fails in $G_3 \wedge \text{NM}_4\wedge \text{Ł}_3$. Deciding that φ fails in $G_3 \wedge \text{NM}_4\wedge \text{Ł}_3$ has complexity NP, as the tautology problem for G_3, NM_4 and $Ł_3$ are known to be in $coNP$. Then WNM$-$Booleanising is in DP. $\square$

Theorem 12. *BL$-$N.C.Extension is in Σ_2^p, while WNM$-$N.C.Extension is in coNP.*

Proof. We notice that for any logic L a formula φ is in $L - $ N.C.Extension iff it is a tautology of some m.m.v. extension of L. Then, as by Theorem 8 the tautology problem for any one of the infinitely many m.m.v. extensions of BL has complexity $coNP$, we readily have that BL$-$N.C.Extension is in Σ_2^p, as one first guesses the m.m.v. extension L of BL such that φ is a tautology of L. This task has complexity NP since by [25] it is sufficient to limit the search space in $\{G_3, \Pi\} \cup \{L_k : k - 1 \text{ prime}\}$ to all k such that $\log k \leq p(size(\varphi))$ for a suitably fixed polynomial p. Finally, we clearly have that WNM$-$N.C.Extension is in $coNP$ as, by Theorem 6, there are exactly three distinct m.m.v. extensions of WNM, each one of them having its tautology problem in $coNP$.   □

6   Conclusions and Future Work

As a future topic, we shall refine the complexity results of Sect. 5. A harder task is the classification of the m.m.v. extensions of MTL, which is probably hopeless in general, but for finite-valued logics the analysis might be easier. Another problem concerns the amalgamation property, which holds for every a.m. variety generated by a finite chain, whilst in general it remains open.

A broader research area concerns bringing together description logics knowledge representation machinery with the ability of describing vague concepts typical of many-valued logics. The second author has proposed a research project aiming at, among other topics, studying application of m.m.v. extensions of MTL to description logics. This will be the first step towards the applications to real-world problems of minimally many-valued logics.

References

1. Aglianò, P., Montagna, F.: Varieties of BL-algebras I: general properties. J. Pure Appl. Algebra **181**(2–3), 105–129 (2003). doi:10.1016/S0022-4049(02)00329-8
2. Aguzzoli, S., Bianchi, M.: On Linear Varieties of MTL-Algebras (2017, in preparation)
3. Aguzzoli, S., Bianchi, M.: Single chain completeness and some related properties. Fuzzy Sets Syst. **301**, 51–63 (2016). doi:10.1016/j.fss.2016.03.008
4. Aguzzoli, S., Bianchi, M., Valota, D.: A note on drastic product logic. In: Laurent, A., Strauss, O., Bouchon-Meunier, B., Yager, R.R. (eds.) IPMU 2014. CCIS, vol. 443, pp. 365–374. Springer, Heidelberg (2014). doi:10.1007/978-3-319-08855-6_37. arXiv:1406.7166
5. Aguzzoli, S., Gerla, B., Haniková, Z.: Complexity issues in basic logic. Soft Comput. **9**(12), 919–934 (2005). doi:10.1007/s00500-004-0443-y
6. Aguzzoli, S., Marra, V.: Two principles in many-valued logic. In: Montagna, F. (ed.) Petr Hájek on Mathematical Fuzzy Logic, pp. 159–174. Springer, Heidelberg (2015). doi:10.1007/978-3-319-06233-4_8
7. Bianchi, M.: The logic of the strongest and the weakest t-norms. Fuzzy Sets Syst. **276**, 31–42 (2015). doi:10.1016/j.fss.2015.01.013
8. Blok, W., Pigozzi, D.: Algebraizable Logics, Memoirs of the American Mathematical Society, vol. 77. American Mathematical Society (1989). tinyurl.com/o89ug5o

9. Burris, S., Sankappanavar, H.: A Course in Universal Algebra, vol. 78. Springer, Heidelberg (1981). 2012 Electronic Edition: http://tinyurl.com/zaxeopo

10. Cai, J., Gundermann, T., Hartmanis, J., Hemachandra, L.A., Sewelson, V., Wagner, K., Wechsung, G.: The boolean hierarchy I: structural properties. Siam J. Comput. **17**(6), 1232–1252 (1988). doi:10.1137/0217078

11. Cignoli, R., D'Ottaviano, I., Mundici, D.: Algebraic Foundations of Many-Valued Reasoning. Trends in Logic, vol. 7. Kluwer Academic Publishers, Dordrecht (1999)

12. Cignoli, R., Esteva, F., Godo, L., Torrens, A.: Basic fuzzy logic is the logic of continuous t-norms and their residua. Soft Comput. **4**(2), 106–112 (2000). doi:10.1007/s005000000044

13. Cignoli, R., Torrens, A.: An algebraic analysis of product logic. Mult. Valued Log. **5**, 45–65 (2000)

14. Cintula, P., Esteva, F., Gispert, J., Godo, L., Montagna, F., Noguera, C.: Distinguished algebraic semantics for t-norm based fuzzy logics: methods and algebraic equivalencies. Ann. Pure Appl. Log. **160**(1), 53–81 (2009). doi:10.1016/j.apal.2009.01.012

15. Cintula, P., Hájek, P., Noguera, C.: Handbook of Mathematical Fuzzy Logic, vols. 1 and 2. College Publications (2011)

16. Esteva, F., Godo, L.: Monoidal t-norm based logic: towards a logic for left-continuous t-norms. Fuzzy Sets Syst. **124**(3), 271–288 (2001). doi:10.1016/S0165-0114(01)00098-7

17. Esteva, F., Godo, L., Hájek, P., Montagna, F.: Hoops and fuzzy logic. J. Log. Comput. **13**(4), 532–555 (2003). doi:10.1093/logcom/13.4.532

18. Galatos, N., Jipsen, P., Kowalski, T., Ono, H.: Residuated Lattices: An Algebraic Glimpse at Substructural Logics. Studies in Logic and the Foundations of Mathematics, vol. 151. Elsevier, Hoboken (2007)

19. Gispert, J.: Axiomatic extensions of the nilpotent minimum logic. Rep. Math. Log. **37**, 113–123 (2003). tinyurl.com/nqsle2f

20. Hájek, P.: Metamathematics of Fuzzy Logic. Trends in Logic, Paperback edn., vol. 4. Kluwer Academic Publishers (1998)

21. Jenei, S., Montagna, F.: A proof of standard completeness for Esteva and Godo's logic MTL. Stud. Log. **70**(2), 183–192 (2002). doi:10.1023/A:1015122331293

22. Katoh, Y., Kowalski, T., Ueda, M.: Almost minimal varieties related to fuzzy logic. Rep. Math. Log., 173–194 (2006). tinyurl.com/h8sc6j3

23. Kowalski, T.: Semisimplicity, EDPC and discriminator varieties of residuated lattices. Stud. Log. **77**(2), 255–265 (2004). doi:10.1023/B:STUD.0000037129.58589.0c

24. Montagna, F.: Completeness with respect to a chain and universal models in fuzzy logic. Arch. Math. Log. **50**(1–2), 161–183 (2011). doi:10.1007/s00153-010-0207-6

25. Mundici, D.: Satisfiability in many-valued sentential logic is NP-complete. Theor. Comput. Sci. **52**(1), 145–153 (1987). doi:10.1016/0304-3975(87)90083-1

26. Noguera, C.: Algebraic study of axiomatic extensions of triangular norm based fuzzy logics. Ph.D. thesis, IIIA-CSIC (2006). tinyurl.com/ncxgolk

27. Noguera, C., Esteva, F., Gispert, J.: On some varieties of MTL-algebras. Log. J. IGPL **13**(4), 443–466 (2005). doi:10.1093/jigpal/jzi034

28. Papadimitriou, C., Yannakakis, M.: The complexity of facets (and some facets of complexity). J. Comput. Syst. Sci. **28**(2), 244–259 (1984). doi:10.1016/0022-0000(84)90068-0

Feature Selection Through Composition of Rough–Fuzzy Sets

Alessio Ferone$^{(\boxtimes)}$ and Alfredo Petrosino

Department of Science and Technology, University of Naples "Parthenope",
Centro Direzionale Isola C4, 80143 Naples, Italy
{alessio.ferone,alfredo.petrosino}@uniparthenope.it

Abstract. The well known principle of curse of dimensionality links both dimensions of a dataset stating that as dimensionality increases samples become too sparse to effectively extract knowledge. Hence dimensionality reduction is essential when there are many features and not sufficient samples. We describe an algorithm for unsupervised dimensionality reduction that exploits a model of the hybridization of rough and fuzzy sets. Rough set theory and fuzzy logic are mathematical frameworks for granular computing forming a theoretical basis for the treatment of uncertainty in many real–world problems. The hybrid notion of rough fuzzy sets comes from the combination of these two models of uncertainty and helps to exploit, at the same time, properties like coarseness and vagueness. Experimental results demonstrated that the proposed approach can effectively reduce dataset dimensionality whilst retaining useful features when class labels are unknown or missing.

Keywords: Rough fuzzy sets · Modelling hierarchies · Unsupervised feature selection

1 Introduction

Feature selection concerns the selection of the most predictive input attributes with respect to a given outcome. The main difference with other dimensionality reduction methods, is that selected features preserve the original meaning of the features after reduction.

In the recent years, granular computing has been extensively employed for feature selection. It is based on the concept of information granule, that is a collection of similar objects which can be considered indistinguishable. Partition of an universe into granules offers a coarse view of the universe where concepts, represented as subsets, can be approximated by means of granules. In this framework, rough set theory can be regarded to as a family of methodologies and techniques that make use of granules [15,16]. Granulation is of particular interest when a problem involves incomplete, uncertain or vague information.

The rough set ideology of using only the supplied data and no other information has many benefits in feature selection, although the requirement that

© Springer International Publishing AG 2017
A. Petrosino et al. (Eds.): WILF 2016, LNAI 10147, pp. 116–125, 2017.
DOI: 10.1007/978-3-319-52962-2_10

all data has to be discrete imposes some limitations. In order to overcome these limitations, the theory of fuzzy sets [24] can be applied to handle uncertainty and vagueness present in information system.

The hybrid notion of rough fuzzy sets comes from the combination of these two models of uncertainty to exploit, at the same time, properties like coarseness, by handling rough sets [15], and vagueness, by handling fuzzy sets [24]. Nevertheless, some considerations are in order. Classical rough set theory is defined over a given equivalence relation, although several equivalence relations, and hence partitions, can be defined over the universe of discourse. In order to exploit different partitions, we propose to refine them in a hierarchical manner, so that partitions at each level of the hierarchy retain all the important information contained into the partitions of the lower levels. The operation employed to perform the hierarchical refinement is called *Rough–Fuzzy product* ($\mathcal{RF}$-product) [18].

The hybridization of rough and fuzzy sets reported here has been observed to possess a viable and effective solution in feature selection. The model exhibits a certain advantage of having a new operator to compose rough fuzzy sets that is able to produce a sequence of composition of rough fuzzy sets in a hierarchical manner.

The article is organized as follows. In Sect. 2 the literature about feature selection using rough and fuzzy theories is reviewed. In Sect. 3 rough–fuzzy sets are introduced along with the rough–fuzzy product operation, while in Sect. 4, application to feature selection is explained. Section 5 presents the experimental results and Sect. 6 concludes the paper.

2   Related Works

Many problems in machine learning involve high dimensional descriptions of input features and therefore much research has been carried out on dimensionality reduction [4]. However, existing approaches tend to destroy the underlying semantic of the features [5] or require apriori information about the data [10]. In order to overcome these limitations, rough set theory is a technique that can reduce dimensionality using only information contained into the dataset, preserving at the same time the semantic of the features. Rough set theory can be used as such a tool to discover data dependencies and to reduce the number of attributes contained in a dataset [15].

In particular, the use of rough set theory to achieve dimensionality reduction has been proved to be a successful approach due to the following aspects:

– only the concepts embedded in data are analysed
– no apriori information about the data is required
– minimal knowledge representation is found

Given a dataset with discretized attribute values, rough set theory finds the most informative subset of the original attributes. This subset is called reduct.

Recently, researchers have focused their attention on reduct and classification algorithms based on rough sets [19] Finding all the reducts has been proved to be NP-Hard [9], but in many applications it is sufficient to compute only one reduct, that is the best reduct with respect to a given cost criterion associated with the selected attributes.

A major drawback when using rough set theory is represented by real value attributes, because it is not possible to say whether two attribute values are similar and to what extent they are indiscernible. A possible solution to this problem consists in discretizing the real valued attributes in order to obtain a dataset composed only by crisp values. This preprocessing step is not always adequate, being the source of potential information loss.

Fuzzy sets provide a framework to handle real value data effectively, by allowing values to belong to more than one class with different degrees of membership, and hence handling vagueness present in data.

In a hybridized approach, rough set theory allows to obtain a linguistic description whereas fuzzy set theory allows to generate numerical values starting from its linguistic description.

In [21] a novel concept of attributes reduction with fuzzy rough sets is proposed and an algorithm using discernibility matrix to compute all the reducts is developed. A solid mathematical foundation is set up for attributes reduction with fuzzy rough sets and a detailed comparison with the Quickreduct algorithm is also presented. Experimental results show that the proposed algorithm is feasible and valid.

Jensen and Shen in [7] proposed an extension of the fuzzy–rough feature selection algorithm, based on interval–valued fuzzy sets, in order to face the problem of missing values. In particular, by exploiting interval–valued fuzzy–rough sets, a new feature selection algorithm is developed that not only handles missing values, but also alleviates the problem of defining overly–specific type–1 fuzzy similarity relations.

In [8] three robust approaches to fuzzy–rough feature selection based on fuzzy similarity relations are proposed. In particular, a fuzzy extension to crisp discernibility matrices is proposed and employed for experimentation, showing that the methods produce small reduct while preserving classification accuracy.

Parthaláin et al. [13] examine a rough set based feature selection technique which exploits information extracted from the lower approximation, from the boundary region and the distance of objects in the boundary region from the lower approximation. This information allows to obtain smaller subset if compared to those obtained using the dependency function alone. The proposed approach demonstrates that information extracted from the boundary region is useful for feature selection.

Supervised feature selection methods evaluate subsets of features using an objective function in order to select only those features related to the decision classes. However, in many applications, class labels are not available or incomplete, and unsupervised feature selection approaches are needed. Approaches to unsupervised feature selection can be divided in two broad classes: those that maximize clustering performance with respect to an index function [3,12], and those that select features based on their relevance. The main idea of the latter methods, is that features with little or no information with respect to the remaining features are redundant and can be eliminated [2,6,11]. The work presented in [14] is based on fuzzy-rough sets and, in particular, employs a fuzzy-rough discernibility measure to compute the discernibility between a single feature and a subset of other features. If the single feature can be discerned by the subset of the other features, than it is considered redundant and removed from the feature set. Features are removed until no further inter-dependency can be found.

In [22] authors propose a new unsupervised quick reduct algorithm based on rough set theory. The proposed algorithm is based on a new definition of positive region for unsupervised subset evaluation measure using rough set theory. The evaluation of degree of dependency value for a features subset leads to each conditional attribute and evaluate mean of dependency values for all conditional attributes.

3   Rough–Fuzzy Sets

Let U be the universe of discourse, X a fuzzy subset of U, such that $\mu_X(u)$ represents the fuzzy membership function of X over U, and R an equivalence relation that induces the partition $U/R = \{Y_1, \ldots, Y_p\}$ over U in p disjoint sets, i.e. $Y_i \bigcap Y_j = \emptyset \ \forall i, j = 1, \ldots, p$ and $\bigcup_{i=1}^{p} Y_i = U$. Considering the lower and upper approximations of the fuzzy subset X as, respectively, the infimum and the supremum of the membership functions of the elements of a class Y_i to the fuzzy set X [18], a rough–fuzzy set can be defined as a triple

$$RF_X = (\mathcal{Y}, \mathcal{I}, \mathcal{S}) \tag{1}$$

where $\mathcal{Y} = U/R$ and $\mathcal{I}, \mathcal{S}$ are mappings of kind $U \to [0,1]$ such that $\forall u \in U$,

$$\mathcal{I}(u) = \sum_{i=1}^{p} \underline{\nu_i} \times \mu_{Y_i}(u) \tag{2}$$

$$\mathcal{S}(u) = \sum_{i=1}^{p} \overline{\nu_i} \times \mu_{Y_i}(u) \tag{3}$$

where

$$\underline{\nu_i} = \inf\{\mu_X(u) | u \in Y_i\} \tag{4}$$
$$\overline{\nu_i} = \sup\{\mu_X(u) | u \in Y_i\} \tag{5}$$

$\mathcal{Y}$ and μ uniquely define a rough–fuzzy set. In order to retain important information contained into different rough–fuzzy sets, it is possible to employ an operation called *Rough–Fuzzy product* ($\mathcal{RF}$-product), defined by:

Definition 1. *Let* $RF^i = (\mathcal{Y}^i, \mathcal{I}^i, \mathcal{S}^i)$ *and* $RF^j = (\mathcal{Y}^j, \mathcal{I}^j, \mathcal{S}^j)$ *be two rough fuzzy sets defined, respectively, over partitions* $\mathcal{Y}^i = (Y_1^i, \ldots, Y_p^i)$ *and* $\mathcal{Y}^j = (Y_1^j, \ldots, Y_p^j)$ *with* $\mathcal{I}^i$ *(resp.* $\mathcal{I}^j$*) and* $\mathcal{S}^i$ *(resp.* $\mathcal{S}^j$*) indicating the measures* expressed in Eqs. (2) and (3). The $\mathcal{RF}$-product between two rough–fuzzy sets, denoted by $\otimes$, is defined as a new rough fuzzy set

$$RF^{i,j} = RF^i \otimes RF^j = (\mathcal{Y}^{i,j}, \mathcal{I}^{i,j}, \mathcal{S}^{i,j}) \tag{6}$$

$$Y_k^{ij} = \begin{cases} \displaystyle\bigcup_{\substack{s=1 \\ q=h}}^{\substack{s=h \\ q=1}} Y_q^i \cap Y_s^j & h = k, \qquad k \leq p \\[2em] \displaystyle\bigcup_{\substack{s=h \\ q=p}}^{\substack{s=p \\ q=h}} Y_q^i \cap Y_s^j & h = k-p+1, \qquad k > p \end{cases} \tag{7}$$

and $\mathcal{I}^{i,j}$ and $\mathcal{S}^{i,j}$ are

$$\mathcal{I}^{i,j}(u) = \sum_{k=1}^{2p-1} \underline{\nu}_k^{i,j} \times \mu_k^{i,j}(u) \tag{8}$$

$$\mathcal{S}^{i,j}(u) = \sum_{k=1}^{2p-1} \overline{\nu}_k^{i,j} \times \mu_k^{i,j}(u) \tag{9}$$

4   Feature Selection by Rough–Fuzzy Product

In this section we describe how the rough–fuzzy product can be exploited in order to find distinctive features in an unsupervised way. In particular, the proposed approach is composed by two steps

1. feature granulation, which consists in partitioning the data considering each single feature and building a rough–fuzzy set for each partition
2. feature selection, which consists in combining rough–fuzzy sets by means of rough–fuzzy product and selecting the most distinctive features

4.1   Feature Granulation

The feature granulation step is based on the principle of justifiable granularity [17], that is concerned with the formation of a meaningful information granule Ω based on some experimental evidence of scalar numeric data, $D = x_1, x_2, \ldots, x_N$. Such construct has to respect two requirements:

1. The numeric evidence accumulated within the bounds of Ω has to be as high as possible, i.e. the existence of the information granule is well motivated, or justified, by the experimental data.
2. The information granule should be as specific as possible meaning that it represents a well-defined semantics, i.e. Ω has to be as specific as possible.

The first requirement is quantified by counting the number of data falling within the bounds of Ω, specifically we consider

$$f_1(card(X_k \in \Omega)) = card(X_k \in \Omega) \tag{10}$$

The specificity of the information granule can be quantified by taking into account its size. The length of the interval Ω can be considered as a measure of specificity, in particular we consider

$$f_2(length(\Omega)) = exp(-\alpha|a - m|) \tag{11}$$

where $m = med(\Omega)$ and a is the lower/upper bound of the interval Ω (we consider symmetric interval centred in m). The lower the value of $f2(length(\Omega))$, the higher the specificity is. The optimal granulation is obtained by maximizing the sum over all the granules Ω_i

$$\sum_i V(\Omega_i) \tag{12}$$

where

$$V(\Omega_i) = f_1(card(X_k \in \Omega_i)) * f_2(length(\Omega_i)) \tag{13}$$

As the requirements of experimental evidence and specificity are in conflict, we consider the maximization of the product $V = f_1 * f_2$.

4.2   Feature Selection

The procedure described in Sect. 4.1 is applied independently to each feature of the dataset, thus yielding many partitions of the same dataset. As explained before, for each partition it is possible to define a rough–fuzzy set which can be composed by means of the rough–fuzzy product. Let $RF^i = (\mathcal{Y}^i, \mathcal{I}^i, \mathcal{S}^i)$ and $RF^j = (\mathcal{Y}^j, \mathcal{I}^j, \mathcal{S}^j)$ be two rough–fuzzy sets relative to feature i and j, and let $RF^{i,j} = RF^i \otimes RF^j = (\mathcal{Y}^{i,j}, \mathcal{I}^{i,j}, \mathcal{S}^{i,j})$ be the rough–fuzzy set obtained by applying the rough–fuzzy product to these rough–fuzzy sets. In order to evaluate the goodness of the newly formed rough–fuzzy set with respect the operands, we propose to exploit again the principle of justifiable granularity, where for each granule $Y_k^{i,j}$ of the new rough–fuzzy set

$$f_1(card(Y_k^{i,j})) = card(Y_k^{i,j}) \tag{14}$$

represents the experimental evidence and

$$f_2(Y_k^{i,j}) = exp(-\beta|\underline{\nu}_k^{i,j} - \overline{\nu}_k^{i,j}|) \tag{15}$$

represents the spread with respect to the membership degrees. The optimal rough–fuzzy set is obtained by maximizing

$$\sum_{k=1}^{p} V(Y_k^{i,j}) \tag{16}$$

where

$$V(Y_k^{i,j}) = card(Y_k^{i,j}) * exp(-\beta|\underline{\nu}_k^{i,j} - \overline{\nu}_k^{i,j}|) \qquad (17)$$

The Rough–Fuzzy Product Feature Selection Algorithm is sketched in Algorithm 1. First each feature is granulized by maximizing Eq. 13 and a rough–fuzzy set is constructed as defined in Eq. 1 (lines 2–5). Then the couple of features that maximize Eq. 17 is found by applying the rough–fuzzy product in Eq. 6 (lines 6–10). At this point the algorithm tries to add, one at time, the remaining features by composing, by means of the rough–fuzzy product (line 14), the new feature with the ones already selected. Only the features that lead to a better solution with respect to Eq. 17 (line 16) are added to the final solution (line 18).

Algorithm 1. RFPFS - Rough–Fuzzy Product Feature Selection

1: $F = \{$set of features$\}$
2: **for all** $c \in F$ **do**
3:    GR=granulation of c
4:    RF^c=Rough–Fuzzy Set of GC
5: **end for**
6: **for all** $(i,j) \in F$ **do**
7:    $VM(i,j) = RF^i \otimes RF^j$
8: **end for**
9: $[i,j] = max(VM)$
10: RFPD=$\{i,j\}$
11: $V_{\max} = 0$
12: **for all** $c \in F \setminus RFPD$ **do**
13:    $TMP = RFPD \cup c$
14:    $RF^{TMP} = RF^{RFPD} \otimes RF^c$
15:    $V(RF^{TMP}) = \sum_{k=1}^{p} V(Y_k^{TMP})$
16:    **if** $V(RF^{TMP}) > V_{\max}$ **then**
17:       $V_{\max} = V(RF^{TMP})$
18:       $RFPD = TMP$
19:    **end if**
20: **end for**

---

5   Experimental Results

In this section, experimental results for the proposed approach are presented. The method is compared with some supervised and unsupervised methods. The comparison with the supervised methods is included to show that despite missing or incomplete labels, RFPFS can effectively reduce dimensionality and discover useful subsets of features. The experimental setup consists of three steps: (1) feature selection, (2) dataset reduction by retaining selected features, (3) classifier learning. Note that the class label have been removed before applying RFPFS algorithm. The classifier used in tests is the J48 classifier that creates decision trees by choosing the most informative features via an entropy measure, and recursively partitions the data into subtables based on their values. Each node in the tree represents a feature with branches from a node representing the alternative values this feature can take according to the current subtable. Partitioning stops when all data items in the subtable have the same classification.

The first test has been performed on three datasets from the UCI repository, namely Wine (178 instances and 13 features), Wisconsin (569 instances and 32 features), and Sonar (208 instances and 60 features). From Table 1 it is possible to note how the proposed algorithm selects approximately half of the features (1/3 in the Sonar dataset) still obtaining good classification accuracy with respect to the unreduced dataset.

Table 1. Results on UCI datasets.

	Unreduced	RFPFS (No. of selected features)
Wine	94.41	93.80 (6)
Wisconsin	72.46	73.62 (14)
Sonar	93.86	95.03 (22)

In the second test, the proposed method has been compared to some unsupervised feature selection methods (fuzzy-rough lower approximation-based (UFRFSs) [14]) and supervised feature selection methods (correlation-based (CFS) [6], consistency-based [23], fuzzy-rough lower approximation-based (FRFSs) [8]). The dataset used in this test are Wine (178 instances and 13 features), Water3 (390 instances and 38 features), Ionosphere (230 instances and 34 features) and Glass (214 instances and 9 features).

Table 2. RFPFS vs. Unsupervised features selection algorithms.

	Unreduced	UFRFS	B–UFRFS	D–UFRFS	RFPFS
Glass	67.29 (9)	65.91 (7)	65.91 (7)	65.91 (7)	70.9 (6)
Ionosphere	67.5 (34)	59.17 (6)	64.17 (6)	59.17 (6)	65.17 (5)
Wine	94.41 (13)	79.74 (7)	81.99 (7)	79.74 (6)	93.80 (6)
Water3	83.08 (38)	81.54 (7)	80.51 (7)	81.54 (7)	80.30 (4)

From Table 2 it is possible to see how RFPFS performs better with respect to the considered unsupervised approaches. The results are even more interesting if compared with those obtained by the supervised approaches, shown in Table 3. Even in this case the proposed method is comparable with the other approaches in terms of accuracy, but without considering the class labels.

Table 3. RFPFS vs. Supervised features selection algorithms.

	Unreduced	CFS	Consis.	FRFS	B–FRFS	D–FRFS	RFPFS
Glass	67.29 (9)	69.98 (6)	67.29 (9)	65.87 (8)	65.87 (8)	65.87 (8)	70.9 (6)
Ionosphere	67.5 (34)	57.5 (15)	62.5 (6)	61.67 (5)	61.67 (5)	61.67 (5)	65.17 (5)
Wine	94.41 (13)	94.41 (11)	97.10 (5)	94.97 (5)	96.08 (5)	94.41 (5)	93.80 (6)
Water3	83.08 (38)	81.54 (11)	81.02 (11)	79.49 (6)	80.26 (6)	80.77 (6)	80.30 (4)

6   Conclusions

In this paper a technique for unsupervised feature selection has been presented.
The proposed approach is based on the rough–fuzzy product operation in order
to retain important information contained in different rough–fuzzy sets. An interesting
characteristic is that no user-defined thresholds or domain-related information
is required. The experimental results demonstrated how the approach
can reduce dataset dimensionality considerably whilst retaining useful features
when class labels are unknown or missing.

References

1. Bellman, R.: Adaptive Control Processes: A Guided Tour. Princeton University Press, Princeton (1961)
2. Das, S.K.: Feature selection with a linear dependence measure. IEEE Trans. Comput. **100**(9), 1106–1109 (1971)
3. Dash, M., Liu, H.: Unsupervised Feature Selection. In: Proceedings of the Pacific and Asia Conference on Knowledge Discovery and Data Mining, pp. 110–121 (2000)
4. Dash, M., Liu, H.: Feature selection for classification. Intell. Data Anal. **1**(3), 131–156 (1997)
5. Devijver, P., Kittler, J.: Pattern Recognition: A Statistical Approach. Prentice Hall, Upper Saddle River (1982)
6. Hall, M.A.: Correlation-based feature selection for discrete and numeric class machine learning. In: Proceedings of the 17th International Conference on Machine Learning, pp. 359–366 (2000)
7. Jensen, R., Shen, Q.: Interval-valued fuzzy-rough feature selection in datasets with missing values. In: IEEE International Conference on Fuzzy Systems, pp. 610–615 (2009)
8. Jensen, R., Shen, Q.: New approaches to fuzzy-rough feature selection. IEEE Trans. Fuzzy Syst. **17**(4), 824–838 (2009)
9. Lin, T.Y., Cercone, N.: Rough sets and Data Mining: Analysis of Imprecise Data. Kluwer Academic Publishers, Berlin (1997)
10. Mitchell, T.: Machine Learning. McGraw-Hill, New York (1997)
11. Mitra, P., Murthy, C.A., Pal, S.K.: Unsupervised feature selection using feature similarity. IEEE Trans. Pattern Anal. Mach. Intell. **24**(4), 1–13 (2002)
12. Pal, S.K., De, R.K., Basak, J.: Unsupervised feature evaluation: a neuro-fuzzy approach. IEEE Trans. Neural Netw. **11**, 366–376 (2000)

13. Parthaláin, N.M., Shen, Q., Jensen, R.: A distance measure approach to exploring the rough set boundary region for attribute reduction. IEEE Trans. Knowl. Data Eng. **22**(3), 305–317 (2010)

14. Parthaláin, N.M., Jensen, R.: Measures for unsupervised fuzzy-rough feature selection. Int. J. Hybrid Intell. Syst. **7**(4), 249–259 (2010)

15. Pawlak, Z.: Rough sets. Int. J. Comput. Inform. Sci. **11**, 341–356 (1982)

16. Pawlak, Z.: Granularity of knowledge, indiscernibility and rough sets. In: Proceedings of IEEE International Conference on Fuzzy Systems, pp. 106–110 (1998)

17. Pedrycz, W., Gomide, F.: Fuzzy Systems Engineering: Toward Human-Centric Computing. Wiley, Hoboken (2007)

18. Petrosino, A., Ferone, A.: Feature discovery through hierarchies of rough fuzzy sets. In: Chen, S.-M., Pedrycz, W. (eds.) Granular Computing and Intelligent Systems: Design with Information Granules of Higher Order and Higher Type, vol. 13, pp. 57–73. Springer, Heidelberg (2011)

19. Shen, Q., Chouchoulas, A.: A modular approach to generating fuzzy rules with reduced attributes for the monitoring of complex systems. Eng. Appl. Artif. Intell. **13**(3), 263–278 (2002)

20. Thangavel, K., Pethalakshmi, A.: Performance analysis of accelerated Quickreduct algorithm. In: Proceedings of International Conference on Computational Intelligence and Multimedia Applications, pp. 318–322 (2007)

21. Tsang, E.C.C., Chen, D., Yeung, D.S., Wang, X.-Z., Lee, J.: Attributes reduction using fuzzy rough sets. IEEE Trans. Fuzzy Syst. **16**(5), 1130–1141 (2008)

22. Velayutham, C., Thangavel, K.: Unsupervised quick reduct algorithm using rough set theory. J. Electron. Sci. Technol. **9**(3), 193–201 (2011)

23. Witten, I.H., Frank, E.: Data Mining: Practical Machine Learning Tools with Java Implementations. Morgan Kaufmann, San Francisco (2000)

24. Zadeh, L.: Fuzzy sets. Inf. Control **8**(3), 338–353 (1964)

A System for Fuzzy Granulation of OWL Ontologies

Francesca A. Lisi and Corrado Mencar[(✉)]

Dipartimento di Informatica, Centro Interdipartimentale di Logica e Applicazioni,
Università degli Studi di Bari, Aldo Moro, Italy
{francesca.lisi,corrado.mencar}@uniba.it

Abstract. In this paper, we describe a preliminary version of
GRANULO, a system for building a granular view of individuals over
an OWL ontology. The system applies granular computing techniques
based on fuzzy clustering and relies on SPARQL for querying the given
ontology and Fuzzy OWL 2 for representing the produced granular view.
The system has been applied on a benchmark ontology in the touristic
domain.

1 Introduction

The Semantic Web is full of *imprecise information* coming from, *e.g.*, perceptual
data, incomplete data, erroneous data, etc. Endowing OWL[1] ontologies with
capabilities of representing and processing imprecise knowledge is therefore a
highly desirable feature. Moreover, even in the case that precise information is
available, imprecise knowledge could be advantageous: tolerance to imprecision
may lead to concrete benefits such as compact knowledge representation, efficient
and robust reasoning, etc. [14].

The integration of fuzzy sets in OWL ontologies can be achieved in different
ways and for different pursuits (see [12] for an overview). However, the definition
of such fuzzy sets could be hard if they must represent some hidden properties
of individuals. In this paper, we take a data-driven approach, where fuzzy sets
are automatically derived from the available individuals in the ontology through
a fuzzy clustering process. The derived fuzzy sets are more compliant to represent similarities among individuals w.r.t. a manual definition by experts. This
approach promotes a *granular view* of individuals that can be exploited to further enrich the knowledge base of an ontology by applying techniques from the
so-called *Granular Computing* (GC) paradigm. According to GC, *information
granules* (such as fuzzy sets) are elementary units of information [2]. In this
capacity, information granules can be represented as individuals in an ontology, eventually belonging to a generic class (e.g. the "Granule" class), and thus
endowed with properties that are specific to information granules and do not
pertain to the original individuals from which the granules have been derived.
An example of granule-specific properties is the granularity of an information

---

[1] http://www.w3.org/TR/2009/REC-owl2-overview-20091027/.

ⓒ Springer International Publishing AG 2017
A. Petrosino et al. (Eds.): WILF 2016, LNAI 10147, pp. 126–135, 2017.
DOI: 10.1007/978-3-319-52962-2_11

granule [2], which can be numerically quantified and then related to fuzzy sets representing linguistic quantifiers, usually expressed with terms such as "most", "few", etc. Thence, a new level of knowledge emerges from information granules, which can be profitably included in the original ontology to express knowledge not pertaining to single individuals, but on (fuzzy) collections of them, being each collection defined by individuals kept together by their similarity.

The paper is structured as follows. After some preliminaries on fuzzy information granulation (Sect. 2) and an overview of related works (Sect. 3), we describe the system GRANULO (Sect. 4). In particular, we provide details of the GC method underlying the system (Sect. 4.1), its implementation with Semantic Web technologies (Sect. 4.2), and its evaluation on a real-world medium-sized ontology from the tourism domain (Sect. 4.3). We report some final considerations and perspectives of future work in Sect. 5.

2   Fuzzy Information Granulation

From the modeling viewpoint, fuzzy sets are very useful to represent perception-based information granules, which are characterized by both *granularity* (*i.e.* concepts that refer to a multiplicity of objects) and *graduality* (*i.e.* the reference of concepts to objects is a matter of degree) [13]. In the case of numerical domains, a simple yet effective way to define such fuzzy information granules is through Strong Fuzzy Partitions (SFP). A SFP can be easily defined by trapezoidal fuzzy sets by properly constraining the characterizing parameters.

Fuzzy clustering, such as Fuzzy C-Means (FCM), is a convenient way to build a SFP by taking into account the available data. FCM, applied to one-dimensional numerical data, can be used to derive a set of c clusters characterized by prototypes $p_1, p_2, \ldots, p_c$, with $p_j \in \mathbb{R}$ and $p_j < p_{j+1}$. These prototypes, along with the range of data, provide enough information to define a SFP with two trapezoidal fuzzy sets and $c - 2$ triangular fuzzy sets.

Fuzzy sets, like crisp sets, can be quantified in terms of their *cardinality*. Several definitions of cardinality of fuzzy sets have been proposed [5], although in this paper we consider only relative scalar cardinalities like the *relative σ-count*, defined for a finite set D in the Universe of Discourse as

$$\sigma(F) = \frac{\sum_{x \in D} F(x)}{|D|} \in [0, 1] \tag{1}$$

where, obviously, $\sigma(\emptyset) = 0$ and $\sigma(D) = 1$.

Since the range of σ is always the unitary interval, a number of fuzzy sets can be defined to represent granular concepts about cardinalities, such as MANY, MOST, etc. These concepts are called *fuzzy quantifiers* [8]. As usual, they can be defined so as to form a SFP; in this way linguistic labels can be easily attached, as illustrated in Fig. 1.

Fuzzy quantifiers can be used to express imprecise properties on fuzzy information granules. More specifically, given a quantifier Q labeled with Q and a

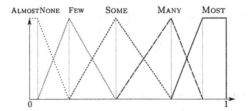

Fig. 1. Examples of fuzzy quantifiers.

fuzzy set F labeled with F, the membership degree $Q\left(\sigma\left(F\right)\right)$ quantifies the truth degree of the proposition

$$Q \text{ x are } F$$

For example, if $Q \equiv$ Many and $F \equiv$ Low, the fuzzy proposition

$$\text{Many x are Low}$$

asserts that many data points have a low value; the truth degree of this proposition is quantified by

$$Q_{\text{Many}}\left(\sigma\left(F_{\text{Low}}\right)\right)$$

By a proper formal representation, these fuzzy propositions can be embodied within an ontology by introducing new individuals corresponding to the information granules.

3 Related Work

The logical foundations of OWL come from the knowledge representation formalism collectively known as *Description Logics* (DLs) [1]. Several fuzzy extensions of DLs can be found in the literature (see the survey in [12]). In particular, we are interested to the extension of $\mathcal{ALC}(\mathbf{D})$ with fuzzy concrete domains [11].

In fuzzy DLs, an *interpretation* $\mathcal{I} = (\Delta^{\mathcal{I}}, \cdot^{\mathcal{I}})$ consist of a nonempty (crisp) set $\Delta^{\mathcal{I}}$ (the *domain*) and of a *fuzzy interpretation function* $\cdot^{\mathcal{I}}$ that, *e.g.*, assigns: *(i)* to each atomic concept A a function $A^{\mathcal{I}} \colon \Delta^{\mathcal{I}} \to [0,1]$; *(ii)* to each object property R a function $R^{\mathcal{I}} \colon \Delta^{\mathcal{I}} \times \Delta^{\mathcal{I}} \to [0,1]$; *(iii)* to each data type property T a function $T^{\mathcal{I}} \colon \Delta^{\mathcal{I}} \times \Delta^{\mathbf{D}} \to [0,1]$; *(iv)* to each individual a an element $a^{\mathcal{I}} \in \Delta^{\mathcal{I}}$; and *(v)* to each concrete value v an element $v^{\mathcal{I}} \in \Delta^{\mathbf{D}}$.

Axioms in a fuzzy $\mathcal{ALC}(\mathbf{D})$ KB $\mathcal{K} = \langle \mathcal{T}, \mathcal{A} \rangle$ are graded, *e.g.* a GCI is of the form $\langle C_1 \sqsubseteq C_2, \alpha \rangle$ (*i.e.* C_1 is a sub-concept of C_2 to degree at least α). (We may omit the truth degree α of an axiom; in this case $\alpha = 1$ is assumed.) An interpretation $\mathcal{I}$ *satisfies* an axiom $\langle \tau, \alpha \rangle$ if $(\tau)^{\mathcal{I}} \geq \alpha$. $\mathcal{I}$ is a *model* of $\mathcal{K}$ iff $\mathcal{I}$ satisfies each axiom in $\mathcal{K}$. We say that $\mathcal{K}$ *entails* an axiom $\langle \tau, \alpha \rangle$, of the form $C \sqsubseteq D$, $a{:}C$ or $(a,b){:}R$, denoted $\mathcal{K} \models \langle \tau, \alpha \rangle$, if any model of $\mathcal{K}$ satisfies $\langle \tau, \alpha \rangle$. Further details of the reasoning procedures for fuzzy DLs can be found in [10].

Fuzzy quantifiers have been also studied in fuzzy DLs. In particular, Sanchez and Tettamanzi [9] define an extension of fuzzy $\mathcal{ALC}(\mathbf{D})$ involving fuzzy quantifiers of the absolute and relative kind, and using qualifiers. They also provide algorithms for performing two important reasoning tasks with their DL: reasoning about instances, and calculating the fuzzy satisfiability of a fuzzy concept.

Some fuzzy DL reasoners have been implemented, such as *fuzzyDL* [3]. Not surprisingly, each reasoner uses its own fuzzy DL language for representing fuzzy ontologies and, thus, there is a need for a standard way to represent such information. In [4], Bobillo and Straccia propose to use OWL 2 itself to represent fuzzy ontologies. More precisely, they use OWL 2 annotation properties to encode fuzzy $\mathcal{SROIQ}(\mathbf{D})$ ontologies. The use of annotation properties makes possible (i) to use current OWL 2 editors for fuzzy ontology representation, and (ii) that OWL 2 reasoners discard the fuzzy part of a fuzzy ontology, producing the same results as if would not exist. Additionally, they identify the syntactic differences that a fuzzy ontology language has to cope with, and show how to address them using OWL 2 annotation properties.

4   The GranulO System

4.1   Method

The GRANULO system described in this paper is based on the refinement of a method previously defined by the authors [7]. Here we outline the method and highlight the differences w.r.t. the previous version.

Let C be a class and T a functional datatype property connecting instances of C to values in a numerical range $\mathbf{d}$. If the class C is populated by n individuals, n data points can be retrieved, which correspond to the respective values of the datatype property. FCM can be applied to this dataset, yielding to c fuzzy sets $F_1, F_2, \ldots, F_c$. For each fuzzy set F_j, the relative cardinality $\sigma(F_j)$ can be computed by means of (1). Given a fuzzy quantifier Q_k, the membership degree

$$q_{jk} = Q_k(\sigma(F_j)) \tag{2}$$

identifies the degree of truth of the fuzzy proposition "$Q_k x$ are F_j". The new granulated view can be integrated in the ontology as follows. The fuzzy sets F_j are the starting point for the definition of new subclasses of C defined as

$$D_j \equiv C \sqcap \exists T.F_j$$

which can be read as: the concept D_j is defined by individuals in C where the property T assumes the granular value F_j.

Also, a new class G is defined, with individuals $g_1, g_2, \ldots, g_c$, where each individual g_i is an information granule corresponding to F_i. Each individual in D_j is then mapped to g_i by means of an object property `mapsTo`. Moreover, for each fuzzy quantifier Q_k, a new class is introduced, which models one of the fuzzy sets over the cardinality of G. The connection between the class G and each class Q_k is established through an object property with a conventional name like `hasCardinality`, with degrees identified as in Eq. (2).

Example 1. For illustrative purposes, we refer to an OWL ontology in the tourism domain, *Hotel*,[2] which encompasses the datatype property `hasPrice` with the class `Hotel` as domain and range in the datatype domain `xsd:double`. Let us suppose that the room price for Hotel Verdi (instance `verdi` of `Hotel`) is 105, *i.e.* the ontology contains the assertion (`verdi, 105`)`:hasPrice`. By applying fuzzy clustering to `hasPrice`, we might obtain three fuzzy sets (with labels `Low`, `Medium`, `High`) from which the following classes are derived:

> `LowPriceHotel` ≡ `Hotel` ⊓ ∃`hasPrice.Low`
> `MidPriceHotel` ≡ `Hotel` ⊓ ∃`hasPrice.Medium`
> `HighPriceHotel` ≡ `Hotel` ⊓ ∃`hasPrice.High`,

With respect to these classes, `verdi` shows different degrees of membership; *e.g.* `verdi` is a low-price hotel at degree 0.8 and a mid-price hotel at degree 0.2 (see Fig. 2 for a graphical representation where fuzzy classes are depicted in gray).

Subsequently, we might be interested in obtaining aggregated information about hotels. Quantified cardinalities allow us, for instance, to represent the fact that "Many hotels are low-priced" with the fuzzy assertion

> `lph : ∃hasCardinality.Many`

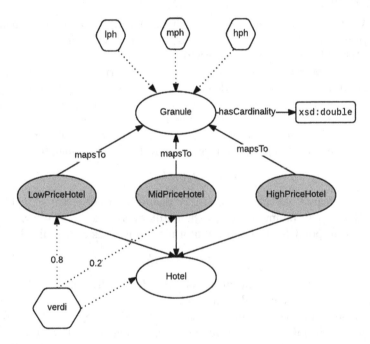

Fig. 2. Graphical representation of the output of GRANULO for the case of Example 1.

---

[2] http://www.umbertostraccia.it/cs/software/FuzzyDL-Learner/download/
FOIL-DL/examples/Hotel/Hotel.owl.

with truth degree 0.8, where `lph` is an instance of the class `Granule` corresponding to `LowPriceHotel` and `Many` is a class representing a fuzzy quantifier.

A natural extension of the proposed granulation method follows when the
class C is specialized in subclasses. With respect to our initial proposal [7], the
GRANULO system is capable of dealing with multiple classes at any level of
inheritance, thanks to the adoption of the SPARQL entailment regime provided
by Apache Jena Fuseki.

A case of particular interest is given by OWL schemes representing *ternary
relations*, involving three domains $C \times D \times \mathbf{d}$ (for our purposes, we will assume
$\mathbf{d}$ a numerical domain). In OWL, ternary relations can be indirectly represented through an auxiliary class E, two object properties R_1 and R_2, and
one datatype property $T.$, as depicted in Fig. 3. Given an individual $a \in C$,
all triplets $(a, b_i, v_i) \in A \times D \times \mathbf{d}$ are retrieved; then fuzzy sets are generated
and quantified as described in the binary case. (The information granules are,
therefore, related to the selected individual of $C.$).

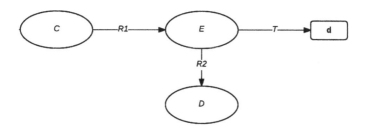

Fig. 3. Graphical representation of the OWL schema modeling a ternary relation.

Example 2. With reference to *Hotel* ontology, we might also consider the distances between hotels and attractions. This is clearly a case of a ternary
relation that needs to be modeled through an auxiliary class `Distance`,
which is connected to the classes `Hotel` and `Attraction` by means of the
object properties `hasDistance` and `isDistanceFor`, respectively, and plays the
role of domain for a datatype property `hasValue` with range `xsd:double`.
The knowledge that "Hotel Verdi has a distance of 100 meters from the
Duomo" can be therefore modeled by means of: (`verdi  hasDistance  d1`),
(`d1  isDistanceFor  duomo`), and (`d1  hasValue  100`).

After fuzzy granulation, the imprecise sentence "Hotel Verdi has a low distance from many attractions" can be considered as a consequence of the previous
and the following axioms and assertions (Fig. 4):

> `LowDistance ≡ Distance ⊓ ∃isDistanceFor.Attraction ⊓ ∃hasValue.Low`
> `d1 : LowDistance` (to some degree), (`d1  mapsTo  ld`), (`ld : Granule`),
> (`ld  hasCardinality  0.5`), `ld : ∃hasCardinality.Many` (to some degree)

where `Many` is defined as mentioned in Example 1.

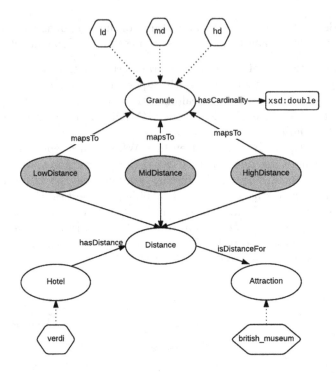

Fig. 4. Graphical representation of the output of GRANUL*O* for the case of Example 2.

4.2   Implementation

The system, implementing the method presented in Sect. 4.1, has been developed with Python. It interfaces a Apache Jena Fuseki server by posing SPARQL queries (under entailment regime) to an OWL ontology in order to extract data in the CSV format. Also, it can be configured by means of a JSON file which specifies the SPARQL Endpoint to be queried, the numerical data property to be fuzzified, the linguistic labels to be used for the fuzzy sets and the linguistic labels for the fuzzy quantifiers together with the prototype values. For instance, the following JSON file tells GRANUL*O* to run the fuzzy granulation process on the data property `hasPrice` by setting the number of clusters to $n = 3$ with `Low`, `Mid`, and `High` as linguistic labels, and the number of fuzzy quantifiers to 5 with `AlmostNone`, `Few`, `Some`, `Many` and `Most` as linguistic labels and the values 0.05, 0.275, 0.5, 0.725 and 0.95 as prototypes.

```
{
    "ontologyName" : "Hotel.owl",
    "ontologyPrefix" : "<http://www.semanticweb.org/ontologies/Hotel.owl#>",
    "SPARQLEndPoint" : "http://localhost:3030/Hotel/query",

    "domainClasses" : ["Hotel"],
    ...

    "dataPropertyToFuzzify" : "hasPrice",
```

```
   "fuzzySetLabels" : ["Low", "Mid", "High"],

   "quantifierLabels" : ["AlmostNone","Few", "Some", "Many", "Most"],
   "quantifierPrototypes" : [0.05, 0.275, 0.5, 0.725, 0.95]
}
```

The resulting granular view is then integrated in the original OWL ontology by means of additional Fuzzy OWL 2 axioms and assertions such as the following:

```
<ClassAssertion>
    <Annotation>
        <AnnotationProperty IRI="#fuzzyLabel"/>
        <Literal datatypeIRI="&rdf;PlainLiteral">
            <fuzzyOwl2 fuzzyType="axiom">
                <Degree value="{degree}"/>
            </fuzzyOwl2>
        </Literal>
    </Annotation>
    <ObjectSomeValuesFrom>
        <ObjectProperty IRI="#hasCardinality"/>
        <Class IRI="#{quantifier}"/>
    </ObjectSomeValuesFrom>
    <NamedIndividual IRI="#{granule}"/>
</ClassAssertion>
```

4.3   Evaluation

Several experiments have been conducted on the OWL ontology *Hotel*, already mentioned in Example 1. Due to the lack of space, we report here only the results obtained for the granulation of hotels w.r.t. their price. The data property `hasPrice` ranges over values between $m = 45$ and $M = 136$. For $n = 3$ clusters have been automatically built around the centroids 54.821, 77.924 and 102.204. They form the SFP shown in Fig. 5(a) which consists of a left-shoulder fuzzy set (labeled as `Low`), a triangular fuzzy set (labeled as `Medium`) and a right-shoulder fuzzy set (labeled as `High`). The results returned by the granulation process are graphically reported in Fig. 6(a). Hotel prices are reported along the x axis, whereas the y axis contains the membership degrees to the fuzzy sets. At most two points having different coloring correspond to a single price value. They

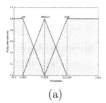   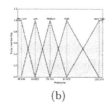   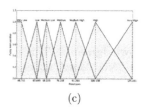

(a)   (b)   (c)

Fig. 5. Fuzzy clustering of hotel prices for $n = 3$ (a), $n = 5$ (b), and $n = 7$ (c).

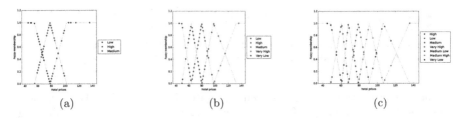

Fig. 6. Fuzzy granulation of hotels w.r.t. price for $n = 3$ (a), $n = 5$ (b), and $n = 7$ (c).

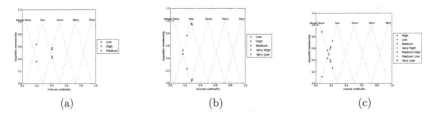

Fig. 7. Fuzzy quantification of hotel granules for $n = 3$ (a), $n = 5$ (b), and $n = 7$ (c).

depict the double membership to two adjacent fuzzy sets. Results for $n = 5$ and $n = 7$ clusters are reported in Figs. 5, 6(b), and Figs. 5, 6(c), respectively.

5 Conclusions

We have presented a system, called GRANULO, able to automatically build granular views over individuals of an OWL ontology. According to these views, obtained by applying Granular Computing techniques, a number of individuals can be replaced by information granules, represented as fuzzy sets. Having such views in alternative to the usual punctual one is a highly desirable feature for many Semantic Web applications in domains pervaded by imprecise information coming from perceptual data, incomplete data, data with errors, etc. An extensive evaluation of GRANULO over several OWL ontologies is ongoing with promising results which we could not include in the paper due to lack of space.

In the future we plan to verify the benefits of information granulation in the context of inductive learning algorithms, such as FOIL-$\mathcal{DL}$ [6], in terms of efficiency and effectiveness of the learning process, as well as in terms of interpretability of the learning results.

Acknowledgments. This work was partially funded by the Università degli Studi di Bari "Aldo Moro" under the IDEA Giovani Ricercatori 2011 grant "Dealing with Vague Knowledge in Ontology Refinement".

References

1. Baader, F., Calvanese, D., McGuinness, D., Nardi, D., Patel-Schneider, P. (eds.): The Description Logic Handbook: Theory, Implementation and Applications, 2nd edn. Cambridge University Press, Cambridge (2007)
2. Bargiela, A., Pedrycz, W.: Granular Computing: An Introduction. Springer, Heidelberg (2003)
3. Bobillo, F., Straccia, U.: fuzzyDL: an expressive fuzzy description logic reasoner. In: FUZZ-IEEE 2008, Proceedings of IEEE International Conference on Fuzzy Systems, Hong Kong, China, pp. 923–930. IEEE (2008)
4. Bobillo, F., Straccia, U.: Representing fuzzy ontologies in OWL 2. In: FUZZ-IEEE 2010, Proceedings of IEEE International Conference on Fuzzy Systems, Barcelona, Spain, pp. 1–6. IEEE (2010)
5. Dubois, D., Prade, H.: Fuzzy cardinality and the modeling of imprecise quantification. Fuzzy Sets Syst. **16**(3), 199–230 (1985)
6. Lisi, F.A., Straccia, U.: Learning in description logics with fuzzy concrete domains. Fundamenta Informaticae **140**(3–4), 373–391 (2015)
7. Lisi, F.A., Mencar, C.: Towards Fuzzy Granulation in OWL Ontologies. In: Proceedings of the 30th Italian Conference on Computational Logic (CILC 2015), vol. 1459, pp. 144–158. CEUR Workshop Proceedings, Genova, Italy (2015)
8. Liu, Y., Kerre, E.E.: An overview of fuzzy quantifiers. (I). Interpretations. Fuzzy Sets Syst. **95**(1), 1–21 (1998)
9. Sanchez, D., Tettamanzi, A.G.: Fuzzy quantification in fuzzy description logics. In: Sanchez, E. (ed.) Fuzzy Logic and the Semantic Web, Capturing Intelligence, vol. 1, pp. 135–159. Elsevier, Amsterdam (2006)
10. Straccia, U.: Reasoning within fuzzy description logics. J. Artif. Intell. Res. **14**, 137–166 (2001)
11. Straccia, U.: Description logics with fuzzy concrete domains. In: UAI 2005, Proceedings of the 21st Conference in Uncertainty in Artificial Intelligence, Edinburgh, Scotland, 26–29 July 2005, pp. 559–567. AUAI Press (2005)
12. Straccia, U.: Foundations of Fuzzy Logic and Semantic Web Languages. CRC Studies in Informatics Series. Chapman & Hall, New York (2013)
13. Zadeh, L.: From computing with numbers to computing with words. From manipulation of measurements to manipulation of perceptions. IEEE Trans. Circ. Syst. I: Fundam. Theory Appl. **46**(1), 105–119 (1999)
14. Zadeh, L.A.: Is there a need for fuzzy logic? Inf. Sci. **178**(13), 2751–2779 (2008)

Clustering and Learning

Graded Possibilistic Clustering of Non-stationary Data Streams

A. Abdullatif[1,2], F. Masulli[1,3(✉)], S. Rovetta[1], and A. Cabri[1]

[1] DIBRIS - Department of Informatics, Bioingengering,
Robotics and Systems Engineering, University of Genoa,
Via Dodecaneso 35, 16146 Genoa, Italy
{amr.abdullatif,francesco.masulli,stefano.rovetta}@unige.it
[2] VEDECOM Institute, Versailles, France
[3] Sbarro Institute for Cancer Research and Molecular Medicine,
College of Science and Technology, Temple University, Philadelphia, PA, USA

Abstract. Multidimensional data streams are a major paradigm in data science. This work focuses on possibilistic clustering algorithms as means to perform clustering of multidimensional streaming data. The proposed approach exploits fuzzy outlier analysis to provide good learning and tracking abilities in both concept shift and concept drift.

1 Introduction

Multidimensional data streams have arisen as a relevant topic in data science during the past decade [1]. They arise in an ever increasing range of fields, from the web, to wearable sensors, to intelligent transportation systems, to smart homes and cities.

Data streams may represent actual time series, or quasi-stationary phenomena that feature longer-term variability, e.g., changes in statistical distribution or a cyclical behavior. In these non-stationary conditions, any model is expected to be appropriate only in a neighborhood of the point in time where it has been learned. Its validity may decrease smoothly with time (*concept drift*), or there may be sudden changes, for instance when switching from one operating condition to a new one (*concept shift*).

This work focuses on possibilistic clustering [3] as a means to perform clustering of multidimensional streaming data. We specifically exploit the ability, provided by the Graded Possibilistic c-Means [2], to learn clustering models iteratively using both batch (sliding-window) and online (by-pattern) strategies that track and adapt to concept drift and shift in a natural way.

2 Clustering Non-stationary Data Streams

Clustering streams requires tackling related, but different, problems: Handling unbounded, possibly large data; detecting model changes (drift and shift); adapting to model changes. These are often treated independently in the literature.

© Springer International Publishing AG 2017
A. Petrosino et al. (Eds.): WILF 2016, LNAI 10147, pp. 139–150, 2017.
DOI: 10.1007/978-3-319-52962-2_12

Model shift, either for rejection or tracking, has been extensively studied, also under the name of change detection or change point identification. The key to detect model differences is measuring the fit of observations. An observation that does not fit a given model is either called novelty [6] or outlier [7] depending on the focus of the inquiry. Individual outliers can be detected or rejected by evaluating their estimated probability or membership and comparing it to a threshold, using distance-based or density-based criteria [8], or simply not taken into account by using robust methods [9]. Alternatively it is possible to gather a new sample and apply sequential statistical tests for comparing empirical distributions [10]. Regarding fuzzy methods, techniques have been proposed mainly for robustness [11], although some approaches directly include measures of *outlierness* [13] that can indicate the inadequacy of a given current model.

Model drift is a much more difficult issue, usually simply tackled by continuous learning. This brings us to the third problem, learning strategies for dynamically changing models. Continuous learning can be done either by recomputing, if affordable, or by incremental updates. Most fuzzy clustering models are fit with a batch training procedure, owing to their derivation from (crisp) k Means. This batch optimization, iterative and prone to local minima, makes them unsuitable for data sets of very large/virtually unbounded cardinality. However, the same prototype-based representation lends itself very well to alternative training methods, e.g., online or "by pattern" [14], which are naturally suitable for incremental updating. Due to the non-stationarity, it is not possible to resort to the extensive literature about convergence of stochastic approximation or related methods [15]. However, a measure of model inadequacy can be used to modulate the required amount of update, avoiding unnecessary waste of computational resources especially when the data are large in size. In addition, it might be appropriate to dynamically change the number of centroids. In [16] this was addressed with a cross-validation approach. However we didn't take this problem into consideration.

We consider the case where the data are represented by numerical feature vectors. In general they will be multi-dimensional, even in the case of scalar-valued time series where time-lag encoding is used. We assume that the data are generated independently by some underlying probability distribution, but on the long run the distribution may change. As long as the underlying distribution doesn't "change much", it can be considered stationary. However, for sufficiently long observation periods this approximation is no longer valid. The extent of validity of a stationary approximation depends on the rate and intensity of change of the source distribution, which is clearly difficult to estimate. In this work we don't consider this problem, but provide methods for two different scenarios.

The goal of the analysis is to summarize these data by (fuzzy) clustering, with a process that should learn continuously from the input patterns as they arrive. The data are stored in a sliding window W of constant size $w \geq 1$ which is updated every s observations by deleting the oldest s patterns and adding the new ones, so that, at each time, the current W has an overlap of $w - s$ patterns with the previous one. We collect the s incoming observations in a probe set S before incorporating them in the window; the probe set is used to adaptively tune the learning parameters.

We assume that, within W, the data can be considered independent and identically distributed i.i.d. As anticipated in the introduction, we will focus on the following two scenarios:

- The source change rate so slow that W is sufficient to infer the clustering structure;
- The rate is so high that W is not sufficient to perform a complete clustering, and an incremental procedure is required.

Accordingly, in this work we consider the cases $w > 1$, $s > 1$ (batch learning, batch density estimate) and $w = s = 1$ (online learning, online density estimate).

3   The Graded Possibilistic $c-$Means Model

In central clustering data objects are points or vectors in data space, and c clusters are represented by means of their "central" points or centroids $\mathbf{y}_j$. The Graded Possibilistic model is a *soft* central clustering method, implying that cluster membership can be partial. This is usually represented by means of cluster indicators (or membership functions) which are real-valued rather than integer.

In many cases methods are derived as the iterative optimization of a constrained objective function [4], usually the mean squared distortion:

$$D = \frac{1}{n} \sum_{l=1}^{n} \sum_{j=1}^{c} u_{lj} ||\mathbf{x}_l - \mathbf{y}_j||^2 \tag{1}$$

Centroids are obtained by imposing $\nabla D = 0$:

$$\mathbf{y}_j = \frac{\sum_{l=1}^{n} u_{lj} x_l}{\sum_{l=1}^{n} u_{lj}}. \tag{2}$$

Usually constraints are placed on the sum $\zeta_l = \sum_{j=1}^{c} u_{lj}$ of all memberships for any given observation x_l. The value ζ_l can be interpreted as the total membership mass of observation x_l. We now survey from this perspective two related soft clustering methods.

The Maximum Entropy (ME) approach [5] imposes $\zeta_l = 1$, so we are in the "probabilistic" case, where memberships are formally equivalent to probabilities.

The objective J_{ME} includes the distortion, plus an entropic penalty with weight β and the normality constraint $\sum_{j=1}^{c} u_{lj} = 1 \ \forall l$. The first-order necessary minimum condition $\nabla J_{\mathrm{ME}} = 0$ then yields

$$u_{lj} = \frac{e^{-||\mathbf{x}_l - \mathbf{y}_j||^2/\beta}}{\zeta_l}. \tag{3}$$

On the other end of the spectrum, the Possibilistic c-Means in its second formulation (PCM-II) [3] does not impose any constraint on ζ_l, so memberships are not formally equivalent to probabilities; they represent degrees of typicality.

The objective J_{PCM-II} includes the distortion plus a penalty term to discourage (but not exclude) extreme solutions. This term contains an individual parameter β_j for each cluster, and $\nabla J_{PCM-II} = 0$ yields

$$u_{lj} = e^{-||x_l - y_j||^2 / \beta_j}. \tag{4}$$

Both Eqs. (4) and (3) can be generalized to a unique, common formulation as was done in [2] as follows:

$$u_{lj} = \frac{v_{lj}}{Z_l}, \tag{5}$$

where the *free membership* $v_{lj} = e^{-||x_l - y_j||^2 / \beta_j}$ is normalized with some term Z_l, a function of $\mathbf{v}_l = [v_{l1}, v_{l2}, \ldots, v_{cl}]$ but not necessarily equal to $\zeta_l = \sum_{j=1}^c v_{lj} = |\mathbf{v}_l|_1$. This allows us to add a continuum of other, intermediate cases to the two extreme models just described, respectively characterized by $Z_l = \zeta_j$ (probabilistic) and $Z_l = 1$ (possibilistic). Here we use the following formulation:

$$Z_l = \zeta_l^{\alpha} = \left(\sum_{j=1}^c v_{lj} \right)^{\alpha}, \quad \alpha \in [0, 1] \subset \mathbb{R} \tag{6}$$

The parameter α controls the "possibility level", from a totally probabilistic ($\alpha = 1$) to a totally possibilistic ($\alpha = 0$) model, with all intermediate cases for $0 < \alpha < 1$.

4   Outlierness Measurement Through Graded Possibilistic Memberships

The nature of membership functions suggests a characterization of outliers similar to Davé's Noise Clustering model that was used in the context of robust clustering [12]. Given a trained clustering model, i.e., a set of centroids and a set of cluster widths β_j, we exploit the properties of the possibilistic memberships to evaluate the degree of *outlierness*. We define outlierness as the membership of an observation to the concept of "being an outlier" with respect to a given clustering model. Differently from other approaches based on analyzing pattern-centroid distances [17], the graded possibilistic model used in this work provides a direct measure of outlierness. We propose to measure the total mass of membership to clusters ζ_l, which, by definition of the graded possibilistic model, does not necessarily equal 1, and measure whether and how much it is less than 1. Quantitatively, we define an index Ω as follows:

$$\Omega(\mathbf{x}_l) = \max\{1 - \zeta_l, 0\}. \tag{7}$$

Outlierness can be modulated by an appropriate choice of α. Low values correspond to sharper outlier rejection, while higher values imply wider cluster regions and therefore lower rejection. For $\alpha = 1$ the model becomes probabilistic and loses any ability to identify or reject outliers.

We observe that $\zeta_l = \sum_j u_{lj} \in (0, c)$. However:

- values $\zeta_l > 1$ are typical of regions well covered by centroids;
- but $\zeta_l \gg 1$ is very unlikely for good clustering solutions without many over-lapping clusters;
- finally, $\zeta_l \ll 1$ characterizes regions not covered by centroids, and any observation occurring there is an outlier.

The index Ω is defined as the complement to one of ζ_l, with negative values clipped out as not interesting.

The outlierness index is a pointwise measure, but it can be integrated to measure the frequency of outliers. For crisp decision-making, a point could be labeled as outlier when Ω exceeds some threshold. It is therefore easy to count the proportion (frequency) of outliers over a given set of probe points S.

However, we take advantage of the fact that Ω expresses a fuzzy concept, and rather than simply counting the frequency we can measure an *outlier density* $\rho \in [0, 1)$ defined in one of the following ways.

$$\rho_M = \frac{1}{|S|} \sum_{l \in S} \Omega(\mathbf{x}_l) \tag{8}$$

The density ρ_M accounts for both frequency and intensity, or degree of anomaly, of outliers. A high number of borderline outliers is equivalent to a lower number of stronger outliers, provided their mean value is the same. To give more emphasis to the case where some observation have a higher outlierness, an alternative definition can be used:

$$\rho_{RMS} = \frac{1}{|S|} \sqrt{\sum_{l \in S} (\Omega(\mathbf{x}_l))^2} \tag{9}$$

The definition $\rho = \rho_{RMS}$ will be adopted in the experiments.

5   Learning Regimes

We distinguish between three different situations, corresponding to three possible *learning regimes*.

1. Concept drift. The source is stationary or changing smoothly and slowly ($\Omega = 0$, density is low). Action to be taken: The model should be incrementally updated to track possible variations. We can call this the *baseline learning regime*.
2. Outliers. One or few isolated observations are clearly not explained by the model, which means that they have outlierness ($\Omega > 0$, density is low). Action to be taken: Incremental learning should be paused to avoid skewing clusters with atypical observations (*no-learning regime*).

3. Concept shift. Several observations have outlierness ($\Omega > 0$, density is high). Action to be taken: The old clustering model should be replaced by a new one. This is the *re-learning regime*.

The learning depends on a parameter θ that balances between stability ($\theta \approx 0$, model stays the same) and plasticity ($\theta \approx 1$, model changes completely), so it is possible to modulate this parameter as a function of outlier density, so that the three learning regimes (baseline, no learning and re-learning) can be implemented.

Since learning regimes are yet another fuzzy concept, rather than splitting them into clear-cut regions, in our experiments we employed a smooth function that is controlled by parameters: The user should interactively select their values to obtain the desired profile. This procedure is similar to defining the membership function for a linguistic variable. The function we used is the following:

$$\theta = 1 + \theta_0 \exp\left(-\frac{\rho}{\tau_1}\right) - \exp\left(-\left(\frac{\rho}{\tau_2}\right)^\gamma\right) \tag{10}$$

- θ_0 is the baseline value of θ, used when new data are well explained by the current model (baseline learning regime).
- τ_1 is a scale constant, analogue to a time constant in linear dynamical system eigenfunctions, that determines the range of values for which the baseline learning regime should hold.
- τ_2 is a scale constant that determines the range of values for which the re-learning regime should hold.
- γ is an exponential gain that controls how quick the relearning regime should go to saturation, i.e., to $\theta \approx 1$.

In the transition between baseline learning and re-learning, this function has a valley that brings the value of θ close to zero, implementing the no-learning regime.

6   Learning Possibilistic Stream Clustering

In the algorithms proposed in this section, the degree of possibility α is assumed to be fixed. This quantity incorporates the a-priori knowledge about the amount of outlier sensitivity desired by the user.

In batch learning, at each time t we train the clustering model on a training set (window) W_t of size w and evaluate ρ on the next $s < w$ observations (set S_t). Then we compute W_{t+1} for the next step by removing the s oldest observations and adding S_t, so that the training set size remains constant. Finally, the amount of learning required is estimated by computing $\theta(\rho)$ according to (10).

From an optimization or learning perspective, we are estimating the true (expected) objective function on the basis of a set W of w observations, that we use to compute a sample average. The batch process is initialized by taking a first sample W_0 and performing a complete deterministic annealing optimization on it with an annealing schedule $B = \{\beta_1, \ldots, \beta_b\}$. In subsequent optimizations, the annealing schedule is shortened proportionally to the computed value of $\theta(\rho) \in [0, 1]$: When $\theta = 1$ the complete B is used ($|B| = b$ optimization steps); when $\theta = 0$ no training is performed (0 steps); when $0 \leq \theta \leq 1$ a corresponding fraction B_θ of the schedule B is used, starting from step number $\lceil b \cdot \theta \rceil$ up to β_b (that is, $\lfloor b \cdot (1 - \theta) \rfloor$ steps in total). The updating rule for a generic centroid (non-stationary data streams) is the following:

$$\mathbf{y}_j(t + 1) = \mathbf{y}_j(t) + \eta \frac{\sum_{l=1}^{w} u_{lj} (\mathbf{x}_l - \mathbf{y}_j(t))}{\sum_{l=1}^{w} u_{lj}} \tag{11}$$

$$= (1 - \eta)\mathbf{y}_j(t) + \eta \frac{\sum_{l=1}^{w} u_{lj}\mathbf{x}_l}{\sum_{l=1}^{w} u_{lj}} \tag{12}$$

For *stationary data streams* the distribution of any sample $W(t)$ is constant w.r.t. t, and therefore its weighted mean $\frac{\sum_{l=1}^{w} u_{lj}\mathbf{x}_l}{\sum_{l=1}^{w} u_{lj}}$ is also constant. In this case

$$\mathbf{y}_j(t \to \infty) \to \frac{\sum_{l=1}^{w} u_{lj}\mathbf{x}_l}{\sum_{l=1}^{w} u_{lj}}. \tag{13}$$

With fixed η, Eq. (11) computes an exponentially discounted moving average.

In summary, for the batch case we use θ to modulate the number of annealing steps and, consequently, the scale parameter β. The optimization is longer, and starts with a higher coverage (higher β), in the re-learning regime; it is shorter and more localized in the baseline learning regime; it does not occur at all (zero steps) in the no-learning regime. Figure 1 outlines the batch learning algorithm.

Now for real time learning we provide an online learning method. This case can be modeled as a limit case of the batch method. At each time t we train the clustering model on a training set W_t of size 1, i.e., one observation, and evaluate ρ on the next $s = 1$ observation forming the "set" S_t. Then we compute W_{t+1} for the next step by replacing the single observation with that in S_t. The amount of learning required is estimated by computing $\theta(\rho)$ according to (10). In this case

---

init: Select α, $B = \{\beta_1, \ldots, \beta_b\}$, $\theta = \theta(\rho)$
      Read first w observations into W_0
      Learn clustering model from W_0 using annealing schedule B

loop: (for each time $t : 1, \ldots, \infty$)
      Read next s observations into S_t
      Use (7) to compute Ω for all observations in S_t
      Use (9) to compute ρ
      Use (10) to compute θ
      Discard s oldest observations from W_{t-1}
      Update $W_t \leftarrow [\, W_{t-1} \quad S \,]$
      Learn clustering model from W_t using annealing schedule B_θ

Fig. 1. Batch possibilistic stream clustering

the updates are incremental and therefore for ρ as well we propose an incremental computation according to the following discounted average formula:

$$\rho_t = \lambda \Omega_t + (1 - \lambda)\rho_{t-1}. \tag{14}$$

In this case as well, the process is initialized by taking a first sample W_0 and performing a complete deterministic annealing optimization on it with an annealing schedule $B = \{\beta_1, \ldots, \beta_b\}$.

However, after the initial phase $w = 1$, and the estimate of the objective function cannot be obtained by approximating an expectation with an average. So we resort to a stochastic approximation procedure [18]. This results in the following iterative update equations:

$$\mathbf{y}_j(t+1) = \mathbf{y}_j(t) + \eta_t u_{lj}(\mathbf{x}_l - \mathbf{y}_j) \tag{15}$$

for each centroid, $j = 1, \ldots, c$, with learning step size η_t. The update equation for the memberships is still given by Eqs. (5) and (6).

Differently from the batch case, after the initialization step a deterministic annealing schedule is not needed; rather, we have a stochastic annealing step size η_t. There are well-known conditions on η_t for convergence in the stationary case [18]:

$$\sum_{t=1}^{\infty} \eta_t = \infty \quad \text{and} \quad \sum_{t=1}^{\infty} \eta_t^2 < \infty \tag{16}$$

However, these conditions obviously do not hold in the nonstationary case, and this topic has not been thoroughly studied in the literature because the conclusions depend on the specific problem setting.

The strategy adopted in this work is to have the step size η be directly proportional to $\rho = \Omega$, i.e., $\eta = \eta_0 \cdot \rho$ for a user-selected constant η_0. An averaging effect is obtained through the stochastic iterative updates. In this way, after initialization, the intensity of updates depends on the degree of outlierness of the current observation.

To avoid premature convergence, the possibility degree α is also made dependent on ρ, so as to increase centroid coverage when outliers are detected. The formula used is:

$$\alpha = \alpha_{\min} + \rho(1 - \alpha_{\min}) \tag{17}$$

Figure 2 sketches the online learning algorithm.

---

init: Select α, $B = \{\beta_1, \ldots, \beta_b\}$, $\theta = \theta(\rho)$, η_0
      Read first w observations into W_0
      Learn clustering model from W_0 using annealing schedule B

loop: (for each time $t : 1, \ldots, \infty$)
      Read next observation
      Use (7) to compute Ω for current observation
      Use (10) to compute θ with $\rho = \Omega$
      Learn clustering model from current observation using Eq. (15) with learning step $\eta_t = eta_0 \cdot \theta$

Fig. 2. Online possibilistic stream clustering

7   Experimental Results

Synthetic data sets containing concept drift (we select the Gaussian and electricity data sets) were generated using the Matlab program ConceptDriftData.m[1] [19]. We also integrated our model in a traffic flow management system [20]. The proposed work was used as a generative model to asses and improve the accuracy of a short term traffic flow forecasting model.

The Gaussian data set already include concept drift, so outliers and concept shift were added by removing a number of observations in two parts of the data sequence. Discontinuities in the sequence are therefore introduced at 50% and 75% of the streams. In addition, the final 25% was shifted by adding an offset to all the data. Results are here shown for the Gaussians dataset. This includes four two-dimensional, evolving Gaussian with equal and known centers and spreads. After introducing discontinuities and shift, the data were remapped into $[0,1] \times [0,1]$. This procedure ensures that a ground truth is available at all times.

The data set contains 2500 observations. The parameters used for the experiments are listed in Table 1.

Figure 3 shows the graphs of outlier density ρ in the batch (upper plot) and online (middle plot) cases. For reference, plots of the data taken at various stages are displayed in the bottom. The upper (batch) plot shows $\rho = \rho_{\text{RMS}}$ computed on the probe set S at each iteration. The middle (online) plot shows $\rho = \Omega$. During training this is evaluated at each pattern. For clearer display, average

Table 1. Parameters used in the experiments

Parameter	Symbol	Batch	Online	Note
Training window size	w	200	—	
Probe window size	s	30	—	
Possibility degree	α	0.7	0.7	(1)
Num. annealing steps	b	20	—	
β schedule		linear	linear	(2)
Starting value for annealing	β_1	0.05	0.05	(2)
Ending value for annealing	β_b	0.002	0.002	(2)
Density estimation function	$\rho(\Omega)$	ρ_{RMS}	Ω	
Coefficient for discounted avg.	λ	—	0.01	
Parameters for computing θ	θ_0	0.3	0.3	
	τ_1	0.01	0.01	
	τ_2	0.5	0.5	
	γ	2	2	

(1) For online: minimum value, maximum is 1.
(2) For online: only in the batch initialization phase.

---

[1] Available under GPL at https://github.com/gditzler/ConceptDriftData.

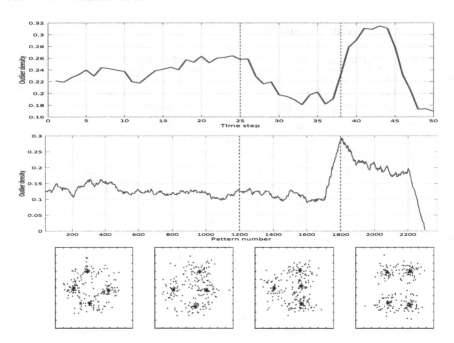

Fig. 3. Outlier density ρ with concept drift and shift. The true model is continuously evolving. The vertical lines mark points where the stream has been cut to create a discontinuity (concept shift). Samples of the training are shown below the graph.

over the past few iterations, rather than instantaneous value, is shown in the figure. In both plots, vertical lines indicate discontinuities (concept shift).

The graphs show the effectiveness of the proposed indexes in indicating the conditions occurring in the stream at each point in time. From the bottom plots it is evident that the first discontinuity occurs between similar configurations, so there is no actual concept shift; in fact, the graph in the batch case highlights that after the discontinuity clusters are closer to each other, so the concentration of data is higher, the clustering task is easier, and the outlier density decreases.

The second discontinuity produces instead a relevant change in centroid configuration, and this is evident in both graphs. However the online version is much quicker to respond to the variation, as indicated by the more steeply increasing plot around the second discontinuity.

A comparative study was performed with two other methods. The first, used as a baseline, is a simple non-tracking k-means, trained once and used without updates. The second comparison is with a method with similar goals, called TRAC-STREAMS [13], which performs a weighted PCM-like clustering while updating by pattern.

To perform a comparative analysis, for each method we measured tracking performance by computing the distortion (mean squared error) with respect to the learned centroids and to the true centroids, that we have available since

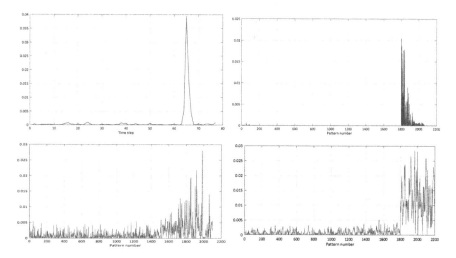

Fig. 4. Tracking error. Absolute difference between distortion w.r.t. true centroids and distortion w.r.t. learned centroids. Top left: batch. Top right: online. Bottom left: TRAC-STREAMS. Bottom right: k means, statically trained.

the data set is synthetic. Figure 4 shows the tracking performance, obtained by plotting the absolute difference of distortions computed for the "true" and learned models. Apart from the abrupt discontinuity, quickly recovered in both the batch and online cases, the difference is extremely limited, and superior to the other two methods considered.

8   Conclusions

We have presented a method that exploits fuzzy outlier analysis to provide good learning and tracking abilities in both concept shift and concept drift. The method builds upon a possibilistic clustering model that naturally offers a fuzzy outlierness measure.

The proposed method is currently being deployed in several applications, ranging from urban traffic forecasting to ambient assisted living. Several aspects are being investigated. Future work will include improvements in automatic setting of model parameters, as well as in the optimization process.

References

1. Aggarwal, C.C.: Data Streams: Models and Algorithms, vol. 31. Springer Science & Business Media, Berlin (2007)
2. Masulli, F., Rovetta, S.: Soft transition from probabilistic to possibilistic fuzzy clustering. IEEE Trans. Fuzzy Syst. **14**(4), 516–527 (2006)
3. Krishnapuram, R., Keller, J.M.: The possibilistic C-means algorithm: insights and recommendations. IEEE Trans. Fuzzy Syst. **4**(3), 385–393 (1996)

4. Bezdek, J.C.: Pattern Recognition with Fuzzy Objective Function Algorithms. Kluwer Academic Publishers, Norwell (1981)
5. Rose, K., Gurewitz, E., Fox, G.: A deterministic annealing approach to clustering. Pattern Recogn. Lett. **11**, 589–594 (1990)
6. Chandola, V., Banerjee, A., Kumar, V.: Anomaly detection: a survey. ACM Comput. Surv. (CSUR) **41**(3), 15 (2009)
7. Hawkins, D.M.: Identification of Outliers, vol. 11. Springer, Heidelberg (1980)
8. Knorr, E.M., Ng, R.T.: Finding intensional knowledge of distance-based outliers. In: VLDB, vol. 99, pp. 211–222 (1999)
9. Huber, P.J.: Robust Statistics. Wiley, New York (1981)
10. Balasubramanian, V., Ho, S.-S., Vovk, V.: Conformal Prediction for Reliable Machine Learning: Theory, Adaptations and Applications. Newnes (2014)
11. Keller, A.: Fuzzy clustering with outliers. In: Fuzzy Information Processing Society, 2000 NAFIPS 19th International Conference of the North American, pp. 143–147 (2000)
12. Davé, R.N., Krishnapuram, R.: Robust clustering methods: a unified view. IEEE Trans. Fuzzy Syst. **5**(2), 270–293 (1997)
13. Nasraoui, O., Rojas, C.: Robust clustering for tracking noisy evolving data streams. In: Proceedings of the 2006 SIAM International Conference on Data Mining, pp. 619–623 (2006)
14. Martinetz, T., Berkovich, S., Schulten, K.: Neural gas' network for vector quantization and its application to time-series prediction. IEEE Trans. Neural Netw. **4**(4), 558–569 (1993)
15. Kushner, H., Yin, G.: Stochastic Approximation and Recursive Algorithms and Applications. Stochastic Modelling and Applied Probability. Springer, New York (2003)
16. Ridella, S., Rovetta, S., Zunino, R.: Plastic algorithm for adaptive vector quantization. Neural Comput. Appl. **7**(1), 37–51 (1998)
17. Yoon, K.-A., Kwon, O.-S., Bae, D.-H.: An approach to outlier detection of software measurement data using the k-means clustering method. In: First International Symposium on Empirical Software Engineering and Measurement, ESEM 2007, pp. 443–445. IEEE (2007)
18. Robbins, H., Monro, S.: A stochastic approximation method. Ann. Math. Stat. **22**(3), 400–407 (1951)
19. Ditzler, G., Polikar, R.: Incremental learning of concept drift from streaming imbalanced data. IEEE Trans. Knowl. Data Eng. **25**(10), 2283–2301 (2013)
20. Abdullatif, A., Masulli, F., Rovetta, S.: Layered ensemble model for short-term traffic flow forecasting with outlier detection. In: 2016 IEEE 2nd International Forum on Research and Technologies for Society and Industry Leveraging a Better Tomorrow (RTSI) (IEEE RTSI 2016), pp. 1–6 (2016)

Unsupervised Analysis of Event-Related Potentials (ERPs) During an Emotional Go/NoGo Task

Paolo Masulli[1]([✉]), Francesco Masulli[2,3], Stefano Rovetta[2], Alessandra Lintas[1], and Alessandro E.P. Villa[1]

[1] NeuroHeuristic Research Group, University of Lausanne,
Internef, Quartier UNIL Dorigny, 1015 Lausanne, Switzerland
{paolo.masulli,alessandra.lintas,alessandro.villa}@unil.ch

[2] Department of Computer Science, Bioengineering,
Robotics and Systems Engineering, DIBRIS, University of Genova,
16146 Genova, Italy
{francesco.masulli,stefano.rovetta}@unige.it

[3] Sbarro Institute for Cancer Research and Molecular Medicine,
Temple University, Philadelphia, PA, USA

Abstract. We propose a framework for an unsupervised analysis of electroencephalography (EEG) data based on possibilistic clustering, including a preliminary noise and artefact rejection. The proposed data flow identifies the existing similarities in a set of segments of EEG signals and their grouping according to relevant experimental conditions. The analysis is applied to a set of event-related potentials (ERPs) recorded during the performance of an emotional Go/NoGo task. We show that the clusterization rate of trials in two experimental conditions is able to characterize the participants. The extension of the method and its generalization is discussed.

Keywords: EEG · ERP · Possibilistic clustering · Decoding · Brain activity

1 Introduction

The recording of human brain activity is generally performed by electroencephalography (EEG), which has the advantage of being a non invasive technique using external electrodes placed over many standard locations determined by skull landmarks. It is recognized that brain circuits that are activated at any occurrence of the same mental or physical stimulation generate transient electric potentials than can be averaged over repeated trials [20]. Such brain signals recorded by EEG and triggered by a specific event are referred to as event-related potentials (ERPs).

The conventional analysis of ERPs rely on a supervised approach, mainly on the experimenter's experience in order to discard outliers and detect wave components that are associated with different neural processes. In order to decrease

© Springer International Publishing AG 2017
A. Petrosino et al. (Eds.): WILF 2016, LNAI 10147, pp. 151–161, 2017.
DOI: 10.1007/978-3-319-52962-2_13

the impact of the bias due to the human supervision, machine learning techniques have been recently presented [15]. Techniques using Echo State Networks [2], or interval features [13], or statistical techniques such as Independent Component Analysis [18,25] have been successfully applied to tackle a specific classification problem following a phase of feature extraction. However, these techniques bear the disadvantage of depending on a careful choice of the feature extraction technique which captures a pre-determined aspect of the variability. In a way the bias of human supervision is moved to another step of the analysis, but the pre-selection of specific features might be prone to learning biases or over-fitting of the training data.

It has been recognized for a long time the value of single-trial analysis [5] and the importance of analyzing all time points to reveal the complete time course of the effect of a triggering event, either mental or physicial [23]. In order to reduce the bias of an *a priori* criteria for a supervised classification we propose an unsupervised data processing flow based on a probabilistic clustering aimed to capture similarities between different trials in the data set, without previously extracting features. Our approach is not intended to provide a direct answer to a specific classification problem. It is rather thought as a general tool to clusterize ERPs based on their internal structure. We present the application of this technique to a set of ERPs recorded during the performance of an emotional Go/NoGo task.

2   Methods

2.1   Graded Possibilistic Clustering

The subject of this study is the analysis of noisy signals, possibly containing anomalies. For the data analysis step a soft clustering method with tunable outlier rejection ability was therefore selected.

The *Graded Possibilistic Clustering* approach [16] is a central clustering model which inherits outlier rejection properties from soft clustering methods based on a possibilistic model (such as Possibilistic C-Means [11,12]), while at the same time avoiding their well-known issues related to convergence and to overlapping clusters, similarly to soft clustering methods with probabilistic constraint (as in Fuzzy C-Means [1,6], and Deterministic Annealing [21,22]). In the current work we propose a new version of the Graded Possibilistic Clustering model (GPC-II). Let us consider a set X of k observations (or instances) $\mathbf{x}_l$, for $l \in \{1, \ldots, n\}$, and a set C of c fuzzy clusters denoted $C_1, \ldots, C_c$. Clusters are represented via their centroids $\mathbf{y}_j$, for $j \in \{1, \ldots, c\}$. Each cluster associates a fuzzy cluster indicator (or membership) function $u_{lj} \in [0,1] \subset \mathbb{R}$ to a given observation $\mathbf{x}_l$. We define the *total membership mass* of an observation $\mathbf{x}_l$ as:

$$\zeta_l = \sum_{j=1}^{c} u_{lj}. \tag{1}$$

The membership of the observation $\mathbf{x}_l$ to the cluster u_j can be expressed as:

$$u_{lj} = \frac{v_{lj}}{Z_l}, \tag{2}$$

where

$$v_{lj} = e^{-d_{lj}/\beta_j} \tag{3}$$

is the *free membership*, and

$$Z_l = \zeta_l^\alpha = \left(\sum_{j=1}^c v_{lj} \right)^\alpha, \quad \alpha \in [0,1] \subset \mathbb{R} \tag{4}$$

is the *generalized partition function*.

In the last two equations, d_{lj} is the distance between the j-th centroid and the observation $\mathbf{x}_l$, coefficients β_j are model parameters playing a role in the representation of data as cluster widths, and the parameter α controls the "possibility level" of the GPC-II, from a totally probabilistic ($\alpha = 1$) to a totally possibilistic ($\alpha = 0$) model, with all intermediate cases for $0 < \alpha < 1$.

In GPC-II cluster centroids are related to membership vectors via the equation:

$$\mathbf{y}_j = \frac{\sum_{l=1}^n u_{lj} \mathbf{x}_l}{\sum_{l=1}^n u_{lj}}. \tag{5}$$

The implementation of the GPC-II is based on a Picard iteration of Eqs. 2 and 5 after a random initialization of centroids.

Note that for $\alpha = 1$ the representation properties of the method coincide with those of Deterministic Annealing [21,22]), where the solution of a regularized minimization problem with constraints yields $u_{lj} = \frac{e^{-d_{lj}/\beta_j}}{\sum_{h=1}^c e^{-d_{lh}/\beta_h}}$, equivalent to (2) when $\alpha = 1$. In turn, this is equivalent to Fuzzy c Means, up to a change of units, as proved in [17]. When $\alpha = 0$, they are equivalent to those of Possibilistic C-Means [12]) which is designed for robust clustering, since its clusters are not influenced by each other or by outliers. This comes at the cost of difficult convergence and collapsing clusters. In the intermediate cases, as soon as $\alpha > 0$, there is a degree of competition between clusters which improves convergence, as in probabilistic models, but memberships eventually vanish for points sufficiently far away from the centroids, as in the possibilistic case, for noise insensitivity. This trade off is under user control, and depends in a complex way on the nature of the data (dimensionality, density, clustering structure, distribution isotropy, distribution uniformity...), so it is usually selected by grid search.

A deterministic annealing version of the Graded Possibilistic Clustering model (DAGPC-II) can be implemented by decomposing the model parameters β_j as $\beta_j = \beta b_j$, where β is the optimization parameter for the deterministic annealing procedure (starting small and enlarging each time the Picard iteration converges) and b_j are the *relative* cluster scales obtained from some heuristic like those proposed in [12]. In this way the Graded Possibilistic Clustering can benefit of the powerful optimization technique proposed in [21,22] that, after starting from fully overlapping clusters, performs a hierarchical clustering by progressively splitting overlapping clusters as β increases. "Natural" aggregations are therefore discovered.

2.2 Measuring Overlap Between Fuzzy Cluster

The Jaccard index is a common measure of overlap between two partitions. It is defined by the following expression:

$$J(A, B) = \frac{|A \cap B|}{|A \cup B|}. \tag{6}$$

For fuzzy clusters, the belonging of an element to a cluster is expressed via its membership indicator, which is a real value in the interval $[0, 1]$. In this case we can use the following definitions of fuzzy cardinality for the fuzzy clusters in Eq. (6).

$$|C_l \cap C_m| = \sum_{j=1}^{m} \min(u_{lj}, u_{mj}), \qquad |C_l \cup C_m| = \sum_{j=1}^{m} \max(u_{lj}, u_{mj}),$$

The resulting fuzzy Jaccard index is still a real value in the interval $[0, 1]$.

2.3 Experimental Data

Nineteen volunteers (5 females, mean age (SD) 28 (6.69)) were fitted with EEG equipment and recorded using 64 scalp active electrodes (ActiveTwo MARK II Biosemi EEG System) at a sampling frequency of 2048 Hz. We present here an analysis limited to brain signals recorded on the electrode Fz.

The behavioral task consisted in an emotional Go/NoGo task [7]. The stimulus presentation and response collection software was programmed using the software EPrime (Psychology Software Tools, Inc., Sharpsburg, PA 15215-2821, USA). The Go-cues of the task required participants to look at a picture with a face presented in the center of a computer screen and respond as fast as possible by pressing a button when a face expressing neutral emotion was displayed. In the NoGo-cues trials the participants withheld responses to non-target stimuli. For each trial the stimulus presentation lasted for a duration of 500 ms, followed by a fixation mark (+) for 1000 ms. The task consisted of four blocks of 30 pseudo-randomized "Go/NoGo" trials. Each block contained 20 "Go" and 10 "NoGo" trials. A neutral emotional expression of a face was paired with emotional expressions (happiness, fear, anger, or sadness) of the same face. The neutral expression served as the "Go" cue, while anyone of the emotional expressions was the "NoGo" cue.

2.4 Within-Participant Data Processing Flow

For each participant the analysis of all Go/NoGo trials is aimed to exclude the trials that contain artefacts, and then identify clusters of the remaining trials, without any information whether the stimulus was a "GO" or a "NoGo" cue. The data processing flow consists of the following phases.

1. **EEG Pre-processing.** The data of each participant are imported in EEGLAB [4] and re-referenced with respect to the two mastoidals M1 and M2. Subsequently, the filter IIR Butterworth (Basic Filter for continuous EEG in the software) is applied, with Half-Power parameters 0.1 Hz for High-Pass and 27.6 Hz for Low-Pass.

2. **Segments extraction.** The resulting signals are read with a Python script using the EDF reader present in the package `eegtools`. The relevant segments of the EEG recording are extracted according to the triggers which identify the experimental trials. We have considered only the correct trials (i.e., the participants either pressed the button after a "Go" cue, or they did not press the button for a "NoGo" cue), on average 108 (out of 120) trials per participant. We considered a time course of 600 ms for the ERPs, starting at the beginning of the trial (the instant when the picture with the face is presented to the participant), in order to avoid the presence of muscular artifacts produced at the time the participant pressed the button. For each trial, we apply a baseline correction by subtracting the average value of amplitude of the signal in the 200 ms immediately before the trial onset. With a sampling rate of 2048 Hz, the segments with a duration of 600 ms correspond to vectors in $\mathbb{R}^{1228}$ and they form a subset of $\mathbb{R}^{1228}$: in other words, the analysis was applied directly to vectors of 1228 signal samples.

3. **Smoothing.** We apply the 1-dimensional Anisotropic Diffusion algorithm by Perona and Malik [19] to smooth each signal segment (parameters: 1000 iterations, `delta_t` $= 1/3$, `kappa` $= 3$).

4. **Artefact rejection.** The peak-to-peak amplitude was computed for each segment. We rejected those segments characterized by a difference larger than twice the median absolute deviation (MAD) between the peak-to-peak value of a trial and the median peak-to-peak value computed over all correct trials. On average this procedure rejected approximately 14 segments per participant. The remaining segments ($n = 94$ on average, out of 120 initial trials) for each participant were considered valid for further analysis. The set of valid data is denoted $X \subset \mathbb{R}^{1228}$. An example is displayed in Fig. 1.

5. **Possibilistic clustering.** We apply the Graded Possibilistic Clustering algorithm with deterministic annealing (Sect. 2.1) with parameter $\alpha = 0.85$ and initial number of clusters equal to $c_0 = 7$, which is chosen heuristically to be between 2 and 3 times the actual number of clusters we expect to find in the data, obtaining, for each cluster C_j, its centroid $\mathbf{y}_j \in X$ and its membership vector $u_{.j} \in \mathbb{R}^{|X|}$, whose components express how much each of the segments belongs to the cluster.

6. **Singleton clusters rejection.** Any cluster with its centroid modeling only a single trial was removed. Such clusters, the singletons, are identified among those with a total membership differing from the average membership of all clusters by more than 1.5 standard deviations. In case there is only one segment, among all the segments, with maximum membership in the cluster,

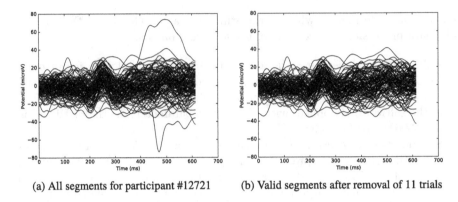

(a) All segments for participant #12721   (b) Valid segments after removal of 11 trials

Fig. 1. Rejection of the ERP segments tagged as artefacts according to the peak-to-peak amplitude criterion.

then the segment is removed and the clustering algorithm of point 5 is applied iteratively. Hence the number of required clusters is decreasing by one and the procedure is repeated until no more segments are removed.

7. **Merging of overlapping clusters.** The Jaccard index, defined in Sect. 2.2, is computed to determine the closeness for each pair of clusters. For each participant we compute independently the distribution of these values, which tends to be a multimodal distribution that can be decomposed as a mixture of distributions. The leftmost mode includes the neighbor clusters, very close to each other given a Jaccard index towards 1. Different methods used to fit this density show that a cut-point separating the neighbor from the remaining clusters tended to be close to a value $J_0 = 0.7$. For this reason we have decided to fix this value for all participants as the significant threshold value characterizing overlapping clusters. Hence, two clusters are merged if their index surpasses the threshold value J_0. After merging, the new centroid is computed as the average of the centroids of the previous clusters weighed by the membership values of their elements and we determine the new membership vector as done in the clustering algorithm of point 5. Then, new Jaccard indexes are recomputed and the procedure is applied iteratively until no neighbor clusters can be merged any more.

2.5   Clusterization Rate

The interest and the validity of the clusters obtained through the procedure described above is dependent on the specific data set. In this study, we evaluate the outcome of the analysis by considering whether the "Go" and "NoGo" trials were clusterized according to the experimental conditions. For each participant we computed a *clusterization rate* of the "Go" trials (respectively, the "NoGo") trials defined as the number of "Go" segments (respectively, "NoGo") characterized by a high membership to one of the identified clusters (i.e., signal segments

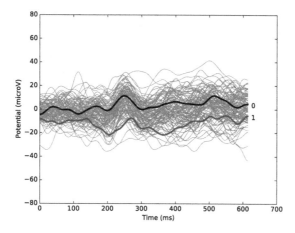

Fig. 2. ERP segments for participant #12721 (thin grey lines) and the two clusters (represented by their centroids with thick blue and black lines). (Color figure online)

with a membership above the 95-th percentile of the cluster) divided by the total number of "Go" trials (respectively, "NoGo").

3   Results

For each participant we observed a number of clusters $2 \leq C \leq 4$. An example is displayed in Fig. 2.

The confusion matrix of the different experimental conditions (Go/NoGo condition, type of emotion) *vs.* the membership to the clusters did not provide relevant differences between the clusters. This suggests that the ERPs features related to high cognitive functions such as affective discrimination are not associated with simple wave components of the brain signal itself and might require the choice of specific features or a different metric of the signals to be detected.

Nevertheless, we observed a significant difference of the clusterization rate of the "Go" and "NoGo" trials for most participants (Fig. 3). The probabilistic clustering is implicitly depending on the random seed used at the begin of the procedure, for we repeated the same analysis with 20 different random seeds in order to evaluate the robustness of the result for each participant. We observed that for 13 participants the average clusterization rates for the Go trials were larger than the corresponding clusterization rates for the NoGo trials (Fig. 3a).

For each participant the variability of the clusterization rates depended on the variability between segments, on the number of correct available trials and on the nonlinear interaction of these factors with the random initialization of the probabilistic clustering. The variability for the Go trials was smaller than for the NoGo trials (Fig. 3b). The average values of clusterization of the Go trials for different participants were more similar to each other, while the average values for the clusterization of the NoGo trials varied more between participants.

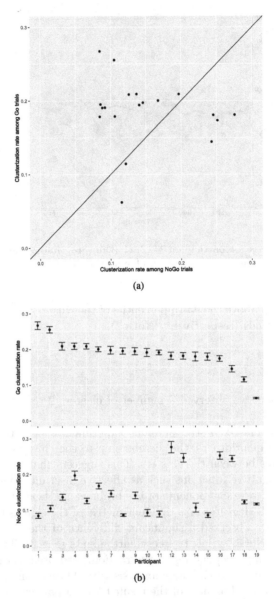

Fig. 3. Clusterization rate comparison of Go and NoGo trials for all participants. (a) scatter plot of the average values over 20 different random seed initializations (each point corresponds to one participant). For the majority of the participants (n = 13), the clusterization rate of the Go trials is higher than the one of the NoGo trials. This expresses the fact that the Go trials tend to be more similar to each other, while the NoGo trials are characterized by higher intrinsic variability which is not necessarily captured in the clusters. (b) average value and standard error of the clusterization rate of Go and NoGo trials for each participant over 20 different random seed initializations. Notice that the clusterization rates of the Go trials are characterized by a similar variation across all the participants, while the NoGo trials have a much broader distribution of the standard error.

Indeed, it was expected that the ERP segments of the "Go" trials, with a neutral expression of the cue, were more similar and hence would be associated with a high membership to one of the identified clusters. Conversely, the NoGo ERPs were associated with different face expressions and tended to be more scattered and differ more between each other.

4 Discussion and Conclusions

The analysis presented here demonstrates a possible approach towards an unsupervised EEG signal analysis. The choices that we have made in the design of the data processing flow make it very general and easy to extend to other datasets, since the user is not required to choose a specific set of features to extract and analyze. The only value used in our processing flow that was derived from a post-hoc analysis is the threshold value J_0 of the Jaccard index used to merge neighbor clusters. We set this value $J_0 = 0.7$ and it we can foresee that for other datasets this value might not be optimal. What is the effect of choosing a value of 0.6, 0.65, or 0.8 is worth to be considered in future work and with other data sets.

The trade-off of our approach is that the clustering algorithm compares the signals directly by computing their distances as vectors, which might lead to the loss of higher order features and to comparison problems between different trials given by time shifts and delayed onset of EEG peaks. The complexity of the information processed by the brain will certainly require a set of techniques able to analyze simultaneously the brain activity at various time scales [3,9,24,26]. Decoding brain states cannot be expected to be achieved by totally unsupervised or partially supervised machine learning techniques [15]. For the analysis of specific features of the ERPs a necessary approach will consider the use of metrics able to detect and compare peaks and wave components and accounting for time warping [8,10,14,27]. Another future development which we shall address is the comparison of the EEG signals of different participants with similar unsupervised techniques, with the aim of grouping the participants according to their mutual similarities. This will require the application of normalization to account for inter-participant variability of the amplitude of the signals.

Beyond brain research and clinical applications, these results will be applicable to end-user applications of EEG recordings and might be integrated in brain-machine interfaces featuring EEG sensors, thereby bringing important societal benefits through applications to the domains of health, wellness and well-being. Additional goals will be the extensions to different types of neurophysiological signals, such as spike trains recorded by intra-cranial electrodes, and to time-series data from different sources.

Acknowledgments. This work was partially supported by the Swiss National Science Foundation grant CR13I1-138032.

References

1. Bezdek, J.C.: Pattern Recognition with Fuzzy Objective Function Algorithms. Springer, Boston (1981)
2. Bozhkov, L., Koprinkova-Hristova, P., Georgieva, P.: Learning to decode human emotions with Echo State Networks. Neural Netw. **78**, 112–119 (2016)
3. Del Prete, V., Martignon, L., Villa, A.E.: Detection of syntonies between multiple spike trains using a coarse-grain binarization of spike count distributions. Network **15**(1), 13–28 (2004)
4. Delorme, A., Makeig, S.: EEGLAB: an open source toolbox for analysis of single-trial EEG dynamics including independent component analysis. J. Neurosci. Methods **134**(1), 9–21 (2004)
5. Donchin, E.: Discriminant analysis in average evoked response studies: the study of single trial data. Electroencephalogr. Clin. Neurophysiol. **27**(3), 311–314 (1969)
6. Dunn, J.C.: Some recent investigations of a new fuzzy partitioning algorithm and its application to pattern classification problems. J. Cybern. **4**(2), 1–15 (1974)
7. Hare, T.A., Tottenham, N., Davidson, M.C., Glover, G.H., Casey, B.J.: Contributions of amygdala and striatal activity in emotion regulation. Biol. Psychiatry **57**(6), 624–632 (2005)
8. Ihrke, M., Schrobsdorff, H., Herrmann, J.M.: Recurrence-based estimation of time-distortion functions for ERP waveform reconstruction. Int. J. Neural Syst. **21**(1), 65–78 (2011)
9. Indic, P.: Time scale dependence of human brain dynamics. Int. J. Neurosci. **99**(1–4), 195–199 (1999)
10. Karamzadeh, N., Medvedev, A., Azari, A., Gandjbakhche, A., Najafizadeh, L.: Capturing dynamic patterns of task-based functional connectivity with EEG. Neuroimage **66**, 311–317 (2013)
11. Krishnapuram, R., Keller, J.M.: A possibilistic approach to clustering. IEEE Trans. Fuzzy Syst. **1**(2), 98–110 (1993)
12. Krishnapuram, R., Keller, J.M.: The possibilistic c-means algorithm: insights and recommendations. IEEE Trans. Fuzzy Syst. **4**(3), 385–393 (1996)
13. Kuncheva, L.I., Rodríguez, J.J.: Interval feature extraction for classification of event-related potentials (ERP) in EEG data analysis. Prog. Artif. Intell. **2**(1), 65–72 (2013)
14. Lederman, D., Tabrikian, J.: Classification of multichannel EEG patterns using parallel hidden Markov models. Med. Biol. Eng. Comput. **50**(4), 319–328 (2012)
15. Lemm, S., Blankertz, B., Dickhaus, T., Müller, K.R.: Introduction to machine learning for brain imaging. Neuroimage **56**(2), 387–399 (2011)
16. Masulli, F., Rovetta, S.: Soft transition from probabilistic to possibilistic fuzzy clustering. IEEE Trans. Fuzzy Syst. **14**(4), 516–527 (2006)
17. Miyamoto, S., Mukaidono, M.: Fuzzy c-means as a regularization and maximum entropy approach. In: Proceedings of 7th IFSA World Congress, Prague, pp. 86–91 (1997)
18. Mueller, A., Candrian, G., Kropotov, J.D., Ponomarev, V.A., Baschera, G.M.: Classification of ADHD patients on the basis of independent ERP components using a machine learning system. Nonlinear Biomed. Phys. **4**(Suppl 1), S1 (2010)
19. Perona, P., Malik, J.: Scale-space and edge detection using anisotropic diffusion. IEEE Trans. Pattern Anal. Mach. Intell. **12**(7), 629–639 (1990)

20. Picton, T.W., Bentin, S., Berg, P., Donchin, E., Hillyard, S.A., Johnson, R., Miller, G.A., Ritter, W., Ruchkin, D.S., Rugg, M.D., Taylor, M.J.: Guidelines for using human event-related potentials to study cognition: recording standards and publication criteria. Psychophysiology **37**(2), 127–152 (2000)

21. Rose, K., Gurewitz, E., Fox, G.: A deterministic annealing approach to clustering. Pattern Recogn. Lett. **11**(9), 589–594 (1990)

22. Rose, K., Gurewitz, E., Fox, G.C.: Statistical mechanics and phase transitions in clustering. Phys. Rev. Lett. **65**(8), 945–948 (1990)

23. Rousselet, G.A., Pernet, C.R.: Quantifying the time course of visual object processing using ERPs: it's time to up the game. Front. Psychol. **2**, 107 (2011)

24. Smith, R.X., Yan, L., Wang, D.J.J.: Multiple time scale complexity analysis of resting state FMRI. Brain Imaging Behav. **8**(2), 284–291 (2014)

25. Stewart, A.X., Nuthmann, A., Sanguinetti, G.: Single-trial classification of EEG in a visual object task using ICA and machine learning. J. Neurosci. Methods **228**, 1–14 (2014)

26. Wohrer, A., Machens, C.K.: On the number of neurons and time scale of integration underlying the formation of percepts in the brain. PLoS Comput. Biol. **11**(3), e1004082 (2015)

27. Zoumpoulaki, A., Alsufyani, A., Filetti, M., Brammer, M., Bowman, H.: Latency as a region contrast: measuring ERP latency differences with dynamic time warping. Psychophysiology **52**(12), 1559–1576 (2015)

Deep Learning Architectures
for DNA Sequence Classification

Giosué Lo Bosco[1,2(✉)] and Mattia Antonino Di Gangi[3,4]

[1] Dipartimento di Matematica e Informatica,
Universitá degli studi di Palermo, Palermo, Italy
giosue.lobosco@unipa.it
[2] Dipartimento di Scienze per l'Innovazione e le Tecnologie Abilitanti,
Istituto Euro Mediterraneo di Scienza e Tecnologia, Palermo, Italy
[3] Fondazione Bruno Kessler, Trento, Italy
[4] ICT International Doctoral School, University of Trento, Trento, Italy

Abstract. DNA sequence classification is a key task in a generic computational framework for biomedical data analysis, and in recent years several machine learning technique have been adopted to successful accomplish with this task. Anyway, the main difficulty behind the problem remains the feature selection process. Sequences do not have explicit features, and the commonly used representations introduce the main drawback of the high dimensionality. For sure, machine learning method devoted to supervised classification tasks are strongly dependent on the feature extraction step, and in order to build a good representation it is necessary to recognize and measure meaningful details of the items to classify. Recently, neural deep learning architectures or deep learning models, were proved to be able to extract automatically useful features from input patterns. In this work we present two different deep learning architectures for the purpose of DNA sequence classification. Their comparison is carried out on a public data-set of DNA sequences, for five different classification tasks.

Keywords: DNA sequence classification · Convolutional Neural Networks · Recurrent Neural Networks · Deep learning networks

1 Introduction

One of the primary goal in biology is the understanding of the relationships between protein structure and function. To understand the structure-function paradigm, particularly useful structural information comes from the primary amino acid sequences. DNA sequence classification is extremely useful for this task, following the principle that sequences having similar structures have also similar functions. Sequence similarity is traditionally established by using

---

G. Lo Bosco and M.A. Di Gangi—Both authors have the same contribution to this paper.

A. Petrosino et al. (Eds.): WILF 2016, LNAI 10147, pp. 162–171, 2017.
DOI: 10.1007/978-3-319-52962-2_14

sequence alignment methods, such as BLAST [1] and FASTA [2]. This choice is motivated by two main assumptions: (1) the functional elements share common sequence features and (2) the relative order of the functional elements is conserved between different sequences. Although these assumptions are valid in a broad range of cases, they are not general. For example, in the case of cis-regulatory elements related sequences, there is little evidence suggesting that the order between different elements would have any significant effect in regulating gene expression. Anyway, despite recent efforts, the key issue that seriously limits the application of the alignment methods still remains their time computational complexity. As such, the recently developed alignment-free methods [3,4] have emerged as a promising approach to investigate the regulatory genome. For sure, these methodologies involve a feature extraction phase such as spectral representation of DNA sequences [5–7]. The performance of conventional machine learning algorithms depends on feature representations, which are typically designed by experts. As a subsequent phase, it is necessary to identify which features are more appropriate to face with the given task, and nowadays this remains a very crucial and difficult step.

Deep learning has recently emerged as a successful paradigm for big data, also because of the technological advances in terms of the low level cost of parallel computing architectures, so that deep learning has given significant contributions in several basic but arduous artificial intelligence tasks. Very important advancements have been made in computer vision [8,9], but also in natural language processing (NLP) [10,11] and machine translation [12–14], where deep learning techniques represent now the state of the art.

The main contribution of deep learning methods to bioinformatics have been in genomic medicine and medical imaging research field. To the best of our knowledge, very few contribution have been provided for the sequence classification problem. For a deep review about deep learning in bioinformatics, see the review by Seonwoo et al. [15].

The taxonomy of deep neural models mainly includes *Convolutional Neural Networks (CNN), Recurrent Neural Networks (RNN)* and *Stacked Auto-encoders (SAE)*. CNNs are characterized by an initial layer of convolutional filters, followed by a non Linearity, a sub-sampling, and a fully connected layer which realized the final classification. LeNet [16] was one of the very first convolutional neural networks which helped propel the field of Deep Learning. Collobert et al. [17] firstly shown that CNNs can be used effectively also for sequence analysis, in the case of a generic text. RNNs are designed to exploit sequential information of input data with cyclic connections among building blocks like perceptrons or long short-term memory units (LSTMs). Differently from other architectures, RNNs are capable of handling sequence information over time, and this is an outstanding property in the case of sequence classification. They are able to incorporate contextual information from past inputs, with the advantage to be robust to localised distortions of the input sequence along the time.

In this study we compare CNNs and RNNs deep learning architectures for the special case of bacterial classification using a publicly available data-set. We will use a variant of the LeNet network as choice for the architecture of the CNN, and a long short-term memory (LSTM) network as choice for RNN. The latter, is a type of recurrent neural networks with a more complex computational unit that leads to better performance. We compare these networks with their variants trained for multi-task learning, which should exploit the relations between the classes at the different levels of the taxonomy. In the next two sections the main components of CNNs and RNNs deep learning architectures are introduced and described, in Sect. 4 data-sets, experiments and results are reported. Conclusions and remarks are discussed in Sect. 5.

2   Deep Learning Architectures

2.1   Convolutional Neural Networks

The convolutional layers are the core components of a CNN. They calculate L one-dimensional convolutions between the kernel vectors w^l, of size $2n + 1$, and the input signal x:

$$q^l(i) = \sum_{u=-n}^{n} w^l(u)x(i-u) \tag{1}$$

In Eq. 1 $q^l(i)$ is the component i of the l-th output vector and $w^l(u)$ is the component u of the l-th kernel vector. After a bias term b^l is added and the nonlinear function g is applied:

$$h^l(i) = g(q^l(i) + b^l). \tag{2}$$

Another important component of a CNN is the max-pooling layer, that usually follows the convolutional layer in the computation flow. It is a non-linear down-sampling layer that partitions the input vector into a set of non-overlapping regions and, for each sub-region, the maximum value is considered as output. This processing layer reduces the complexity for the following layers and operates a sort of translational invariance. Convolution and max-pooling are usually considered together as two highly connected blocks.

2.2   Recurrent Neural Networks

Recurrent Neural Networks are generally used for processing sequences of data which evolves along the time axis. The simpler version of a RNN owns an internal state h_t which is a summary of the sequence seen until the previous time step $(t-1)$ and it is used together with the new input x_t:

$$h_t = \sigma(W_h x_t + U_h h_{t-1} + b_h)$$

$$y_t = \sigma(W_y h_t + b_y)$$

where W_h and U_h are respectively the weight matrices for the input and the internal state, W_y is the weight matrix for producing the output from the internal state, and the two b are bias vectors.

One problem that limits the usage of RNNs that respect this formulation, is that all the time steps have the same weight and, consequently, the contribution of an input in the hidden state is subjected to exponential decay. A variant of RNN has been introduced in [18] with the name of *Long Short-Term Memory (LSTM)*. LSTM is a recurrent neural unit consisting of four internal gates, each computing a different function of the input, that allow the layer to exploit long range relations. The four gates provide the unit with the capability of choosing the weight to give to every single input, and to forget the current hidden state for the computation of the next output.

2.3   Softmax Layer

Both the two previous architectures needs a further layer in order to compute a classification task. Softmax layer [19] is composed by K units, where K is the number of different classes. Each unit is densely connected with the previous layer and computes the probability that an element is of class k by means of the formula:

$$softmax_k(x) = \frac{e^{W_k x + b_k}}{\sum_{l=1}^{k} e^{W_l x + b_l}}$$

where W_l is the weight matrix connecting the l-th unit to the previous layer, x is the output of the previous layer and b_l is the bias for the l-th unit. Softmax is widely used by deep learning practitioners as a classification layer because of the normalized probability distribution it outputs, which proves particularly useful during back-propagation.

2.4   Character Embedding

In this work we evaluate models for classifying genomic sequences without providing a-priori information by means of feature engineering. One method for achieving this is represented by the use of *character-level one-hot encoding*. This representation takes into account each character i of the alphabet by a vector of length equal to the alphabet size, having all zero entries except for a single one in position i. This method leads to a sparse representation of the input, which is tackled in the NLP literature by means of an embedding layer.

Word Embedding [20, 21] is used to give a continuous vector representation to each word of a sequence, and in the vector space obtained with this process words with a similar meaning are close in distance.

The same idea has been also applied at character-level [22], and while in this case there is no semantic similarity to find, it represents a first step for fully-automated feature learning.

The embedding representation can be implemented by a single feed-forward layer with a non-linear activation, where each row of the weight matrix W gives the representation for the corresponding symbol and it can be jointly trained with the rest of the network.

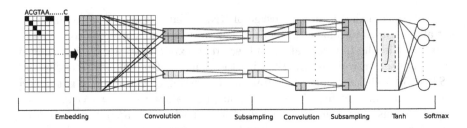

Fig. 1. The architecture of the proposed CNN

3 Deep Learning Networks for DNA Sequence Classification

As the used data-set encodes the sequences according to *IUPAC* nucleotide codes, the vectors that we consider are of size 16. Note that the IUPAC alphabet is composed by 15 characters, so we have used a padding character in order to have only equally sized sequences.

3.1 Convolutional Neural Network

The Convolutional Neural Network (CNN) used in this study is inspired by Lecun's LeNet-5 [16]. The first layer is an embedding layer, which takes as input 16-dimensional one-hot encoding of sequence characters, and produces a 10 dimensional continuous vector. On top of the input we have two convolutional layers, each followed by a max pooling layer. The two convolutional layers use filters of size 5 and have a growing number of filters, respectively 10 and 20. The width and the stride of the pooling layers are both of 5 time steps, considering a one-hot vector as a single unit.

The convolutional layers are stacked with two fully connected layers. The first one, consisting of 500 units, uses a *tanh* activation function, which is the same activation used by the lower levels. The second one is the classification layer and uses the *softmax* activation.

This network is similar to that proposed in [23], except for the addition of an embedding layer and the absence of preprocessing of the sequences that is achieved by means of the one-hot encoding. Figure 1 shows the architecture of the proposed CNN.

3.2 Recurrent Neural Network

The recurrent network is a 6-layered network whose input is in the form of one-hot encoding vectors, analogously to the previous section's model. The first layer is an embedding layer and it is followed by a max pooling layer of size 2. The max pooling reduces the computation for the following layer, and at the same time gives some capability of translational invariance to the network. After the

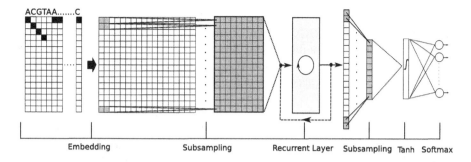

Fig. 2. The architecture of the proposed LSTM

max pooling there is a recurrent layer, implemented in this work as an LSTM, which processes its input from left to right and produces an output vector of size 20 at each time step. The LSTM layer is followed by another max pooling layer and in the top we have two fully-connected layers like those explained in the previous section (Fig. 2).

3.3 Multi-task Learning

Multi-task learning, in its original formulation, is the joint training of a machine learning system on multiple related tasks [24] in order to share the internal representation and achieve a better regularization. The used data-set is organized into five nested taxonomic ranks, i.e. each of the labels at a given rank of the taxonomy exists within one of the categories of the more coarse rank. For this reason it appears natural to jointly train a deep learning model for the five ranks at the same time.

We propose a multi-task learning variant for both networks by simply replicating the dense layers. The first dense layer is replicated five times, once for each taxon, each replica having 500 units, exactly like the layer in the original architecture. The softmax layers are also five, and each of them is stacked to only one of the fully-connected layers. We expect these networks to be able to perform better than their single task version, while requiring less time to train. In fact, even though the resulting network is bigger, the train is performed just once instead of five times, causing a sensitive gain in terms of time.

4 Experimental Results

4.1 Dataset Description

We have tested the effectiveness of the two proposed deep learning models in the case of the so called 16S data-set.

Studies about bacteria species are based on the analysis of their 16S ribosomal RNA (rRNA for short) housekeeping gene, as biologically validated by Drancourt et al. [25]. Ribosomes are the complex structures that acts the protein synthesis in each living organism. They consists of two subunits, each of which is composed of protein and a type of RNA, known as ribosomal RNA (rRNA). Prokaryotic ribosomes consist of 30S subunit and 50S subunit. The first subunit is also composed of 16S ribosomal RNA. The presence of hyper variable regions in the 16S rRNA gene provides a species specific signature sequence which is useful for bacterial identification process. Because the 16S rRNA gene is very short, just around 1500 nucleotide bases, it can be easily copied and sequenced. The use of 16S rRNA gene sequences to study bacterial phylogeny and taxonomy has been by far the most common housekeeping genetic marker used for several reasons that include its presence in almost all bacteria, its unchanged role function over evolution, and its versatility, in terms of length, for computational purposes.

The 16S rRNA data-set used in this study was downloaded from the RDP Ribosomal Database Project II (RDP-II) [26] and the selected sequences were chosen according to the type strain, uncultured and isolates source, average length of about 1200–1400 bps, good RDP system quality and *class* taxonomy by NCBI. By a subsequent filtering phase, a total amount of 3000 sequences have been selected, and can be grouped into 5 ordered taxonomic ranks, named *Phylum* (most coarse grained), *Class*, *Order*, *Family* and *Genus* (more fine grained). The resulting number of classes is increasing from a minimum of 3 classes (Phylum) to a maximum of 393 (Genus). Between them there are *Class*, *Order* and *Family* with 6, 22, 65 classes.

4.2 Experiments

For each taxonomic rank a 10-Fold cross-validation has been performed. We have chosen 15 epochs for each fold, without using early stopping validation. This choice has been motivated by our experimental observations.

In Table 1 we report the results of mean (μ) and standard deviation (σ) of the accuracy of the two networks computed on 10 test folds, for both single-task (CNN and LSTM) and multi-task (CNN-MT and LSTM-MT) learning.

Table 1. Average accuracies of the proposed models over 10-fold.

	Phylum		Class		Order		Family		Genus	
	μ	σ	μ	σ	μ	σ	μ	σ	μ	σ
CNN	0,995	0,003	0,993	0,006	0,937	0,012	0,893	0,019	0,676	0,065
CNN-MT	0,981	0,007	0,978	0,008	0,908	0,021	0,851	0,024	0,692	0,024
LSTM	0,982	0,028	0,977	0,022	0,902	0,028	0,857	0,034	0,728	0,030
LSTM-MT	0,992	0,007	0,990	0,008	0,941	0,029	0,897	0,023	0,733	0,030

As far as we know, the deep learning model state of the art for classifying this data-set is represented by a CNN that uses k-mers representation, that has been recently introduced by Rizzo et al. [23]. The comparison between the

Table 2. Approximate time in minutes for training one step of 10-fold cross-validation for the single-task (first five columns) and multi-task architecture (last column).

	Phylum	Class	Order	Family	Genus	Total	Mtask
CNN	6.5	6.5	6.7	6.7	6.8	33.2	21.5
LSTM	13.0	14.7	14.7	15.0	17.5	74.9	44.0

single-task models is absolutely favorable for the CNN, also in terms of training time, indeed the CNN can be trained in half the time than the LSTM for the same number of epochs and minibatch size. multi-task learning affects the two models in different ways, for instance it makes the CNN worse in the first four taxonomy ranks, while slightly improving the last, but improves the performance of LSTM, making it comparable to that of the single-task CNN on the first two categories and slightly better for the rest. Also in this case the time for the CNN is a bit less than the half respect to the time of the LSTM (see Table 2). Our results are anyway far from the state of the art, with the exception of the first two taxonomy ranks, which doesn't use positional information. We expect that this will not be the general case, and we plan to investigate on this by applying our networks to other sequence data sets related to other biological properties.

5 Conclusion

In this study, two deep learning architectures were compared for an automatic classification of bacteria species with no steps of sequence preprocessing. In particular, we have proposed a long short-time memory network (LSTM) that exploits the nucleotides positions of a sequence, in comparison with a classical convolutional neural network (CNN). Results show a superiority of the CNNs in the four simplest classification tasks, while its performance poorly degrades for the last where the LSTM works better. Moreover, we have studied how the multi-task learning affects the models, both in terms of performance and training time. Our results showed that it helps the LSTM for all the tasks, while it harms CNN performance. Further analysis is needed in order to explain this difference. Taking into consideration the results of the state of the art, we are going to expand the two models with positional information, coming from both convolutional and recurrent layers, trying to further exploit contextual information, for example with bidirectional RNNs.

References

1. Altschul, S., Gish, W., Miller, W., et al.: Basic local alignment search tool. J. Mol. Biol. **25**(3), 403–410 (1990)
2. Lipman, D., Pearson, W.: Rapid and sensitive protein similarity searches. Science **227**(4693), 1435–1441 (1985)

3. Vinga, S., Almeida, J.: Alignment-free sequence comparison a review. Bioinformatics **19**(4), 513–523 (2003)
4. Pinello, L., Lo Bosco, G., Yuan, G.-C.: Applications of alignment-free methods in epigenomics. Brief. Bioinform. **15**(3), 419–430 (2014)
5. Pinello, L., Lo Bosco, G., Hanlon, B., Yuan, G.-C.: A motif-independent metric for DNA sequence specificity. BMC Bioinform. **12**, 1–9 (2011)
6. Lo Bosco, G., Pinello, L.: A new feature selection methodology for K-mers representation of DNA sequences. In: Serio, C., Liò, P., Nonis, A., Tagliaferri, R. (eds.) CIBB 2014. LNCS, vol. 8623, pp. 99–108. Springer, Heidelberg (2015). doi:10.1007/978-3-319-24462-4_9
7. Lo Bosco, G.: Alignment free dissimilarities for nucleosome classification. In: Angelini, C., Rancoita, P.M.V., Rovetta, S. (eds.) CIBB 2015. LNCS, vol. 9874, pp. 114–128. Springer, Heidelberg (2016). doi:10.1007/978-3-319-44332-4_9
8. Farabet, C., Couprie, C., Najman, L., et al.: Learning hierarchical features for scene labeling. IEEE Trans. Pattern Anal. Mach. Intell. **35**(8), 1915–1929 (2013)
9. Tompson, J.J., Jain, A., LeCun, Y., et al.: Joint training of a convolutional network and a graphical model for human pose estimation. In: Advances in Neural Information Processing Systems, pp. 1799–1807 (2014)
10. Kiros, R., Zhu, Y., Salakhutdinov, R.R., et al.: Skip-thought vectors. In: Advances in Neural Information Processing Systems, pp. 3276–3284 (2015)
11. Li, J., Luong, M.-T., Jurafsky, D.: A hierarchical neural autoencoder for paragraphs and documents. In: Proceedings of 53rd Annual Meeting of the Association for Computational Linguistics and the 7th International Joint Conference on Natural Language Processing, pp. 1106–1115 (2015)
12. Luong, M.-T., Pham, H., Manning, C.D.: Effective approaches attention-based neural machine translation. In: Proceedings of Conference on Empirical Methods in Natural Language Processing, pp. 1412–1421 (2015)
13. Cho, K., Van Merrienboer, B., Gulcehre, C., et al.: Learning phrase representations using RNN encoder-decoder for statistical machine translation. In: Proceedings of Conference on Empirical Methods in Natural Language Processing (EMNLP), pp. 1724–1734 (2014)
14. Chatterjee, R., Farajian, M.A., Conforti, C., Jalalvand, S., Balaraman, V., Di Gangi, M.A., Ataman, D., Turchi, M., Negri, M., Federico, M.: FBK's neural machine translation systems for IWSLT 2016. In: Proceedings of 13th International Workshop on Spoken Language Translation (IWSLT 2016) (2016)
15. Seonwoo, M., Byunghan, L., Sungroh, Y.: Deep learning in bioinformatics. In: Briefings in Bioinformatics (2016)
16. LeCun, Y., Bottou, L., Bengio, Y., Haffner, P.: Gradient-based learning applied to document recognition. Proc. IEEE **86**(11), 2278–2324 (1998)
17. Collobert, R., Weston, J., Bottou, L., Karlen, M., Kavukcuoglu, K., Kuksa, P.: Natural language processing (almost) from scratch. J. Mach. Learn. Res. **12**, 2493–2537 (2011)
18. Hochreiter, S., Schmidhuber, J.: Long short-term memory. Neural Comput. **9**(8), 1735–1780 (1997)
19. Bridle, J.S.: Probabilistic interpretation of feedforward classification network outputs, with relationships to statistical pattern recognition. In: Soulié, F.F., Hérault, J. (eds.) Neurocomputing, pp. 227–236. Springer, Heidelberg (1990)
20. Mikolov, T., Chen, K., Corrado, G., Dean, J.: Efficient estimation of word representations in vector space. CoRR, abs/1301.3781 (2013)

21. Pennington, J., Socher, R., Manning, C.D.: Glove: global vectors for word representation. In: Proceedings of the 2014 Conference on Empirical Methods in Natural Language Processing (EMNLP), pp. 1532–1543, October 2014

22. Dos Santos, C.N., Zadrozny, B.: Learning character-level representations for part-of-speech tagging. In: Proceedings of the 31st International Conference on Machine Learning (ICML), pp. 1818–1826 (2014)

23. Rizzo, R., Fiannaca, A., La Rosa, M., Urso, A.: A deep learning approach to dna sequence classification. In: Angelini, C., Rancoita, P.M.V., Rovetta, S. (eds.) CIBB 2015. LNCS, vol. 9874, pp. 129–140. Springer, Heidelberg (2016). doi:10. 1007/978-3-319-44332-4_10

24. Caruana, R.: Multi-task learning: a knowledge-based source of inductive bias. Mach. Learn. **28**, 41–75 (1997)

25. Drancourt, M., Berger, P., Raoult, D.: Systematic 16S rRNA gene sequencing of atypical clinical isolates identified 27 new bacterial species associated with humans. J. Clin. Microbiol. **42**(5), 2197–2202 (2004)

26. https://rdp.cme.msu.edu/

A Deep Learning Approach to Deal with Data Uncertainty in Sentiment Analysis

Michele Di Capua[1] and Alfredo Petrosino[2(✉)]

[1] Department of Computer Science, University of Milan, Milan, Italy
michele.dicapua@unimi.it
[2] Department of Science and Technology,
University of Naples "Parthenope", Naples, Italy
petrosino@uniparthenope.it

Abstract. Sentiment Analysis refers to the process of computationally identifying and categorizing opinions expressed in a piece of text, in order to determine whether the writer's attitude towards a particular topic or product is positive, negative, or even neutral. Recently, deep learning approaches emerge as powerful computational models that discover intricate semantic representations of texts automatically from data without hand-made feature engineering. These approaches have improved the state-of-the-art in many Sentiment Analysis tasks including sentiment classification of sentences or documents. In this paper we propose a semi-supervised neural network model, based on Deep Belief Networks, able to deal with data uncertainty for text sentences and adopting the Italian language as a reference language. We test this model against some datasets from literature related to movie reviews, adopting a vectorized representation of text and exploiting methods from Natural Language Processing (NLP) pre-processing.

Keywords: Deep learning · Sentiment analysis · Deep Belief Networks

1 Introduction

Sentiment analysis attempts to identify and analyze opinions and emotions from sentences providing a classification of text. The ability to analyze and measure the "feeling" of users respect to a generic product or service, expressed in textual comments, can be an interesting element of evaluation that can guide business dynamics, both in the industry area and also in marketing. Automatic text classification of texts into pre-defined categories has witnessed an increasing interest in the last years, due to the huge availability of documents in digital form.

Some examples of sentences that express a feeling or an opinion are: *"This movie is really ugly and the direction is very bad"*, or *"The story is pretty but the director could do more"*.

In the research community the recent dominant approach to general text classification and Sentiment Analysis is increasingly based on machine learning

© Springer International Publishing AG 2017
A. Petrosino et al. (Eds.): WILF 2016, LNAI 10147, pp. 172–184, 2017.
DOI: 10.1007/978-3-319-52962-2_15

techniques: a general inductive process that automatically builds a classifier by learning the main features of the categories from a set of pre-classified documents. The advantages of this approach over the knowledge engineering classical approach (consisting in the manual definition of a classifier by domain experts) are a very good effectiveness, a considerable savings in terms of expert manpower, but also a straightforward portability to different domains. Depending on the problem statement, in Sentiment Analysis tasks, classifiers like SVM (Support Vector Machine) or Naive Bayes, have shown good performance accuracy, provided proper feature engineering and also dedicated pre-processing steps, to be executed before the classification process.

Also, these traditional approaches are lacking in face of structural and cultural subtleties in the written language. For instance, negating a highly positive phrase can completely reverse its sentiment, but unless we can efficiently present the structure of the sentence in the feature set, we will not be able to capture this effect. On a more abstract level, it will be quite challenging for a machine to understand sarcasm in a review. The classic approaches to sentiment analysis and natural language processing are heavily based on engineered features, but it is very difficult to hand-craft features to extract properties mentioned above. But also, due to the dynamic nature of the language, those features might become obsolete in a very short amount of time and so a different approach to overcame these problems is needed.

1.1   Terms and Definitions

In order to obtain a formal definition of Sentiment Analysis let's first consider the definition of the single word *sentiment*. From the Cambridge vocabulary we can read that a sentiment is defined as *"a thought, opinion, or idea based on a feeling about a situation, or a way of thinking about something"*. It is clear from this definition that a sentiment is strictly related to a personal or subjective (not factual or objective) opinion. Wiebe et al. in [1] more formally defines the subjectivity of a sentence as *"the set of all the elements that describe the emotional state of the author"*. Typical subjectivity clues can be considered: assumptions, beliefs, thoughts, experiences, and also opinions. In summary, we can say that a *feeling* or *sentiment*, related to a given text, can be defined as the set of subjective expressions. These expression can commonly be measured in terms of positive, neutral or negative orientation.

With this premise Sentiment Analysis, introduced first in 2003 in [2], describes the process of evaluating the *polarity* expressed by a set of documents in an automatic way. The term polarity here still refers to the orientation of the author expressed in the sentence.

Definitions and terms in this field have been subsequently refined. In [3], Liu defines formally an opinion as a binomial expression made of two fundamental parts:

- a **target g**, also called topic, which represent an entity (or one of its aspects);

– a **sentiment s**, expressed on the target, which can assume positive, negative or even neutral values. Normally the sentiment can be expressed also using a number (score), or by a kind of ranking (e.g. from 1 to 5 stars). Positive or negative terms represent also the so called polarity of the expressed sentiment;

Considering the following product review, made by different sentences, related to a camera:

1. *"I bought this camera six months ago."*
2. *"It's really wonderful."*
3. *"The picture quality is really amazing."*
4. *"The battery life is good."*
5. *"However, my wife thinks it's too heavy for her."*

Looking at these sentences, that belong to the same review, it's possible to note some basic features. The review contains both positive and negative sentences on the same product. The phrase number 2 expresses a very positive opinion on the whole product. The phrase number 3 expresses still a very positive opinion, but only on one aspect of the product. The phrase number 4 expresses a good opinion but always on a single aspect of the product. The phrase number 5 expresses instead a negative opinion on a precise aspect (the weight) of the product.

We can now formally extends the definition of opinion as a quadruple:

$$Op = (g, s, h, t) \tag{1}$$

where g is the opinion, s is the sentiment, h is the opinion holder and t is a time stamp related to the time at which the opinion has been expressed.

To ensure that the text classification task could be really feasible and significant when applied, existent literature usually adopts an important assumption, sometimes implicitly, that has been defined by Liu in [3]: in the classification activity of a document d it is assumed that this document expresses an opinion (e.g., a product review) only on a single subject s or entity, and this opinion is expressed by a single author h. In practice, if a document contains opinions on different entities, then these opinions can potentially be different, i.e. positive on certain entities and negatives on other entities. Therefore, it makes no sense to assign a unique feeling or sentiment to the entire document. This assumption is still valid in the case of product or service reviews, that are generally wrote by a single author and are usually related to a single target.

While it's possible to approach Sentiment Analysis in a few ways, commonly in literature 3 typical levels of analysis are defined:

– **Document Level:** at this level we analyze the overall sentiment expressed in the text. This level of analysis is based on the assumption that the whole document discusses only one topic and thus cannot be applied to documents that contain opinions on more than one entity.

- **Sentence Level:** at this detail we examine the sentiment expressed in single sentences. This kind of analysis determines whether each sentence contains a positive or negative opinion. This level of analysis can be also considered as a subjectivity classification [1], which aims to distinguish objective sentences, that simply express factual information, from subjective sentences, that express personal views and opinions.
- **Entity and Aspect Level:** this much more granular analysis takes into consideration each opinion expressed in the content and it is generally based on the idea that an opinion consists of both a sentiment (positive or negative) and a target of opinion.

In this paper we will focus with classical Sentiment Analysis tasks, adopting binary polarities (positive, negative) of text, and analyzing text at a sentence level detail.

2   Related Works

A more general Sentiment Analysis research activity, dealing with interpretation of metaphors, sentiment adjectives, subjectivity, view points, and affects, mainly started from early 2000, with some earlier works done by Hatzivassiloglou and McKeown [4]; Hearst [5], Wiebe [6,7], Bruce and Wiebe [8].

Specific works that contain the term Sentiment Analysis or Opinion Mining appeared in the same years, with Turney [9], Das and Chen [10], Morinaga et al. [11], Pang et al. [12], Tong [13], Wiebe et al. [14].

The work done by Turney [9], on review classification, presents an approach based on the distance measure of adjectives found in text, using preselected words with known polarity as *excellent* and *poor*. The author presents an algorithm based on three steps which processes documents without human supervision. First, the adjectives are extracted along with a word that provides contextual information. Words to extract are identified by applying predefined patterns (for instance: adjective-noun or adverb-noun etc.). Next, the semantic orientation is measured. This is done by measuring the distance from words of known polarity. The mutual dependence between two words is found by analysis of hit count with the AltaVista search engine for documents that contain two words in a certain proximity of each other. At the end the algorithm counts the average semantic orientation for all word pairs and classifies a review as recommended or not.

In contrast, Pang et al. [12] present a work based on classic topic classification techniques. The proposed approach aims to test whether a selected group of machine learning algorithms can produce good result when Sentiment Analysis is perceived as document topic analysis with two topics: positive and negative. Authors present results for experiments with: Naive Bayes, Maximum Entropy and Support Vector Machine algorithms that will be discussed in the next paragraph. Interestingly the performed tests have shown results comparable to other solutions ranging from 71 % to 85 % depending on the method and test data sets.

Riloff and Wiebe [15] put most of impact in their work on the task of subjective sentences identification. They propose a method that at bootstrap uses
a high precision (and low recall) classifiers to extract a number of subjective
sentences. During this phase sentences are labeled by two classifiers: first for
high confidence subjective sentences, second for high confidence objective sentences. The sentences that are not clearly classified into any category are left
unlabeled and omitted at this stage. Both of the classifiers are based on preset
list of words that indicate sentence subjectivity. The subjective classifier looks
for the presence of words from the list, while the objective classifier tries to locate
sentences without those words. According to the results presented by authors
their classifiers achieve around 90% accuracy during the tests. In the second
step, the gathered data is used a for training an extraction algorithm that generates patterns for subjective sentences. The patterns are used to extract more
sentences in the same text. The presented method has such split in order to
increase recall after the initial bootstrap phase (however, as expected, author
report the precision to fall between 70–80%).

In opposition to it, the work done by Yu and Hatzivassiloglou [16] discusses both sentence classification (subjective/objective) and orientation (pos-
itive/negative/neutral). For the first step of sentence classification, authors
present test results for three different algorithms: sentence similarity detection,
Naive Bayes classification and Multiple Naive Bayes classification. In the second
step of sentence orientation recognition authors use a technique similar to the
one used by Turney [9] for document level sentiment analysis. The main different is that the algorithm is extended to use more then two (excellent/poor) base
words to which all others are compared.

Sometimes authors go even further and present methods for specific text
format, for instance reviews where positive and negative features are explicitly
separated is different areas.

A similar approach is presented by Hu and Liu in their work about customer
reviews analysis [17]. In their research authors present opinion mining based
on feature frequency. Only the most frequent features, recognized by precessing
many review, are taken into consideration during summary generation.

During last years a very active research group in Sentiment Analysis has
been the Stanford NLP group. From this group several reference papers has been
produced. In [18], Socher et al. proposed a semi-supervised approach based on
recursive autoencoders for predicting sentiment distributions. The method learns
vector space representation for multi-word phrases and exploits the recursive
nature of sentences. In [19], Socher et al. proposed a matrix-vector recursive
neural network model for semantic compositionality, which has the ability to
learn compositional vector representations for phrases and sentences of arbitrary
length. The vector captures the inherent meaning of the constituent, while the
matrix captures how the meaning of neighboring words and phrases are changed.
Then in [20] the same authors propose the Recursive Neural Tensor Network
(RNTN) architecture, which represents a phrase through word vectors and a

parse tree and then compute vectors for higher nodes in the tree using the same tensor-based composition function.

From the literature review is almost clear that the approaches towards this task, based on machine learning, are manly based on supervised techniques together with Natural Language Processing methods. There is an evident lack in experiencing unsupervised methods or semi-supervised methods, always based on machine learning, to address Sentiment Analysis problems. It is also clear that there is a lack of works and resources in specific languages other than English, for Sentiment Analysis tasks, and so in the next part of this work, we will also propose, in the generic case of polarity detection, a model applied to the Italian language.

3 Current Limitations

From what we have seen so far, many issues are evident in Sentiment Analysis tasks. First of all we must consider the complexity of human language (both written or spoken). In fact, it would be too naive to oversimplify language thinking that its underlying sentiment can always be accurately examined by a machine or an algorithm. Mainly there are four main factors that currently can make sentiment analysis a complex task:

1. **Context:** a positive or negative sentiment word can have the opposite connotation depending on context.
2. **Sentiment Ambiguity:** a sentence with a positive or negative word doesn't necessarily express any sentiment although it uses the positive or negative sentiment words. Likewise, sentences without sentiment words can express sentiment too.
3. **Sarcasm:** a positive or negative sentiment word can switch sentiment if there is sarcasm in the sentence.
4. **Language:** a word can change sentiment and meaning depending on the language used. This is often seen in slang, dialects, and language variations.

A basic consideration related to the complexity of text classification can also be done considering that an automatic sentiment analysis tool's accuracy is merely the percentage of times that human judgment agrees with the tool's judgment. This degree of agreement among humans is also known as human concordance. There have been various studies in this field and they concluded that the rate of human concordance is between 70% and 79%. Taking that into consideration, we can safely say that a good accuracy for sentiment analysis tools is around 70%. The problem is: a tool that is accurate less than 70% of the time is not accurate, and a "perfect" tool that is accurate 100% of the time will draw data that we disagree with roughly 30% of the time.

4 The Proposed Model

Generally pattern classification tasks can be roughly divided into two groups: the so-called *supervised* methods and the *unsupervised* methods. In supervised learning, the class labels in the dataset, which is used to build the classification model, are a priori known. In contrast, unsupervised learning tasks deal with unlabeled dataset, and the classes have to be a posteriori inferred from the unstructured dataset. Typically, unsupervised learning employs a clustering technique in order to group the unlabeled samples based on certain similarity (or distance) measures. But recently another approach, that we can found in the middle, is the so called *semi-supervised* approach [21], that represents a class of supervised learning tasks and techniques that make use of unlabeled data for training, using typically a small amount of labeled data with a large amount of unlabeled data. Semi-supervised learning falls between unsupervised learning (without any labeled training data) and supervised learning (with completely labeled training data).

4.1   Deep Belief Networks

In the 2006 Hinton et al. [22] introduced DBNs networks, that uses a deep architecture that is in capable to learn features representation from labeled and unlabeled data. This kind of network integrates both a unsupervised learning step, and a strategy to supervised fine-tuning, to build a more robust and efficient model. Unsupervised step is used to learn data distribution without a their knowledge, instead supervised step execute a local search to obtain an optimization of results.

In Fig. 1 is showed a typical architecture for a DBN, it is composed of a stack of Restricted Boltzmann Machines (RBM), and an additional layer dedicated to specific task, such as classification.

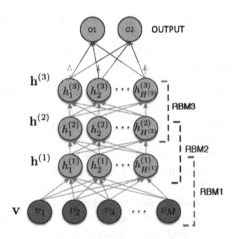

Fig. 1. Example of a deep belief network

An RBM is a generative stochastic artificial neural network that can learn a probability distribution over its set of inputs. RBMs are a variant of Boltzmann machines, with the restriction that their neurons must form a bipartite graph: a pair of nodes from each of the two groups of units (commonly referred to as the "visible" and "hidden" units respectively) may have a symmetric connection between them; and there are no connections between nodes within a group. This restriction allows for more efficient training algorithms than are available for the general class of Boltzmann machines, in particular the gradient-based contrastive divergence algorithm.

4.2   The Network Model

The neural network model designed for our Sentiment Analysis activities is based on Deep Belief Network, obtained by stacking some Restricted Boltzmann Machines. Different structural models have been tested, varying the shape of the network, or by varying the size of the input level, the number of hidden levels as well as the number of neurons of each level. In the Fig. 2 below we present a general outline of the proposed model of our deep network.

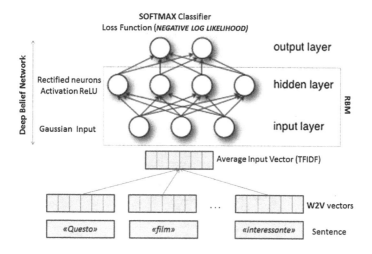

Fig. 2. Architecture of the proposed deep network.

Some choices of our model are represented by classifier type SoftMax, with error function based on Negative Log Likelihood loss function. SoftMax is a function used as the output layer of a neural network that classifies input. It converts vectors into class probabilities. SoftMax normalizes the vector of scores by first exponentiating and then dividing by a constant.

At the visible unit layer, i.e. the layer of nodes where input goes in, we adopted Gaussian neurons, which have an activation function based on the Gaussian function.

Our activation function of the hidden layer nodes choice is the Rectified Linear Unit (ReLU) [23].

The initialization of the input level weights were made through the Xavier algorithm, proposed by Glorot and Bengio in [24]. In short, this kind of initialization helps signals reach deep levels of the network. If the weights in a network start with too small values, then the signal shrinks as it passes through each layer until it is too tiny to be useful. On the contrary, if the weights in a network start too large, then the signal grows as it passes through each layer until it is too massive to be useful.

4.3   The Processing Architecture

The realization of the whole architectural solution is composed of several processing steps that define a processing pipeline that will be detailed here, evaluating the each individual steps from a qualitative point of view (see Fig. 3).

In order to achieve efficient representation of the text used by our deep network, it was necessary to study and subsequently adopt an algorithm able to convert sentences in numerical values (i.e. vectors in our case), that pushed semantic properties of text, necessary for the correct realization of a Sentiment Analysis solution. Our choice has been Word2Vec, the recent algorithm proposed by Mikolov et al. [25] in 2013. We then select a representative Italian corpus (Paisá from CNR)[1] that was compatible with context, to be used with Word2Vec algorithm, in order to produce an effective vector representation of Italian terms. Considering the size of the corpus (about 250 millions of words), we obtained a vocabulary of about 250.000 Italian terms.

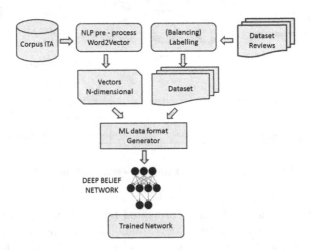

Fig. 3. The processing architecture.

Once obtained a cleaned corpus it was possible to apply the Word2Vec algorithm to obtain a vector representations of the terms on the generated dictionary.

---

[1] The Paisá corpus is made available at http://www.corpusitaliano.it through a Creative Commons license.

Last, before submitting input data to the deep network, we built a simple software component to transform cleaned and balanced textual data in a vectorized representation, that can be processed by the DBN.

5   Datasets and Results

In order to train the Deep Belief Network two different datasets have been used, both in Italian language and both relative to movie reviews, publicly available on the web.

The first dataset (A) has been built upon the website cinemadelsilenzio.it, it comprises about 7.000 reviews written both by experts and common registered users. The second dataset (B), for the training stage of the Deep Belief Network, is still made of about 10.000 movie reviews, taken from the website filmup.com, and it comprises reviews written by common users (not experts).

Experimentally, once fixed the dataset size, we evaluate the response of the network in terms of classification *accuracy*. In order to also validate the correctness of the representation of the input vectors dimension some comparisons were made during testing.

In line with what occurred in the literature, from the tests that were carried out it was found that a first hidden layer, configured so as to be larger (e.g. double) than the dimensionality of the input data, actually works better than one with a reduced dimensionality.

In Table 1 we report a summary of the most significant cases of our testing activities done with the proposed model. We emphasize that the training dataset were all validated by K cross validation [26], with $K = 10$.

Table 1. Summary of the main results obtained from the experiments carried out varying the DBN configuration parameters.

Network structure	Batch size	Epochs	Dataset	Accuracy
100-200-2	500	2	A	0,88
100-200-2	1000	20	A	0,8243
100-300-2	1000	1	A	0,8229
100-400-2	1000	5	A	0,82
200-400-2	1000	30	A	0,8757
300-600-2	1000	5	A	0,8714
300-600-2	1000	5	B	**0,9171**
300-600-2	5000	30	B	0,7849
200-400-800-2	3000	5	A	0,7448
300-600-900-2	1000	10	B	0,7829

In order to evaluate the correctness of our approach it's worth now to compare these results with also different dataset and similar models taken from literature

Table 2. Comparison of our DBN with other models [27], on MOV dataset.

Network type	Accuracy
Spectral	67,3
TSVM	68,7
HDBN	72,2
DBN	**76,2**

that address the same objective. To accomplish this evaluation we take into account a classical reference work *"Thumbs up?: Sentiment classification using machine learning techniques"* published by Pang and Vaithyanathan in 2002 [12]. In this work authors adopted a widely-used movie review dataset called MOV, with 2.000 labeled reviews (1.000 positive reviews and 1.000 negative reviews). A recent model comparison, using also DBN networks, is available in the work done by Zhou et al. in [27]. In Table 2 it's possible to see four different representative semi-supervised learning methods to which we compared our DBN model. The first model is the Spectral Clustering [28], introduced by Kamvar et al. in 2003. The TSVM model refers to the Transductive SVM [29] introduced by Collobert et al. in 2006. In the HDBN [27] model, Hybrid means that this architecture use a modified RBM layer (a convolutional layer), and the input vectors are represented using a BoW approach, and with a network layers configuration of 100-100-4-2 neurons. Our DBN model, applied on the same MOV dataset, with a network structure of 100-200-2 neurons, performs quite well also when compared to these other models and datasets from literature.

6 Conclusions

In this paper we have investigated the issue of defining and evaluating novel methods for Sentiment Analysis, having as a target the Italian language, and focusing on machine learning techniques, based on semi supervised approaches. We proposed to adopt a Deep Belief Network, together with a vectorized representation of textual input using the recent Word2Vec algorithm. We also analyzed and suggested a possible structure of the neural network in terms of its configuration layers, neurons type, error functions, classifier, etc. We built from scratch a vectorized representation of the data sets in Italian language and provides also 2 new different datasets of movie reviews, still in Italian language. From our studies we can observe that deep learning can be considered an effective approach to general Sentiment Analysis tasks and polarity detection activity, even if it's really important to achieve an efficient text representation for input data, also able to keep semantic properties of textual data in the final vectorized representation. Nevertheless, the proposed semi-supervised approach has obtained promising results respect to other methods from literature.

References

1. Wiebe, J., Wilson, T., Bell, M.: Identifying collocations for recognizing opinions. In: Proceedings of ACL-2001 Workshop on Collocation: Computational Extraction, Analysis, and Exploitation, pp. 24–31 (2001)
2. Nasukawa, T., Yi, J.: Sentiment analysis: capturing favorability using natural language processing. In: Proceedings of 2nd International Conference on Knowledge Capture, K-CAP 2003, New York, NY, USA, pp. 70–77. ACM (2003)
3. Liu, B.: Sentiment analysis and opinion mining. Synth. Lect. Hum. Lang. Technol. **5**(1), 1–167 (2012)
4. Hatzivassiloglou, V., McKeown, K.R.: Predicting the semantic orientation of adjectives. In: Proceedings of 35th Annual Meeting of the Association for Computational Linguistics and 8th Conference of the European Chapter of the Association for Computational Linguistics, ACL 1998, Stroudsburg, PA, USA, pp. 174–181. Association for Computational Linguistics (1997)
5. Hearst, M.A.: Automatic acquisition of hyponyms from large text corpora. In: Proceedings of 14th Conference on Computational Linguistics, COLING 1992, Stroudsburg, PA, USA, vol. 2, pp. 539–545. Association for Computational Linguistics (1992)
6. Wiebe, J.M.: Identifying subjective characters in narrative. In: Proceedings of 13th Conference on Computational Linguistics, COLING 1990, Stroudsburg, PA, USA, vol. 2, pp. 401–406. Association for Computational Linguistics (1990)
7. Wiebe, J.M.: Tracking point of view in narrative. Comput. Linguist. **20**, 233–287 (1994)
8. Bruce, R.F., Wiebe, J.M.: Recognizing subjectivity: a case study in manual tagging. Nat. Lang. Eng. **5**, 187–205 (1999)
9. Turney, P.D.: Thumbs up or thumbs down? Semantic orientation applied to unsupervised classification of reviews. In: Proceedings of 40th Annual Meeting on Association for Computational Linguistics, ACL 2002, Stroudsburg, PA, USA, pp. 417–424. Association for Computational Linguistics (2002)
10. Das, S., Chen, M.: Yahoo! for Amazon: extracting market sentiment from stock message boards. In: Asia Pacific Finance Association Annual Conference (APFA) (2001)
11. Morinaga, S., Yamanishi, K., Tateishi, K., Fukushima, T.: Mining product reputations on the web. In: Proceedings of 8th ACM SIGKDD International Conference on Knowledge Discovery and Data Mining, KDD 2002, New York, NY, USA, pp. 341–349. ACM (2002)
12. Pang, B., Lee, L., Vaithyanathan, S.: Thumbs up? Sentiment classification using machine learning techniques. In: Proceedings of ACL-2002 Conference on Empirical Methods in Natural Language Processing, vol. 10, pp. 79–86. Association for Computational Linguistics (2002)
13. Tong, R.: An operational system for detecting and tracking opinions in on-line discussions. In: Working Notes of the SIGIR Workshop on Operational Text Classification, New Orleans, Louisianna, pp. 1–6 (2001)
14. Wiebe, J., Wilson, T., Bruce, R., Bell, M., Martin, M.: Learning subjective language. Comput. Linguist. **30**, 277–308 (2004)
15. Riloff, E., Wiebe, J.: Learning extraction patterns for subjective expressions. In: Proceedings of Conference on Empirical Methods in Natural Language Processing, EMNLP 2003, Stroudsburg, PA, USA, pp. 105–112. Association for Computational Linguistics (2003)

16. Yu, H., Hatzivassiloglou, V.: Towards answering opinion questions: separating facts from opinions and identifying the polarity of opinion sentences. In: Proceedings of Conference on Empirical Methods in Natural Language Processing, EMNLP 2003, Stroudsburg, PA, USA, pp. 129–136. Association for Computational Linguistics (2003)

17. Hu, M., Liu, B.: Mining and summarizing customer reviews. In: Proceedings of 10th ACM SIGKDD International Conference on Knowledge Discovery and Data Mining, KDD 2004, New York, NY, USA, pp. 168–177. ACM (2004)

18. Socher, R., Pennington, J., Huang, E.H., Ng, A.Y., Manning, C.D.: Semi-supervised recursive autoencoders for predicting sentiment distributions. In: Proceedings of Conference on Empirical Methods in Natural Language Processing, EMNLP 2011, Stroudsburg, PA, USA, pp. 151–161. Association for Computational Linguistics (2011)

19. Socher, R., Huval, B., Manning, C.D., Ng, A.Y.: Semantic compositionality through recursive matrix-vector spaces. In: Proceedings of Joint Conference on Empirical Methods in Natural Language Processing and Computational Natural Language Learning, EMNLP-CoNLL 2012, Stroudsburg, PA, USA, pp. 1201–1211. Association for Computational Linguistics (2012)

20. Socher, R., Perelygin, A., Wu, J.Y., Chuang, J., Manning, C.D., Ng, A.Y., Potts, C.P.: Recursive deep models for semantic compositionality over a sentiment TreeBank. In: EMNLP (2013)

21. Chapelle, O., Schlkopf, B., Zien, A.: Semi-Supervised Learning, 1st edn. The MIT Press, Cambridge (2010)

22. Hinton, G.E., Osindero, S., Teh, Y.-W.: A fast learning algorithm for deep belief nets. Neural Comput. 18(7), 1527–1554 (2006)

23. Krizhevsky, A., Sutskever, I., Hinton, G.E.: ImageNet classification with deep convolutional neural networks. In: Pereira, F., Burges, C.J.C., Bottou, L., Weinberger, K.Q. (eds.) Advances in Neural Information Processing Systems, vol. 25, pp. 1097–1105. Curran Associates Inc. (2012)

24. Glorot, X., Bengio, Y.: Understanding the difficulty of training deep feedforward neural networks. In: Proceedings of International Conference on Artificial Intelligence and Statistics (AISTATS 2010). Society for Artificial Intelligence and Statistics (2010)

25. Mikolov, T., Chen, K., Corrado, G., Dean, J.: Efficient estimation of word representations in vector space. CoRR, abs/1301.3781 (2013)

26. Stone, M.: Cross-validatory choice and assessment of statistical predictions. J. Roy. Stat. Soc. Ser. B (Methodol.) 36, 111–147 (1974)

27. Zhou, S., Chen, Q., Wang, X., Li, X.: Hybrid deep belief networks for semi-supervised sentiment classification (2014)

28. Kamvar, K., Sepandar, S., Klein, K., Dan, D., Manning, M., Christopher, C.: Spectral learning. In: International Joint Conference of Artificial Intelligence. Stanford InfoLab (2003)

29. Collobert, R., Sinz, F., Weston, J., Bottou, L.: Large scale transductive SVMs. J. Mach. Learn. Res. 7(August), 1687–1712 (2006)

Classification of Data Streams by Incremental Semi-supervised Fuzzy Clustering

G. Castellano$^{(\boxtimes)}$ and A.M. Fanelli

Computer Science Department, Università degli Studi di Bari "A. Moro",
Via E. Orabona 4, 70126 Bari, Italy
{giovanna.castellano,annamaria.fanelli}@uniba.it

Abstract. Data stream mining refers to methods able to mine continuously arriving and evolving data sequences or even large scale static databases. Mining data streams has attracted much attention recently. Many data stream classification methods are supervised, hence they require labeled samples that are more difficult and expensive to obtain than unlabeled ones. This paper proposes an incremental semi-supervised clustering approach for data stream classification. Preliminary experimental results on the benchmark data set KDD-CUP'99 show the effectiveness of the proposed algorithm.

1 Introduction

Recent technological advances in many disciplines have seen an increase in the amount of data that are not static but are often continuously coming as data chunks in a data stream. The term "data stream" is defined as a sequence of data that arrives at a system in a continuous and changing manner. Data streams are produced for example by sensors, emails, online transactions, network traffic, weather forecasting and health monitoring. Also several real world phenomena produce data streams, such as climate sensing and deforestation analysis.

Unlike static datasets with a fixed number of data points, in data stream of chunks, the number of data points is continuously increasing. A data stream system may constantly produce huge amounts of data. Hence managing and processing data streams raises new challenges and research problems. Indeed, it is usually not feasible to simply store the arriving data in a traditional database management system in order to process data. Rather, stream data must generally be processed in an online manner. Traditional algorithms are not suitable for such type of data because they extract patterns from data by considering the global properties, rather than undertaking the local ones. Moreover, they require the whole training data set.

The mining of data streams has been attracting much attention in the recent years. Similar to data mining, data stream mining includes classification, clustering, frequent pattern mining etc. In particular the analysis and classification of data streams has become increasingly important [3]. The focus of this paper is on classification of data streams.

© Springer International Publishing AG 2017
A. Petrosino et al. (Eds.): WILF 2016, LNAI 10147, pp. 185–194, 2017.
DOI: 10.1007/978-3-319-52962-2_16

Most of the existing work relevant to classification on data streams [10,12] assume that all arrived streaming data are completely labeled and these labels could be utilized at hand. These methods have been proved to be effective in some applications. However in many applications, labeled samples are difficult or expensive to obtain, meanwhile unlabeled data are relatively easy to collect. Semi-supervised algorithms solve this problem by using large amount of unlabeled samples, together with a few labeled ones, to build models for prediction or classification. In particular, semi-supervised classification is the task of learning from unlabeled data, along with a few labeled data, to predict class labels of test data.

In this paper, we consider the problem of semi-supervised classification in data streams. Our focus is on providing an effective method for incremental update of the classification model. To this aim we propose an approach for data stream classification that works in an incremental way based on a semi-supervised clustering process applied to subsequent, non-overlapping chunks of data. Every time a new chunk arrives, the cluster prototypes resulting from the last chunk are aggregated as labeled data points with the new chunk.

The organization of the rest of the paper is as follows. In Sect. 2 a brief review of some related works for classification of data stream is provided. Section 3 presents the proposed approach for data stream classification. In Sect. 4 the effectiveness of the approach is evaluated by comparing it with other methods on a benchmark dataset. The last section draws the conclusion and outlines future work.

2   Related Works

Several semi-supervised learning methods have been developed in the literature [9,19]. Here we focus just on some well-known algorithms that have been applied to classify data streams.

One famous semi-supervised algorithm is co-training [5]. This method splits the set of features into two subsets that represent different views of an example and each subset is used to train a classifier. Initially two classifiers are trained separately on the two subsets respectively. Then, each classifier is retrained on a training set including unlabeled samples and additional samples confidently labeled by the other classifier. Despite the effectiveness of co-training, its assumption can hardly be met in most scenarios. Indeed it is not always possible to partition the set of features into subsets that can in principle be used on their own to make the classification.

An extension of co-training is Tri-training [18], which trains three classifiers from the original labeled training samples using bootstrap sampling. These classifiers are then refined using unlabeled examples. In each step of Tri-training, an unlabeled sample is labeled for a classifier if the other two classifiers agree on the labeling. Tri-training does not require several views of samples as co-training, however it does not use the attribute values of one sample sufficiently as it uses bootstrap sampling from the original sample.

Self-training is a commonly used technique for semi-supervised learning [19]. In self-training a classifier is first trained with a small amount of labeled data. The classifier is then used to classify the unlabeled data. Typically the most confident unlabeled points, together with their predicted labels, are added to the training set. The classifier is re-trained and the procedure repeated.

All the above algorithms are not incremental in nature, namely they are not designed to process data streams. One incremental semi-supervised algorithm is Clustering-training [17]. The idea underlying Clustering-training is somehow similar to the incremental mechanism of our method. First of all, the labeled samples are used to train an initial classifier. Next, the data are processed in an incremental way, hence they are regarded as a data stream. In each step, a number of samples are extracted from the data stream and the K-prototype clustering [11] is applied to these samples (where K is fixed the same as the number of data classes) and their clustering labels are adjusted by the current classifier, which keeps the two labels by the two methods having the same meanings. Moreover, some of the labeling samples whose labels given by the two methods are identical are selected as confident ones to re-train the classifier, and others are discarded.

3   The Proposed Approach

The proposed approach for data stream classification is based on the assumption that data belonging to C different classes are continuously available during time and processed as chunks. Namely, a chunk of N_1 data points is available at time t_1, a chunk of N_2 data points is available at t_2 and so on[1]. We denote by X_t the data chunk available at time t. Our approach is semi-supervised, hence we assume that in each chunk only some data are eventually pre-labeled from a label set $L = 1, \ldots, C$ and the remaining samples are unlabeled. We further assume that the algorithm creates a fixed number of clusters K that is set equal to the number of classes C that is assumed to be known in advance.

The data chunks are processed as they are available by applying incrementally the Semi-Supervised FCM (SSFCM) clustering algorithm originally proposed in [15]. The SSFCM algorithm works in the same manner as FCM (Fuzzy C-Means) [4], i.e. it iteratively derives K clusters by minimizing an objective function. Unlike FCM, that performs a completely unsupervised clustering, SSFCM is a semi-supervised clustering algorithm, i.e. it uses a set of pre-labeled data to improve clustering results. To embed partial supervision in the clustering process, the objective function of SSFCM includes a supervised learning component, as follows:

$$J = \sum_{k=1}^{K} \sum_{j=1}^{N_t} u_{jk}^m d_{jk}^2 + \alpha \sum_{k=1}^{K} \sum_{j=1}^{N_t} (u_{jk} - b_j f_{jk})^m d_{jk}^2 \tag{1}$$

---

[1] Any stream can be turned into a chunked stream by simply waiting for enough data points to arrive.

where

$$b_j = \begin{cases} 1 \text{ if } \mathbf{x}_j \text{ is pre-labeled} \\ 0 \text{ otherwise} \end{cases} \tag{2}$$

f_{jk} denotes the true membership value of $\mathbf{x}_j$ to the k-th cluster (class), d_{jk} represents the Euclidean distance between $\mathbf{x}_j$ and the center of the k-th cluster, m is the fuzzification coefficient ($m \geq 2$) and α is a parameter that serves as a weight to balance the supervised and unsupervised components of the objective function. The higher the value of α, the higher the impact coming from the supervised component is. The second term of J captures the difference between the true membership f_{jk} and the computed membership u_{jk}. The aim to be reached is that, for the pre-labeled data, these values should coincide. As described in [15], the problem of optimizing the objective function J is converted into the form of unconstrained minimization using the standard technique of Lagrange multipliers. By setting the fuzzification coefficient $m = 2$, the objective function is minimized by updating membership values u_{jk} according to:

$$u_{jk} = \frac{1}{1+\alpha} \left[\frac{1 + \alpha(1 - b_j \sum_{h=1}^{K} f_{jh})}{\sum_{h=1}^{K} d_{jk}^2 / d_{jh}^2} \right] + \alpha b_j f_{jk} \tag{3}$$

and the centers of clusters according to:

$$\mathbf{c}_k = \frac{\sum_{j=1}^{N_t} u_{jk}^2 \mathbf{x}_j}{\sum_{j=1}^{N} u_{jk}^2} \tag{4}$$

The clustering process ends when the difference between the values of J in two consecutive iterations drops below a prefixed threshold or when the established maximum number of iterations is achieved.

Our idea is to apply incrementally the SSFCM algorithm so as to enable continuous update of clusters based on new data chunks. To take into account the evolution of the data in the incremental clustering process, the cluster prototypes discovered from one chunk are added as pre-labeled data to the next chunk.

When the first chunk is available, it is clustered into K clusters by SSFCM and each cluster is represented by a prototype. The prototype of each cluster is chosen to be the data-point belonging with maximal membership to that cluster. To start the process, it is necessary to include in the first data chunk a significant amount of pre-labeled data for each class, so that the resulting prototypes correspond to pre-labeled data and a class label is automatically associated to each prototype. If not enough pre-labeled data are contained in the first chunk, then the prototype may result in an unlabeled data point. In this case the prototype is manually associated to a proper class label. Hence, the output of SSFCM applied to the data chunk is a set of K clusters represented by K labeled prototypes $P = \{\mathbf{p}_1, \mathbf{p}_2, \ldots, \mathbf{p}_K\}$. These prototypes are used for classification, namely all data belonging to cluster k are associated with the class label assigned to prototype $\mathbf{p}_k$.

Next, each time a new data chunk is available, previously derived cluster prototypes are added as pre-labeled data samples to the current chunk, and SSFCM is applied again to derive K clusters and K labeled prototypes. At the end of each SSFCM clustering run, data belonging to the k-th cluster are classified as belonging to the class of the k-th prototype. Moreover, the derived labeled proto-types can be used to classify all data accumulated in a database via a matching mechanism. Namely each data sample is matched against all prototypes and assigned to the class label of the best-matching prototype. Matching is based on simple Euclidean distance. The overall scheme of the proposed incremental approach is show in Fig. 1 and described in Algorithm 1.

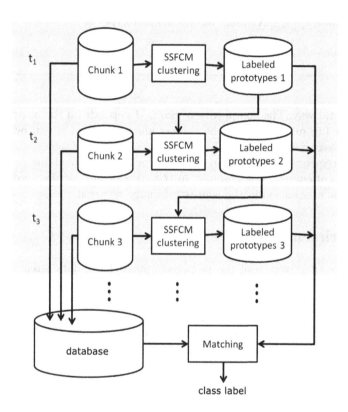

Fig. 1. Scheme of the incremental approach for data stream classification.

Summarizing, based on a sequence of chunks, our incremental scheme gener-ates cluster prototypes which capture the incoming of new data during time and reflect evolution of the data stream. The clustering mechanism is incremental in the sense that the cluster prototypes derived from one chunk are added as labeled data to the next chunk. At the end of the process, the derived prototypes offer a mechanism that enables automatic classification of data. Namely, at any given time an unlabeled data point can be classified by means of prototype-matching. The computational complexity of our approach is mainly based on the complexity of the SSFCM algorithm, which is applied to each data chunk in

---

Algorithm 1. Incremental data stream classification

Require: Stream of data chunks $X_1, X_2, \ldots$ belonging to C classes
Ensure: P: set of labeled prototypes
 1: $P \leftarrow \emptyset$
 2: $t \leftarrow 1$ /* Initialization of time step */
 3: **while** $\exists$ non empty chunk X_t **do**
 4: $X_t \leftarrow X_t \cup P$ /* Add previous prototypes to current chunk */
 5: Cluster X_t into K clusters using SSFCM
 6: Derive the set P of K cluster prototypes
 7: Associate a class label to each prototype in P
 8: Classify data in $\bigcup_{\tau=1}^{t} X_\tau$ using the labeled prototypes in P
 9: $t := t + 1$
10: **end while**
11: **return** P

---

an incremental way. The complexity of SSFCM depends on the computation of the membership matrix $\mathbf{U}$ and the centers matrix $\mathbf{c}$ hence it can be estimated as the complexity of FCM [2].

This incremental approach can be useful for any task involving processing of data in form of chunks. For example, in [7,8] it was successfully applied for the task of shape annotation in the context of image retrieval.

4   Experimental Results

To check the effectiveness of the proposed approach for data stream classification we need a real data set that evolve significantly over time. We considered the KDD-CUP'99 Network Intrusion Detection (KDDCup99) dataset [13] which was firstly used for the KDD - Knowledge Discovery and Data Mining - Tools Conference Cup 99 Competition and later used to evaluate several stream classification algorithms. This dataset refers to the important problem of automatic and real-time detection of cyber attacks. It consists of a series of TCP connection records from two weeks of LAN network traffic managed by MIT Lincoln Labs simulating network intrusion attacks on a military installation. Each sample can either correspond to a normal connection or an intrusion or attack. The KDDCUP99 is a hard dataset for the sake of classification because of its large size and high number of input features, some of them with unbalanced values.

The original dataset contains information to classify the data into four categories of attack types - Denial of Service (DoS), Probe, Remote-to-local (R2L) and User-to-root (U2R) - in addition to the normal network traffic, hence there are 5 classes. Around 20% of the dataset are normal patterns (no attacks). However, in general, the main interest of Intrusion Detection System is to distinguish between attack and no-attack situations. The type of attack being performed is just one more step in the detection. Hence the KDDCup99 dataset can be treated as a binary classification problem, that consists of distinguishing attacks versus normal samples. In this work, as in other works in the literature [6], the binary

Table 1. Data distribution of the KDDCup99 dataset.

Classes	# training samples	# test samples
Normal	97278	60588
Error	396743	250441
total	494021	311029

classification case is considered. There are approximately 5 million samples in the full dataset and each sample represents a TCP/IP connection that is composed of 41 features that are both qualitative and quantitative in nature, such as the duration of the connection, the number of data bytes transmitted from source to destination (and vice versa), percentile of connections that have "SYN" errors, the number of "root" accesses, etc. The dataset used in our simulations is a smaller subset (10% of the original training set), it has 494021 samples and it was employed as the training set in the original competition. As test set we used the KDDCup99 test dataset which is an additional set of 311029 samples. Table 1 shows the distribution of data samples from the two classes in the training and test set.

To test the proposed approach for data stream classification, the KDDCup99 dataset was divided in chunks and presented sequentially to mimic an online data stream. Precisely, the original KDDCup99 training set (494021) was divided in 48 equally-sized chunks. Each chunk contains 10291 data samples. To test the accuracy of the classifier during the online learning, the KDDCup99 test set was used. To evaluate the classification results we used the accuracy measure:

$$Acc = \frac{|\{\mathbf{x}_j | t_j = a_j\}|}{N_t}$$

where $\mathbf{x}_j$ is the j-th data point, t_j is the true class label and a_j is the predicted class label, N_t is the number of data points in the test set. After processing a single data chunk we computed the classification accuracy on the test set. At the end of the online process an average accuracy was evaluated.

For the simulations we considered two cases of partial supervision: 30% of pre-labeled data (Case A) and 50% of pre-labeled data (Case B). In Case A the average accuracy of our classification method is 82.44%. In Case B our method achieves an average accuracy of 95.29%. This last result is comparable with the semi-supervised algorithm SUN [16] that achieves 94.68% with 50% of pre-labeled data.

In the previous simulations each sample in the data set is represented by 41 features. A preliminary study on this dataset [14] suggests that some features have no relevance in intrusion detection. Following the feature selection described in [14] we considered a reduced set of 14 features, listed in Table 2. Using the reduced set of features the average accuracy of our method is 83.76% in Case A and 95.30% in Case B. Hence the performance of the classifier improves slightly by introducing feature selection.

Table 2. The 14 most relevant features used in the KDDCup99 dataset.

Feature ID	Feature name
3	Service
5	Source_bytes
6	Destination bytes
7	Land
8	Wrong fragment
11	Failed login
14	Root sheel
23	Count
24	Srv count
28	Srv error rate
29	Same srv error rate
30	Diff srv rate
36	Dst host name src port rate
39	Srv count

Finally, to better assess the effectiveness of the proposed approach, it was compared with other semi-supervised methods for data stream classification, namely Clustering-training [17], Self-training [19], Co-training [5] and Tri-training [18]. The results are shown in Fig. 2. It can be seen that our method creates a quite good initial classifier (created on the first chunk with 50% of labeled data) in comparison to other methods. The accuracy of the final classifier

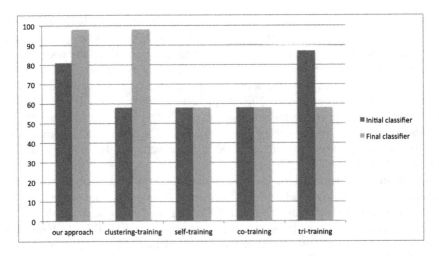

Fig. 2. Comparative results taken from [17].

(updated upon presentation of the last data chunk) improves up to about 98%. In comparison to other semi-supervised methods, our method performs as well as Clustering-training [17], which works incrementally as ours, and outperforms non-incremental methods such as Self-training, Co-training and Tri-training.

5   Conclusions

This paper introduced an approach for data stream classification based on an incremental semi-supervised fuzzy clustering algorithm. According to preliminary results on a benchmark dataset, the proposed approach compares favorably with other well-known state-of-the-art data stream classification methods. However, several issues have still to be investigated. As well known, the evolution of the stream [1] significantly affects the behavior of the classifier. The composition of the data stream over consecutive chunks influences the incremental building of the classifier and requires a deep analysis of the clusters behavior. In this preliminary work for convenience we used fixed sized data chunks to construct the classifier. Moreover we assumed that the distribution of data was unchanged in the consecutive chunks. However, in real-world contexts the incoming chunks may vary in size and the underlying distribution of data may change over the time intervals. We are currently studying the effects of varying the size of the chunks and the data distribution during the incremental clustering. Another step forward is to better take into account the evolution of data. There are two kinds of evolution in data: concept drifting and concept evolution. Concept drifting occurs whenever there is a change in data distribution with respect to class labels. Concept evolution occurs when one or more new class labels emerge. As a future work we plan to make more flexible our approach so as to handle both concept drifting and concept evolution. This requires the design of a dynamic component able to add new cluster prototypes dynamically when new classes appear and remove unnecessary prototypes during the online clustering process. Defining such a dynamic module is the direction for our future study.

References

1. Aggarwal, C.C.: A framework for diagnosing changes in evolving data streams. In: Proceedings of ACM SIGMOD Conference, pp. 575–586 (2003)
2. Almeida, R.J., Sousa, J.M.C.: Comparison of fuzzy clustering algorithms for classification. In: Proceedings of International Symposium on Evolving Fuzzy Systems, pp. 112–117 (2006)
3. Babcock, B., Babu, S., Datar, M., Motwani, R., Widom, J.: Models and issues in data stream systems. In: Proceedings of 21st ACM SIGMOD-SIGACT-SIGART Symposium on Principles of Database Systems, pp. 1–16 (2002)
4. Bezdek, J.C.: Pattern Recognition with Fuzzy Objective Function Algorithms. Plenum Press, New York (1981)
5. Blum, A., Mitchell, T.: Combining labeled and unlabeled data with co-training. In: Proceedings of Annual Conference on Computational Learning Theory, pp. 92–100 (1998)

6. Bolon-Canedo, V., Sanchez-Marono, N., Alonso-Betanzos, A.: A combination of discretization and filter methods for improving classification performance in KDD Cup 99 dataset. In: Proceedings of International Joint Conference on Neural Networks, pp. 359–366 (2009)

7. Castellano, G., Fanelli, A.M., Torsello, M.A.: Shape annotation by semi-supervised fuzzy clustering. Inf. Sci. **289**(24), 148–161 (2014)

8. Castellano, G., Fanelli, A.M., Torsello, M.A.: Incremental semi-supervised fuzzy clustering for shape annotation. In: Proceedings of 2014 IEEE Symposium on Computational Intelligence for Multimedia, Signal and Vision Processing (SSCI-CIMSIVP 2014), Orlando, Florida, USA, pp. 190–194, December 2014

9. Chapelle, O., Scholkopf, B., Zien, A. (eds.): Semi-Supervised Learning. MIT Press, Cambridge (2006)

10. Domingos, P., Hulten, G.: Mining high-speed data streams. In: Proceedings of KDD, pp. 71–80 (2000)

11. Huang, Z.: Clustering large data sets with mixed numeric and categorical values. In: Proceeding of PAKDD, pp. 21–34 (1997)

12. Hulten, G., Spencer, L., Domingos, P.: Mining time-changing data streams. In: Proceedings of KDD, pp. 97–106 (2006)

13. K. C. 1999, KDDCup 1999, Technical report (1999). http://kdd.ics.uci.edu/data bases/kddcup99/kddcup.data_10_percent.gz

14. Olusola, A.A., Oladele, A.S., Abosede, D.O.: Analysis of KDD '99 intrusion detection dataset for selection of relevance features. In: Proceedings of World Congress on Engineering and Computer Science, vol. I (2010)

15. Pedrycz, W., Waletzky, J.: Fuzzy clustering with partial supervision. IEEE Trans. Syst. Man Cybern. **27**(5), 787–795 (1997)

16. Wu, X., Li, P., Hu, X.: Learning from concept drifting data streams with unlabeled data. Neurocomputing **92**, 145–155 (2012)

17. Wu, S., Yang, C., Zhou, J.: Clustering-training for data stream mining. In: Proceedings of ICDMW 2006, pp. 653–656 (2006)

18. Zhou, Z.-H., Li, M.: Tri-training: exploiting unlabeled data using three classifiers. IEEE Trans. Knowl. Data Eng. **17**(11), 1529–1541 (2005)

19. Zhu, X.: Semi-supervised learning literature survey. Report No. 1530, University of Wisconsin (2001)

Knowledge Systems

A Multi-criteria Decision Making Approach for the Assessment of Information Credibility in Social Media

Marco Viviani[(⊠)] and Gabriella Pasi

Dipartimento di Informatica, Sistemistica e Comunicazione (DISCo),
Università degli Studi di Milano-Bicocca,
Viale Sarca, 336 – Edificio U14, 20127 Milano, Italy
{marco.viviani,pasi}@disco.unimib.it
http://www.ir.disco.unimib.it

Abstract. In Social Media, large amounts of User Generated Content (UGC) generally diffuse without any form of trusted external control. In this context, the risk of running into misinformation is not negligible. For this reason, assessing the credibility of both information and its sources in Social Media platforms constitutes nowadays a fundamental issue for users. In the last years, several approaches have been proposed to address this issue. Most of them employ machine learning techniques to classify information and misinformation. Other approaches exploit multiple kinds of relationships connecting entities in Social Media applications, focusing on credibility and trust propagation. Unlike previous approaches, in this paper we propose a model-driven approach based on Multi-Criteria Decision Making (MCDM) and quantifier guided aggregation. An overall credibility estimate for each piece of information is obtained based on multiple criteria connected to both UGC and users generating it. The proposed model is evaluated in the context of opinion spam detection in review sites, on a real-world dataset crawled from Yelp, and it is compared with well-known supervised machine learning techniques.

Keywords: Credibility assessment · Opinion spam detection · Social Media · Multi-Criteria Decision Making · OWA aggregation operators

1 Introduction

The Oxford Dictionary has recently declared *post-truth* to be its international word of the year for 2016. The dictionary defines it as an adjective "relating to or denoting circumstances in which objective facts are less influential in shaping public opinion than appeals to emotion and personal belief". Social Media, which have been considered since their diffusion a democratic instrument for exchanging and sharing information, have become nowadays the vehicle of the spread of misinformation and hoaxes, contributing to the emergence of post-truth.

© Springer International Publishing AG 2017
A. Petrosino et al. (Eds.): WILF 2016, LNAI 10147, pp. 197–207, 2017.
DOI: 10.1007/978-3-319-52962-2_17

Individuals have only limited cognitive abilities to assess credibility of pieces and sources of information with which they come in contact. This may have serious consequences when consumers choose products and services based on fake reviews, or when fake news influence the public opinion, or again when patients run into inaccurate medical information. For this reason, the interest in studying possible ways of helping users in assessing the level of credibility of online information is particularly important, especially in the Social Web scenario.

Due to the above reasons, in the last years numerous approaches have appeared to tackle the issue of the automatic assessment of information credibility, in particular in Social Media. Individuals' credibility perceptions can be evaluated in terms of multiple characteristics, which may be associated with sources of information, shared contents, and media across which information diffuses. State-of-the-art proposals are usually based on data-driven approaches, i.e., employing machine learning techniques that classify pieces and/or sources of information as credible or not credible, or on propagation-based approaches that exploit the underlying graph-based structure connecting entities in social platforms to propagate credibility and trust.

In this paper, a Multi-Criteria Decision Making (MCDM) approach is presented; the approach assesses information credibility in Social Media based on the use of aggregation operators to estimate an overall credibility score associated with the considered User Generated Content. We illustrate in particular the use of *Ordered Weighted Averaging* aggregation operators associated with *linguistic quantifiers* to tune the number of (important) features that allows to correctly identify misinformation. The proposed approach is presented and evaluated in the context of opinion spam detection in review sites, by identifying fake reviews on a crawled Yelp dataset.

The paper is organized as follows. Section 2 introduces the concept of credibility and illustrates its main characteristics. Section 3 briefly summarizes the approaches that in the literature have tackled the issue of the assessment of information credibility in Social Media. In Sect. 4 we describe our MCDM approach, and we present its effectiveness on a real-world scenario. Finally, Sect. 5 concludes the paper and discusses some further research.

2   Credibility

The concept of credibility has been studied since Plato and Aristotle (4th century BC), and more recently in many different research fields, such as Communication, Psychology, and Social Sciences. The research undertaken by Hovland and colleagues [8] in the 1950s constitutes the first systematic work about credibility and mass media.

According to Fogg and Tseng [5], credibility is a *perceived* quality of the *information receiver*, and it is composed of multiple dimensions. In everyday life, people usually reduce the uncertainty about credibility based on (*i*) the reputation of the source of information, (*ii*) the presence of trusted intermediaries such as experts and/or opinion leaders, and (*iii*) personal trust based on first-hand experiences. Conversely, the process of 'disintermediation' that characterizes the digital realm, makes credibility assessment a more complex task with respect to 'offline' media. In this sense, the process of assessing *perceived credibility* involves different *characteristics*, which can be connected to: (*i*) the *source* of the message, (*ii*) the *message* itself, i.e., its structure and its content, and (*iii*) the *media* used to diffuse the message [17].

In particular, evaluating information credibility in the Social Web deals with the analysis of both UGC and their authors' characteristics, and to the intrinsic nature of Social Media platforms. Generally speaking, this means to take into account characteristics of credibility both connected to information (i.e., User Generated Content) and to information sources, as well as to the social relationships connecting the involved entities.

In this scenario, where the complexity of the features to be taken into account increases, several approaches have been proposed to help users in automatically or semi-automatically assess information credibility. They will be discussed in the next section.

3   Related Work

Depending on both the context and the aim to which a given Social Media application is developed, several different characteristics can concur to the assessment of information credibility. They can be simple *linguistic features* associated with the text of the User Generated Content, they can be additional *meta-data features* associated with the content of a review or a tweet, they can also be extracted from the behavior of the users in Social Media, i.e., *behavioral features* or they can be connected to the user's profile (if available). Furthermore, different approaches have taken into consideration *product-based features*, connected with the product or service reviewed, or have considered *social features*, which exploit the network-structure and the relationships connecting entities in Social Media platforms [7].

Over the years, several approaches have been developed for example to detect opinion spam in review sites [7], to identify fake news and spam in microblogging sites [4], or for assessing the credibility of online health information [16]. Some studies focus on survey-based studies that aim at identifying the perception that users have of certain features with respect to credibility [11,15,25], sometimes considering contextual aspects [21]. In general, the majority of approaches that propose automatic or semi-automatic techniques for assessing information credibility focus on data-driven techniques, i.e., machine learning. Many of them propose supervised [2,22] or semi-supervised approaches [13,23] focusing on linguistic features. Other works propose more effective multiple-feature-based approaches, which employ distinct features of different nature in addition to simple linguistic features [4,6,10,12,19]. Recent approaches focus more on the identification of spammers or group spammers to detect misinformation, in some cases using features connected to the behavior of users [14,18], and in many other cases by exploiting the underlying network structure and credibility/trust propagation mechanisms [9,24,26,29]. With respect to fake review and fake news detection, the issue of the assessment of credibility in the healthcare scenario by computer scientist has received less attention. Only few approaches have been proposed so far [1,20], although we feel this is a context in which the problem of identification of false information is particularly significant.

In the next section, we will describe our approach focusing on Multi-Criteria Decision Making, which is based on a model-driven paradigm that has not been deeply investigated yet in the scenario of credibility assessment.

4   An MCDM Approach for Assessing Information Credibility

As already outlined in the literature, several features connected to both the content and the user who generated it can concur at the information credibility assessment process. In machine learning, these features are considered to define an optimal solution in both supervised and semi-supervised contexts. While in these data-driven approaches the process of tuning the interplay of the credibility features is in most cases invisible to the user, in an MCDM context this can be analyzed by selecting and comparing different aggregation schemes over the evaluation scores associated with the considered features for a given piece of information. Then, these scores can be aggregated through a function that synthesizes them into an overall credibility estimate. In the literature a variety of *aggregation operators* [3] has been proposed, belonging to distinct classes characterized by distinct and well defined mathematical properties. Our approach is based on two quantifier-guided OWA-based aggregation schemes [28], which also allow to take into consideration in the credibility assessment process the unequal *importance* of features.

4.1   Quantifier Guided Aggregation

In this section we remind the definition of the parameterized family of operators called *Ordered Weighted Averaging* (OWA) operators, [27] and the concept of *quantifier guided aggregation*.

Definition 1. *An* aggregation operator F

$$F : I^n \to I$$

is called an Ordered Weighted Averaging (OWA) *operator of dimension n if it has associated a weighting vector $W : [w_1, w_2, \ldots, w_n]$, such that (i) $w_j \in [0,1]$ and (ii) $\sum_{i=1}^{n} w_j = 1$, and where, given n values denoted by $a_1, a_2, \ldots, a_n$, their aggregated value is computed as follows:*

$$F(a_1, a_2, \ldots, a_n) = \sum_{j=i}^{n} w_j b_j \tag{1}$$

in which $b_j \in B$ is the jth largest of the a_i values.

The w_j values of the vector W can be automatically inferred based on the notion of *quantifier guided aggregation*. Yager in [28] has defined an elegant solution to the problem of computing the w_j weights starting by an analytic definition of a function associated with a *linguistic quantifier*. In this paper we consider *relative* quantifiers such as *most* and *around k*.

Let Q be a function denoting a relative quantifier; the weights w_j of a weighting vector W of dimension n (n values to be aggregated) can be defined as follows:

$$w_j = Q\left(\frac{j}{n}\right) - Q\left(\frac{j-1}{n}\right), \quad \text{for } j = 1 \ldots n \tag{2}$$

This way of defining the weighting vector W assumes that all the considered criteria are equally important. However, in real scenarios it is crucial to discriminate the importance of the criteria that concur in a decision making process. For this reason, let us consider the importance of the n criteria denoted by $V_1, V_2, \ldots, V_n$, each $V_i \in [0,1]$. In the reordering process of the a_i values, it is important to maintain the reordered values correctly associated with the importance of the criteria that originated them. For this reason, formula (3) can be applied, were u_j denotes the importance originally associated with the criterion that has the jth largest satisfaction degree. To obtain the weight w_j of the weighting vector with weighted criteria, we compute it for each alternative as follows [28]:

$$w_j = Q\left(\frac{\sum_{k=1}^{j} u_k}{T}\right) - Q\left(\frac{\sum_{k=1}^{j-1} u_k}{T}\right) \tag{3}$$

where $T = \sum_{k=1}^{n} u_k$ is the sum of the importance values u_js.

4.2 A Real-Case Study

In this section we define and instantiate our MCDM approach to the Yelp scenario, by selecting a set of appropriate criteria for assessing the credibility of reviews associated with restaurants, to identify fake reviews. We then illustrate two different aggregation schemes: (i) the OWA associated with a *linguistic quantifier* (OWA + LQ), and (ii) the OWA associated with a linguistic quantifier with unequal *importance* of criteria (OWA + LQ + I).

Yelp contains several data and metadata related to restaurants, reviews and users. In our approach, we consider the following features: (i) n_f, the number of friends of a user; (ii) n_r, the number of the reviews written by a user; (iii) l_r, the length of the review; (iv) m_r, the evaluation (in terms of assigned stars) provided for a restaurant connected to a review; (v) d_e, the distance between the evaluation (in terms of stars) given by a user to a restaurant with respect to the global evaluation of the restaurant itself; (vi) s_i, the presence of the standard image rather then a customized one.

For each review r_i, a function ϕ_γ associated with each feature γ (i.e., criterion) produces a binary score 0/1 that represents the degree of satisfaction of the considered constraints. Formally,

- $\phi_{n_f}(r_i) = 0$, if $n_f = 0$, $\phi_{n_f}(r_i) = 1$, otherwise
- $\phi_{n_r}(r_i) = 0$, if $n_r \leq 5$, $\phi_{n_r}(r_i) = 1$, otherwise
- $\phi_{l_r}(r_i) = 0$, if $l_r \leq 120$, $\phi_{l_r}(r_i) = 1$, otherwise
- $\phi_{m_r}(r_i) = 0$, if $m_r = 1$ or $m_r = 5$, $\phi_{m_r}(r_i) = 1$, otherwise
- $\phi_{d_e}(r_i) = 0$, if $d_e \geq 2$, $\phi_{d_e}(r_i) = 1$, otherwise
- $\phi_{s_i}(r_i) = 0$, if $s_i = $ standard, $\phi_{s_i}(r_i) = 1$, if $s_i = $ customized

The obtained evaluation scores must be aggregated into an overall credibility score $v(r_i)$, where $v(r_i) = F(\phi_{n_f}(r_i), \phi_{n_r}(r_i), \phi_{l_r}(r_i), \phi_{m_r}(r_i), \phi_{d_e}(r_i), \phi_{s_i}(r_i))$. The aggregated score must be interpreted in the following way: the lower the value (close to 0) the higher the probability of being in the presence of a fake review. The higher the value (close to 1), the higher the probability of trustworthiness of the review.

The OWA + LQ Aggregation Scheme. In the definition of an OWA operator guided by a linguistic quantifier (LQ), different shapes of Q can be selected, as illustrated in Fig. 1.

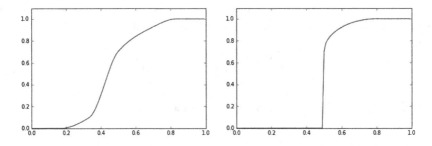

Fig. 1. Possible representations of the Q function for the OWA + LQ scheme.

In this paper, we adopt the softer definition reported in Fig. 2, corresponding to the linguistic quantifier *around* 50%. By observing the function, the maximum increase in satisfaction over 6 criteria is obtained when around 50% of the criteria (in our case 3 over 6 criteria) are satisfied. The fact that the overall satisfaction is not saturated when 3 over 6 criteria are satisfied means that more criteria increase the satisfaction, up to reaching a full satisfaction when 5 of the 6 criteria are satisfied. The weighting vector obtained by Eq. (2) associated with this shape of the Q function is $W = [0 \ \ 0.1 \ \ 0.6 \ \ 0.2 \ \ 0.1 \ \ 0]$.

The OWA + LQ + I Aggregation Scheme. In this scheme, we want to aggregate by ensuring that *around* 50% of the *important* criteria are considered in the aggregation process[1]. This implies to construct a weighting vector for each

---

[1] For the sake of conciseness, we omit the details concerning the definition of the function Q for the OWA + LQ + I scheme.

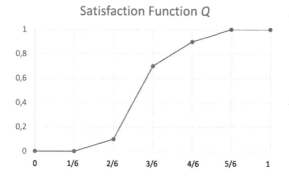

Fig. 2. Graphical representation of the Q function for the OWA + LQ scheme.

of the considered reviews, based on the reordering of the satisfaction degrees of the 6 criteria by maintaining the association with the importance of the criteria that generated them. As importance values associated with criteria, we consider the following ones: $V_1 = 1, V_2 = 0.75, V_3 = 0.35, V_4 = 0.25, V_5 = 0.15, V_6 = 0.5$, where V_1 is associated with n_f, V_2 with n_r, V_3 with l_r, V_4 with m_r, V_5 with d_e and V_6 with s_i. Let us consider for example three reviews r_1, r_2, r_2, whose values associated with criteria are the following:

- r_1: $n_f = 50$, $n_r = 318$, $l_r = 214$, $m_r = 5$, $d_e = 3$, $s_i = $ customized image
- r_2: $n_f = 61$, $n_r = 4$, $l_r = 71$, $m_r = 4$, $d_e = 1$, $s_i = $ standard image
- r_3: $n_f = 0$, $n_r = 2$, $l_r = 539$, $m_r = 3$, $d_e = 1$, $s_i = $ customized image

For each review, the credibility scores associated with each criterion and obtained by applying the binary evaluation functions ϕ_γ previously defined are: $r_1 = [1\ 1\ 1\ 0\ 0\ 1]$, $r_2 = [1\ 0\ 0\ 1\ 1\ 0]$, and $r_3 = [0\ 0\ 1\ 1\ 1\ 1]$; thus, the u_j values obtained after the reordering are:

- r_1: $u_1 = 1, u_2 = 0.75, u_3 = 0.5, u_4 = 0.35, u_5 = 0.25, u_6 = 0.15$
- r_2: $u_1 = 1, u_2 = 0.25, u_3 = 0.15, u_4 = 0.75, u_5 = 0.5, u_6 = 0.35$
- r_3: $u_1 = 0.5, u_2 = 0.35, u_3 = 0.25, u_4 = 0.15, u_5 = 1, u_6 = 0.75$

Finally, by applying Eq. (3), we have: $W_{r_1} = [0.1\ \ 0.6\ \ 0.3\ \ 0\ \ 0\ \ 0]$, $W_{r_2} = [0.1\ \ 0.6\ \ 0\ \ 0.2\ \ 0.1\ \ 0]$, and $W_{r_3} = [0\ \ 0.1\ \ 0.6\ \ 0\ \ 0.3\ \ 0]$.

Evaluation. For evaluation purposes, we crawled the Yelp Website by obtaining a total number of about 140,000 'true' (recommended) reviews and 20,000 'fake' (non-recommended) reviews. This dataset is automatically labeled by Yelp, and it allows us to verify the number of review correctly assigned to the proper category by our approach (with respect to Yelp). For each review r_i, we have produced an overall veracity score $v(r_i)$ by applying the proposed aggregation schemes to the performance scores of the criteria associated with the review itself. The aggregation operators produce a score in the interval $[0, 1]$; in our scenario, 0 indicates the complete *unreliability* of the review, while 1 indicates its complete *reliability*. To classify reviews as 'true' or 'fake' (based on its overall veracity score), we selected a threshold value of 0.75. The reviews having an overall veracity score lower that the veracity threshold are classified as fake, and viceversa. The following evaluation measures have been considered: *precision*, *recall*, *accuracy*, and *f1-score* [19]. Table 1 summarizes the results obtained for the considered aggregation schemes, and it provides a comparison with some state-of-the-art machine learning baselines (i.e., Support Vector Machines and Random Forests). For each ML technique, the presented results have been obtained by a *5-fold cross-validation*; to take into account the imbalanced dataset issue, we applied *undersampling*.

Table 1. Summarization of the evaluation measures.

Approach	Precision	Recall	Accuracy	f1-score
OWA + LQ	0.80	0.76	0.79	0.78
OWA + LQ + I	0.82	0.84	0.82	0.82
SVM	0.74	0.90	0.79	0.81
Random Forests	0.83	0.83	0.83	0.82

By analyzing the results from Table 1, it emerges that the OWA + LQ + I aggregation scheme produce better results, as expected, compared to the simple OWA + LQ scheme. Furthermore, with respect to the use of machine learning techniques, it emerges that dealing with a manageable number of criteria, a Multi-Criteria Decision Making approach associated with a good knowledge of the domain can lead to better (w.r.t. SVM) and comparable (w.r.t. Random Forests) results, preserving at the same time the control over the involved criteria. It is worth to be underlined that the use of the Yelp automatic classification of reviews into recommended and non-recommended has been a practical way (already employed in the literature) to evaluate our model and the diverse schemes of aggregation on a real-case study. Nevertheless, our results (as well as the results of the related work using the same classification as a gold standard) are subjected to this particular choice, which has its limitations.

5   Conclusions

In the *post-truth* era, individuals are in constant contact with misinformation spread across Social Media platforms. In most cases, due to the human limited cognitive skills, facing this huge amount of User Generated Content (UGC) is difficult, if not impossible. For this reason, in the last years several techniques have been proposed to help users to discern in an automatic or semi-automatic way genuine from fake information. Even if Social Media applications are developed for different aims and in different contexts (e.g., review sites, microblogging, health-related communities, etc.), most of the approaches for the credibility assessment of pieces and/or sources of information focus on different kinds of characteristics associated with the source of information, the spread messages, and the network-structure characterizing social applications.

The majority of the approaches proposed so far are based on supervised or semi-supervised machine learning techniques, or on credibility and trust propagation-based models. In this paper, unlike previous approaches, we focus on a model-driven approach, based on Multi-Criteria Decision Making and quantifier guided aggregation. For each piece of UGC considered (alternative), several characteristics (criteria) connected both to users and the content they generate are taken into account. This way, it is possible to obtain distinct scores representing the satisfaction degrees of each alternative w.r.t. criteria. By aggregating these scores by means of quantifier-guided Ordered Weighted Averaging operators, an overall credibility estimate is obtained, measuring the credibility level of each piece of information.

By evaluating our approach on a real-case scenario, i.e., opinion spam detection in review sites, we illustrated the effectiveness of our approach, by outlining that our method reaches results comparable to those of well-known machine learning techniques. In the future, a higher number of characteristics will be considered, focusing for example on social features. In addition to this, the approach will be improved by considering interdependence among criteria, by using suitable aggregation operators.

References

1. Abbasi, A., Fu, T., Zeng, D., Adjeroh, D.: Crawling credible online medical sentiments for social intelligence. In: 2013 International Conference on Social Computing (SocialCom), pp. 254–263. IEEE (2013)
2. Banerjee, S., Chua, A.Y.: Applauses in hotel reviews: genuine or deceptive? In: 2014 Science and Information Conference (SAI), pp. 938–942. IEEE (2014)
3. Calvo, T., Mayor, G., Mesiar, R. (eds.): Aggregation Operators: New Trends and Applications. Physica-Verlag GmbH, Heidelberg (2002)
4. Castillo, C., Mendoza, M., Poblete, B.: Predicting information credibility in time-sensitive social media. Internet Res. 23(5), 560–588 (2012)
5. Fogg, B.J., Tseng, H.: The elements of computer credibility. In: Proceeding of SIGCHI Conference on Human Factors in Computing Systems, pp. 80–87. ACM (1999)
6. Gupta, A., Kumaraguru, P., Castillo, C., Meier, P.: TweetCred: a real-time web-based system for assessing credibility of content on Twitter. In: Proceedings of 6th International Conference on Social Informatics (SocInfo), Barcelona, Spain (2014)
7. Heydari, A., ali Tavakoli, M., Salim, N., Heydari, Z.: Detection of review spam: a survey. Expert Syst. Appl. 42(7), 3634–3642 (2015)
8. Hovland, C.I., Janis, I.L., Kelley, H.H.: Communication and Persuasion. Yale University Press, New Haven (1953)
9. Jin, Z., Cao, J., Jiang, Y.G., Zhang, Y.: News credibility evaluation on microblog with a hierarchical propagation model. In: 2014 IEEE International Conference on Data Mining, pp. 230–239. IEEE (2014)

10. Jindal, N., Liu, B.: Opinion spam and analysis. In: Proceedings of 2008 International Conference on Web Search and Data Mining, pp. 219–230. ACM (2008)

11. Kang, B., Höllerer, T., O'Donovan, J.: Believe it or not? Analyzing information credibility in microblogs. In: Proceedings of 2015 IEEE/ACM International Conference on Advances in Social Networks Analysis and Mining 2015, pp. 611–616. ACM (2015)

12. Li, H., Liu, B., Mukherjee, A., Shao, J.: Spotting fake reviews using positive-unlabeled learning. Computación y Sistemas 18(3), 467–475 (2014)

13. Li, X., Liu, B.: Learning to classify texts using positive and unlabeled data. In: IJCAI, vol. 3, pp. 587–592 (2003)

14. Lim, E.P., Nguyen, V.A., Jindal, N., Liu, B., Lauw, H.W.: Detecting product review spammers using rating behaviors. In: Proceedings of 19th ACM International Conference on Information and Knowledge Management, pp. 939–948. ACM (2010)

15. Luca, M., Zervas, G.: Fake it till you make it: reputation, competition, and Yelp review fraud. Manag. Sci. 62(12), 3412–3427 (2016)

16. Ma, T.J., Atkin, D.: User generated content and credibility evaluation of online health information: a meta analytic study. Telemat. Informat. (2016). In Press

17. Metzger, M.J., Flanagin, A.J., Eyal, K., Lemus, D.R., McCann, R.M.: Credibility for the 21st century: integrating perspectives on source, message, and media credibility in the contemporary media environment. Ann. Int. Commun. Assoc. 27(1), 293–335 (2003)

18. Mukherjee, A., Liu, B., Glance, N.: Spotting fake reviewer groups in consumer reviews. In: Proceedings of 21st International Conference on World Wide Web, pp. 191–200. ACM (2012)

19. Mukherjee, A., Venkataraman, V., Liu, B., Glance, N.S.: What Yelp fake review filter might be doing? In: Proceedings of ICWSM (2013)

20. Mukherjee, S., Weikum, G., Danescu-Niculescu-Mizil, C.: People on drugs: credibility of user statements in health communities. In: Proceedings of 20th ACM SIGKDD International Conference, pp. 65–74. ACM (2014)

21. O'Donovan, J., Kang, B., Meyer, G., Höllerer, T., Adalii, S.: Credibility in context: an analysis of feature distributions in Twitter. In: Privacy, Security, Risk and Trust (PASSAT), 2012 International Conference on Social Computing, pp. 293–301. IEEE (2012)

22. Ott, M., Cardie, C., Hancock, J.: Estimating the prevalence of deception in online review communities. In: Proceedings of 21st International Conference on World Wide Web, pp. 201–210. ACM (2012)

23. Ren, Y., Ji, D., Zhang, H.: Positive unlabeled learning for deceptive reviews detection. In: EMNLP, pp. 488–498 (2014)

24. Seo, E., Mohapatra, P., Abdelzaher, T.: Identifying rumors and their sources in social networks. In: SPIE Defense, Security, and Sensing, p. 83891I. International Society for Optics and Photonics (2012)

25. Sikdar, S., Kang, B., ODonovan, J., Höllerer, T., Adah, S.: Understanding information credibility on Twitter. In: 2013 International Conference on Social Computing (SocialCom), pp. 19–24. IEEE (2013)

26. Wang, G., Xie, S., Liu, B., Yu, P.S.: Identify online store review spammers via social review graph. ACM Trans. Intell. Syst. Technol. (TIST) 3(4), 61 (2012)

27. Yager, R.R.: On ordered weighted averaging aggregation operators in multicriteria decisionmaking. IEEE Trans. Syst. Man Cybern. 18(1), 183–190 (1988)

28. Yager, R.R.: Quantifier guided aggregation using OWA operators. Int. J. Intell. Syst. **11**(1), 49–73 (1996)
29. Ye, J., Akoglu, L.: Discovering opinion spammer groups by network footprints. In: Appice, A., Rodrigues, P.P., Santos Costa, V., Soares, C., Gama, J., Jorge, A. (eds.) ECML PKDD 2015. LNCS (LNAI), vol. 9284, pp. 267–282. Springer, Heidelberg (2015). doi:10.1007/978-3-319-23528-8_17

Fuzzy Consensus Model in Collective Knowledge Systems: An Application for Fighting Food Frauds

Maria Vincenza Ciasullo[1], Giuseppe D'Aniello[2(✉)], and Matteo Gaeta[2]

[1] Dipartimento di Scienze Aziendali, Management and Innovation Systems,
University of Salerno, 84084 Fisciano, SA, Italy
mciasullo@unisa.it
[2] Dipartimento di Ingegneria dell'Informazione ed Elettrica e Matematica Applicata,
University of Salerno, 84084 Fisciano, SA, Italy
{gidaniello,mgaeta}@unisa.it

Abstract. Food fraud is related to different illicit conducts which aim at gaining economic benefit from counterfeiting food and ignoring the damage they cause to public economy and health. Consumers use the new technologies, like social networks, in order to share their worries about food frauds and to stay informed about them. But, in such a complex and dynamic context, it is important to ensure the reliability of news about food frauds in order to avoid misinformation and general panic phenomena. In this context, we propose an extension of a Collective Knowledge System aiming at verifying the reliability of news about food frauds and to decide whether to publish and spread information on the food frauds in the society. A Fuzzy Consensus Model has been proposed for helping the experts in achieving a shared decision about the reliability of each news and about its publication and diffusion. An illustrative example demonstrates the feasibility and the usefulness of the proposed approach.

Keywords: Collective Knowledge System · Fuzzy Consensus Model · Food frauds

1 Introduction and Motivation

In our society, consumers are globally concerned about food safety, asking for very diverse products in terms of quality, certifications, safety and information [10]. It has to be noted that the complexity of current food markets represents for consumers a source of emerging and challenging risks of frauds or adulterations [11]. Food fraud is a wide concept without a statutory definition, being related to several different illicit conducts aiming to gain economic benefit counterfeiting food. This concept is often "used to encompass the deliberate and intentional substitution, addition, tampering, or misrepresentation of food, food ingredients, or food packaging; or false or misleading statements made about a

© Springer International Publishing AG 2017
A. Petrosino et al. (Eds.): WILF 2016, LNAI 10147, pp. 208–217, 2017.
DOI: 10.1007/978-3-319-52962-2_18

product" [11]. In this context, there is a collective responsibility for providing information about food safeness; consequently, emergent technologies can contribute to facilitate not only food quality traceability, but also data collection, elaboration, and sharing about real or possible food frauds or risks.

It is now evident that citizens can spontaneously feed the lively debate on food security and the related frauds, through those web and social channels (e.g. social media, online communities, blogs etc.) that consumers usually use to share information. Consequently, they can also support the institutional struggle against food fraud, acting as "bell-ringers" in order to report to the authorities real or potential risks concerning food purchasing or consumption.

Such collaborative approaches and information sharing processes that, in the Knowledge society, are really successful in many cases, suffer of clear difficulties in the agri-food sector. Think, for instance, to the potential risk of misinformation campaigns that some companies or producers may carry on by publishing false news on social web sites in order to discredit potential competitors.

In such context, we propose an extension of the traditional model of Collective Knowledge System (CKS) [6] which aims at involving citizens in the identification, checking and communication of food frauds' news, with the intent of avoiding to communicate false news which may produce misinformation, panic and may cause also economic damages to the society. Indeed, in the classic models of CKS, users are free to communicate and introduce any information in the systems, usually by answering to questions asked by other users, thus sharing their knowledge to a wider audience. Such approach, however, lacks an assessment and verification process regarding the truthfulness of the information autonomously introduced by the users. Although such an approach works well in certain domains, it can be really dangerous in sensitive and critical sectors like public health and food adulteration. Accordingly, in this paper we extend the traditional model of CKS with two phases: verification of the news on food frauds by a pool of experts and approval of the publication process of the news to the consumers, after having ascertained its truthfulness. A Fuzzy Consensus process is proposed for supporting the experts in the verification and publication of the food frauds' news. An illustrative example is proposed in order to demonstrate the feasibility of the approach.

2   Theoretical Background

The term "Collective Intelligence" is attributed to the French philosopher Pierre Lévy in 1994, although similar concepts and thoughts can be found in previous works. By simplifying, Collective Intelligence can be defined as a kind of intelligence distributed everywhere and continuously enhanced by the collaboration among individuals. From the point of view of ICT, the Collective Intelligence can be seen as the possibility for a user to employ applications, services and ICT systems to generate contents and to communicate and compare such contents with other users, by using pervasive Web 2.0/3.0 technologies, which de facto enables the creation and diffusion of new knowledge. From a merely technological

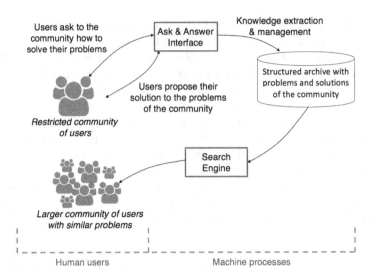

Fig. 1. Example of a Collective Knowledge System (our elaboration from [6])

viewpoint, the collective intelligence can be sustained by the so-called Collective Knowledge System (CKS) [6], depicted in Fig. 1. Such frameworks enable effective and efficient access to information. They represent human-machine systems able to sustain the Collective Intelligence, in which the machine allows for collecting huge amount of information generated by the humans. Generally, such systems consist of three subsystems: (i) A social system supported by information and communication technologies, which generate a problem that can be solved by means of discussion in the community; (ii) A search engine which is capable to find questions and answers in the contents generated by the social system; (iii) Intelligent users, which are capable of formulate their problems by means of queries for the search engine. A typical example of CKS is the Faq-O-Sphere proposed by Gruber in his work [6]. One of the key characteristic of the CKS is the presence of user-generated contents. The system is also able to make inference from the user-generated contents, by means of knowledge extraction approaches [5], thus producing answers and results that can not be found explicitly in such contents, which represents emerging knowledge enabling the shift from gathered and individual intelligence to collective intelligence.

In such collaborative processes, users may be involved in decision making problems in which they have to reach a shared decision. Such kind of problem is defined as Group Decision Making (GDM) [2] problem for indicating that a group of individuals have to find a shared decision. A GDM problem can be conceived as composed of two processes: a consensus process and a selection process (see Fig. 2). The former allows to reach the higher level of consensus among the experts on the available alternatives. Such process is driven by a moderator that, by evaluating the current level of consensus among the experts, is able to supervise the consensus process and to drive it to the success. The

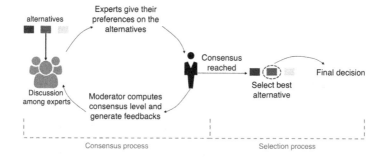

Fig. 2. A group decision making process

process is iterative: at each round, the level of consensus is measured. If it is below a given threshold, the moderator generates some feedback asking to the experts to revise their decisions and to discuss with the others in order to change their opinions and finally find an agreement on the decision. When a consensus is reached, the second process (Selection) starts and a final solution is selected among the available alternatives. The consensus process has been applied in many domains, from digital libraries [9], to logistics [1], sensor networks [4], control of dynamic systems [12], workplace environments [3], demonstrating its value in solving heterogeneous group decision making problems.

3   Overall Approach

The proposed approach leverages on the theoretical approaches of Collective Intelligence and Collective Knowledge Systems (CKS) and proposes an extension of the processes of a CKS by means of an additional phase of verification of the news reliability and another phase in which the experts decide whether to publish the news in the community. In this way, the proposed approach is able to solve one of the main issue of a CKS, i.e., to ensure the certainty and reliability of the information therein contained, thanks to meticulous and precise checks performed by experts of several institutions.

The approach is depicted in Fig. 3. Notice that, with respect to the traditional CKS proposed by Gruber [6] (Fig. 1), we do not distinguish among a small group of users that propose and create the contents and a wider community that only use the provided solutions. We consider a unique, large community of consumers, associations, law enforcement agencies' officers (LEAs), producers and so on, which report and share information on possible food frauds. Such information, shared by the community, are further elaborated in order to extract and aggregate the knowledge coming from different sources and reports, so to organize the information on each potential food fraud into an archive.

After that (step 3 in Fig. 3), a group of experts evaluates each potential news on food fraud stored in the archive, performs further investigations about it, and discusses its truthfulness and dangerousness. In this phase, a consensus process (like the one in Fig. 2) is applied in order to reach a final decision about the food fraud. Further details on the consensus process are reported in next section.

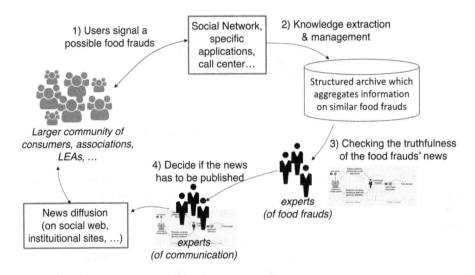

Fig. 3. Overall approach

For the food frauds that have been verified by the experts, a process regarding the publishing of the news is executed by another group of experts (in some cases, it can be also the previous group). These experts discuss on the impact that the news may have if published. Even in this case, a consensus process is applied in order to achieve a shared decision among the experts.

The consensus processes are implemented by means of ICT tools and systems, thus to allow the experts to discuss and express their preferences also in an asynchronous way and remotely. This eases the participation of the experts and speeds up the decision about the food frauds.

4   A Fuzzy Consensus Approach for Assessing News Reliability

In the consensus processes of the proposed approach, we adopt the *utility functions* method for representing the preferences of the experts [7]. Essentially, as regarding the verification phase (step 3 in Fig. 3), each expert express his/her preferences with respect to 5 alternatives $S_v = \{s_1 = Completely\ False, s_2 = Partially\ False, s_3 = Neutral, s_4 = Partially\ True, s_5 = Completely\ True\}$. With respect to the publishing phase (step 4 in Fig. 3), instead, the set of alternatives is $S_p = \{s_1 = Strongly\ Avoid\ Publish, s_2 = Avoid\ Publish, s_3 = Neutral, s_4 = Publish, s_5 = Absolutely\ Publish\}$. Notice that we adopt ordinal linguistic values as the set of alternatives. In other traditional approaches, instead, the linguistic values are used to express the preference degrees on a given set of alternatives [9].

Each expert assigns an utility value to each alternative, resulting in a set of 5 utility values $V = \{v_1, \ldots v_5\}$ for each expert, with $v_i \in [0,1]$. The higher is the value v_i for alternative x_i, the better it satisfies the experts' objective. Moreover, we set an additional constraint to the utility values in order to avoid that, if an expert assigns a high preference to an alternative (e.g., $v_1 = 0.9$), he/she can not assign high values also to the other alternatives, especially to the opposite ones (eg., $v_5 = 0.8$). For doing so, we impose that $\sum_{i=1}^{5} v_i = 1$.

Starting from the preferences expressed by the experts, in each of the two phases of the proposed approach, we adopt a consensus process similar to the one proposed in [2,4,8]. Such consensus process adopts fuzzy preference relations in order to represent the preferences of a set of experts $E = \{e_1, \ldots, e_m\}$ and to compute the consensus measures. Consequently, we transform the utility values used by the expert $e_h \in E$ in preference relations by means of the following formula [7]:

$$p_{ik}^h = \frac{(v_i^h)^2}{(v_i^h)^2 + (v_k^h)^2} \tag{1}$$

In this way, we obtain a fuzzy preference relation $P^h = (p_{ik}^h)$ for each expert $e_h \in E$. Then, the consensus process starts. First, for each pair of experts (e_h, e_l), a similarity matrix $SM^{hl} = (sm_{ik}^{hl})$ is defined.

$$sm_{ik}^{hl} = 1 - |p_{ik}^h - p_{ik}^l| \tag{2}$$

Then, a global similarity matrix $SM = (sm_{ik})$ is obtained by aggregating all the similarity matrices by using an aggregation function ϕ. The similarity matrices are used for computing the consensus degree:

– First, we consider as the consensus degrees on pairs of alternatives the values of the similarity matrix: $cop_{ik} = sm_{ik}$ which represents the agreement amongst all the experts on the pair of alternatives (x_i, x_k).
– Then, the consensus degree on alternatives is calculated:
$ca_i = \frac{\sum_{k=1; k \neq i}^{n}(cop_{ik} + cop_{ki})}{2n-2}$.
– Lastly, the consensus on the relation, used for controlling the consensus process, is obtained: $cr = \frac{\sum_{i=1}^{n} ca_i}{n}$.

The consensus on the relation is compared with the consensus threshold γ. If it is greater than the threshold, the consensus process ends. Otherwise, the moderator generates feedback for each expert in order to increase the overall consensus. The feedback generation process is based on the proximity matrix. The proximity matrices for each expert are computed with respect to the collective preference relation, $P^c = (p_{ik}^c)$ which represents an aggregation of all the preferences given by all the experts:

$$p_{ik}^c = \Phi(p_{ik}^1, \ldots, p_{ik}^m) \tag{3}$$

where Φ is an aggregation operator. The proximity measures are then computed with respect to the global preference:

– For each expert e_h, the proximity matrix $PM^h = (pm_{ik}^h)$, with $pm_{ik}^h = 1 - |p_{ik}^h - p_{ik}^c|$, is obtained.
– The proximity measure on pairs of alternative, between an expert e_h and the group, is given by the matrix $PM^h = (pp_{ik}^h)$, where $pp_{ik}^h = pm_{ik}^h$.
– The proximity measure on alternative x_i for expert e_h is obtained by adding all the proximity measures on pairs of the alternative x_i in this way: $pa_i^h = \frac{\sum_{k=1,k\neq i}^n pp_{ik}^h}{n-1}$.
– Finally, the proximity measure of the expert e_h on the relation is given by the average of all the proximity measures on the alternatives, i.e., $pr^h = \frac{\sum_{i=1}^n pa_i^h}{n}$.

Proximity measures are useful to send feedback to the experts on how to change their opinions in order to increase the consensus degree. The moderator generates the feedback after having identified the experts that have to change their opinions, the alternatives that should be changed and also the specific pairs of alternatives that are contributing less to the consensus. Such identification is performed in three steps by considering the desired consensus degree threshold γ:

1. The set of experts that should receive feedback is: $EXP = \{e_h | pr^h < \gamma\}$
2. The alternatives that the set of experts EXP should consider in changing their preferences are: $ALT = \{x_i \in X | pa_i < \gamma\}$
3. The specific pairs of alternatives (x_i, x_k) that the experts e^h in the set EXP should change is given by: $PALT^h = \{(x_i, x_k) | x_i \in ALT \wedge e_h \in EXP \wedge pp_{ik}^h < \gamma\}$

Thanks to this rules, the moderator can send feedback to each expert $e_h \in EXP$ for each preference $p_{ik}^h \in PALT$ in the following ways:

– if $p_{ik}^h > p_{ik}^c$, the expert e^h should decrease the preference associated to the pair of alternatives (x_i, x_k)
– if $p_{ik}^h < p_{ik}^c$, the expert e^h should increase the preference associated to the pair of alternatives (x_i, x_k)

After a discussion among the experts, by following the advices sent by the moderator, a new iteration of the consensus process above described starts, and the consensus degree is computed again.

5   Illustrative Example

In order to demonstrate the applicability of the proposed approach, we propose the following illustrative example. We consider a news story[1] appeared on "The Telegraph" on October 2015 regarding the probable use of "disappeared" horses for making food sold in Britain. As written in the article, a Government inquiry is still ongoing. Many experts share the opinion that such possibility is very plausible. We have involved five experts and we asked them to say if such news

---

[1] http://www.telegraph.co.uk/news/uknews/law-and-order/11938438/Up-to-50000-mi ssing-horses-may-have-ended-up-as-food-sold-in-Britain.html.

V	s_1	s_2	s_3	s_4	s_5
e_1	0	0	0.1	0.2	0.7
e_2	0	0	0	0.1	0.9
e_3	0	0	0	0	1
e_4	0	0	0.1	0.5	0.4
e_5	0	0	0.6	0.3	0.1

a)

e_1	s_1	s_2	s_3	s_4	s_5
s_1	0.5	0.5	0	0	0
s_2	0.5	0.5	0	0	0
s_3	1	1	0.5	0.2	0.02
s_4	1	1	0.8	0.5	0.08
s_5	1	1	0.98	0.92	0.5

b)

Fig. 4. Preferences of the experts. (a) The utility values of each experts e_h with respect to each alternative in S_v. (b) Preference relation of expert e_1.

can be considered reliable or not, according to their knowledge and experience, by using the consensus process described in previous section. In such consensus process, the moderator establishes that the consensus degree level that the experts should reach is $\gamma = 0.8$. Figure 4a shows the utility values that each experts assigned to the 5 alternatives of the set S_v described in previous section. Figure 4b shows an example of the preference relation for one expert (e_1), calculated form the values of Fig. 4a by means of Eq. 1. Starting from these preference relations, we compute the similarity among the experts by using Eq. 2. Then, by aggregating all the similarity matrices, we obtain the global similarity matrix, in Fig. 5a, from which it is possible to compute the consensus among the experts. In particular, Fig. 5b shows the consensus on alternatives, from which we obtain the consensus on relation that is $c_r = 0.77 < \gamma$. Thus, the consensus reached by the experts is below the threshold, and the moderator generates the feedback for the experts before a new round of discussion and consensus evaluation may start. The collective preference matrix is obtained by applying Eq. 3 to the preference relation. Such matrix is shown in Fig. 6a. By considering this matrix, it is possible to calculate the proximity matrices for each expert PM^h. Figure 6b shows an example of such matrix for expert e_1.

The proximity measures on alternatives for each expert, PA^h, are shown in Fig. 6c. From these, it is possible to calculate the proximity measure on the relation for each expert (Fig. 6d), useful for identifying the set of experts EXP that should change their opinions.

The moderator, by observing the matrix in Fig. 6d, identifies the set EXP by selecting the expert with a proximity on the relation below the threshold $\gamma = 0.8$, i.e. $EXP = \{e_3, e_5\}$. Then, by considering the proximity measure on the alternative for the experts e_3 and e_5 in Fig. 6c, it is possible to notice that e_3 should change the preference on alternative s_3, while e_5 should change the preferences on alternatives s_3, s_4, s_5. The moderator sends the feedback to the two experts by also considering the proximity matrix PM^3 and PM^5.

Following such feedback and after having discussed among them, the experts change their opinions. In particular:

- expert e_3 changes the utility value of s_4 from 0 to 0.1 and of s_5 form 1.0 to 0.9.
- expert e_5 changes the utility value of s_3 from 0.6 to 0.5 and s_4 from 0.3 to 0.4.

SM	s_1	s_2	s_3	s_4	s_5
s_1	1	1	0.7	0.8	1
s_2	1	1	0.7	0.8	1
s_3	0.7	0.7	1	0.59	0.6
s_4	0.8	0.8	0.59	1	0.52
s_5	1	1	0.6	0.52	1

	ca_i
s_1	0.87
s_2	0.87
s_3	0.65
s_4	0.68
s_5	0.78

a)   b)

Fig. 5. (a) Global similarity matrix. (b) Consensus on alternatives.

P^c	s_1	s_2	s_3	s_4	s_5
s_1	0.5	0.5	0.2	0.1	0
s_2	0.5	0.5	0.2	0.1	0
s_3	0.8	0.8	0.5	0.31	0.21
s_4	0.9	0.9	0.69	0.5	0.32
s_5	1	1	0.79	0.68	0.5

PM^1	s_1	s_2	s_3	s_4	s_5
s_1	1	1	0.8	0.9	1
s_2	1	1	0.8	0.9	1
s_3	0.8	0.8	1	0.89	0.81
s_4	0.9	0.9	0.89	1	0.75
s_5	1	1	0.81	0.75	1

PA^h	s_1	s_2	s_3	s_4	s_5
PA^1	0.925	0.925	0.825	0.862	0.891
PA^2	0.9	0.9	0.72	0.796	0.87
PA^3	0.825	0.825	0.749	0.672	0.867
PA^4	0.925	0.925	0.795	0.81	0.889
PA^5	0.925	0.925	0.586	0.682	0.664

	PR^h
e_1	0.886
e_2	0.837
e_3	0.788
e_4	0.869
e_5	0.756

a)   b)   c)   d)

Fig. 6. (a) Collective preference matrix P^c. (b) Proximity matrix PM^h for each expert $e_h, h = 1..5$. (c) Proximity measures on alternatives for each expert PA^h. (d) Proximity measure on the relation for each expert PR^h

By applying again the entire process for computing the consensus measures with this new set of utility values, we obtain a consensus degree on the relation $c_r = 0.823 > \gamma$, which is greater than the threshold. Accordingly, the consensus process ends, and it is possible to select the final decision. By applying a selection process and considering the collective preference relation, the experts agree on considering the news reliable. Notice that in this example the consensus has been reached with just a slightly modification of the preferences of two experts with respect to only two alternatives. This demonstrates that the consensus process helps in reaching a shared decision while respecting, as much as possible, the opinions of each expert.

6   Conclusions and Future Works

In this work we have proposed an extension of the Collective Knowledge System by means of a verification and publishing phases, implemented by means of a consensus process that uses utility values for representing experts' opinions. An illustrative example has been proposed to demonstrate the applicability of the approach in evaluating the reliability of food frauds' news. Future works aim at experimenting the approach in different domains with more experts and a wider community. We will also consider the presence of experts with different experience and importance in the decisional process [8]. Lastly, we will consider and evaluate the role of different aggregation functions in the consensus reaching process. Specifically, we will evaluate the effect of the chosen aggregation function on the consensus reaching process with respect to the number of different experts in order to verify also the sensitivity of the approach to the outliers (experts with very different opinions from the group). On the other hand, we will evaluate the computational complexity resulting from the used data structures for implementing the method. A jointly optimization between the outlier sensitivity and the size complexity of the approach will be addressed in future works.

References

1. Bassano, C., Ciasullo, M.V., Gaeta, M., Rarità, L.: A consensus-based approach for team allocations: the case of logistics in Campania region. Complex Syst. Inform. Model. Q. **6**, 12–30 (2016)
2. Cabrerizo, F.J., Alonso, S., Pérez, I.J., Herrera-Viedma, E.: On consensus measures in fuzzy group decision making. In: Torra, V., Narukawa, Y. (eds.) MDAI 2008. LNCS (LNAI), vol. 5285, pp. 86–97. Springer, Heidelberg (2008). doi:10.1007/978-3-540-88269-5_9
3. D'Aniello, G., Gaeta, M., Tomasiello, S., Rarità, L.: A fuzzy consensus approach for group decision making with variable importance of experts. In: 2016 IEEE International Conference on Fuzzy Systems (FUZZ-IEEE), pp. 1693–1700 (2016)
4. D'Aniello, G., Loia, V., Orciuoli, F.: A multi-agent fuzzy consensus model in a situation awareness framework. Appl. Soft Comput. **30**, 430–440 (2015)
5. Gaeta, A., Gaeta, M., Piciocchi, P., Ritrovato, P., Vollero, A.: Evaluation of the human resources relevance in organisations via knowledge technologies and semantic social network analysis. Int. J. Knowl. Learn. **9**(3), 219–241 (2014)
6. Gruber, T.: Collective knowledge systems: where the social web meets the semantic web. Web Semant. **6**(1), 4–13 (2008)
7. Herrera-Viedma, E., Herrera, F., Chiclana, F.: A consensus model for multiperson decision making with different preference structures. Trans. Syst. Man Cybern. Part A **32**(3), 394–402 (2002)
8. Pérez, I.J., Cabrerizo, F.J., Alonso, S., Herrera-Viedma, E.: A new consensus model for group decision making problems with non-homogeneous experts. IEEE Trans. Syst. Man Cybern.: Syst. **44**(4), 494–498 (2014)
9. Pérez, I., Cabrerizo, F., Morente-Molinera, J., Urea, R., Herrera-Viedma, E.: Reaching consensus in digital libraries: a linguistic approach. Procedia Comput. Sci. **31**, 449–458 (2014)
10. Pulina, P., Timpanaro, G.: Ethics, sustainability and logistics in agricultural and agri-food economics research. Ital. J. Agron. **7**(3), 33 (2012)
11. Spink, J., Moyer, D.C.: Defining the public health threat of food fraud. J. Food Sci. **76**(9), R157–R163 (2011)
12. Tomasiello, S., Gaeta, M., Loia, V.: Quasi-consensus in second-order multi-agent systems with sampled data through fuzzy transform. J. Uncertain Syst. **10**(4), 243–250 (2016)

Innovative Methods for the Development of a Notoriety System

Massimiliano Giacalone[1(✉)], Antonio Buondonno[2], Angelo Romano[2], and Vito Santarcangelo[3]

[1] Department of Economics and Statistics,
University of Naples "Federico II", Naples, Italy
massimiliano.giacalone@unina.it
[2] iInformatica S.r.l.s., Corso Italia, 77, 91100 Trapani, Italy
ant.buondonno@gmail.com, angelo.rom06@gmail.com
[3] Department of Mathematics and Computer Science,
University of Catania, Catania, Italy
santarcangelo@dmi.unict.it

Abstract. The role of internet in our society is growing day by day and is becoming more and more the only way for getting information, exchange opinions and for improving our personal culture. So, an huge mole of data, in all fields, is today easily accessible and everybody can express and exchange ideas. This represents the greatness of the web. But at the same time, to this huge amount of data does not always correspond an appropriate quality of information that we are reading, and nowadays this represents the biggest weakness of the web. Aim this the work is to analyze the approach based on marked chain used by Pagliarani et al. as we explain in the introduction, showing the relation of this method with BigData analysis.

1 Introduction

The problem of data information is identified by these two phenomena called respectively disinformation and misinformation. Disinformation understood like an intentional inaccurate or false information, and misinformation instead like an unintentional inaccurate information. Some authors of this paper have already opened this debate in other conferences and papers as [1,3]. These previous approaches consider a crawler, a parser, a text similarity analyzer and a notoriety analyzer based on a knowledge base of website manually managed by system maintainers. Scope of this work is to improve the approach and the system already proposed through the use of transfer learning method already based on Markov Chain used by Domeniconi et al. in [2]. This approach is applied on OSINT data, then this is strictly related to BigData analysis and its usage is also useful for other purposes. The system obtained and shown in our paper could be an important instrument to examine preventively the reliability of the information that comes from the web, in such a way as to select just the news that present a high degree of certainty and indisputability to be considered as 'notorious fact'.

© Springer International Publishing AG 2017
A. Petrosino et al. (Eds.): WILF 2016, LNAI 10147, pp. 218–225, 2017.
DOI: 10.1007/978-3-319-52962-2_19

2 The Legal Importance of the Notoriety System: The Article 115 of the Italian Civil Procedure Code and the Notion of 'Notorious Fact'

Thanks to this work it is possible to see how the relations between three different areas, such as the field of law, the engineering and the statistics ones, can be strong. We can particularly observe how the two last fields can help the first. To fully understand the analysis, it is very necessary to explain some legal concepts. One of them is contained in the article 2697 of the Italian civil code [4] which states, at the first paragraph, that: '*Chi vuol far valere un diritto in giudizio deve provare i fatti che ne costituiscono il fondamento.*' This article introduces the principle of 'burden of proof' or 'onus probandi' which implies that if a person wants to see recognized in a trial a certain right by the judge, he/she has to prove the facts where that right comes from. More precisely, in order to establish who is the owner of a certain right, it is necessary to refer to 'evidence' because just with a proof we can ascertain the existence or not of the facts. With the term 'evidence' we refer to any suitable means to determine the conviction of the judge on the existence or not of the facts. In the Italian legal system we have different classifications of the proofs. The most important classification is between 'prove precostituite' and 'prove costituende'. The first term refers to the evidence that can be formed before and outside of a process such as a document. The second term is related to the evidence that can be formed just inside of a process such as witness or confession. The article 2697 of the Italian civil code is connected to another important rule that is contained into the Italian civil Procedure code, that is the article 115. This article is composed by two paragraphs. The first paragraph contains the general rule while the second paragraph represents the exception. In particular the first paragraph states that: '*Salvo i casi previsti dalla legge, il giudice deve porre a fondamento della decisione le prove proposte dalle parti o dal pubblico ministero, nonche' i fatti non specificatamente contestati dalla parte costituita*'. This paragraph introduces the 'dispositive principle' which involves that, in a trial, the judge must reconstruct the truth of the facts, and base its decision, just using the proofs that are adduced by the parties of the dispute (the plaintiff and the defendant). This implies, moreover, that the judge cannot use in a trial his personal knowledge to reconstruct a certain fact because there is a real ban for this. For example, if the judge saw a car accident between two subjects, he could not use his personal knowledge about that accident because in this way he would not be an impartial organ and the first paragraph of the article 115 of the Italian civil procedure code would be violated. However the second paragraph of the article 115 contains an exception to the dispositive principle because it states: '*Il giudice puo' tuttavia, senza bisogno di prova, porre a fondamento della decisione le nozioni di fatto che rientrano nella comune esperienza*'. With this paragraph we introduce the concept of 'notorious fact'. The notorious fact is a fact that a community is able to know at a given time and in a given place [5] (eg. A natural disaster or an act of war). As stated by the second paragraph of the article 115, the judge can found his decision on a notorious fact without the parties of the trial have proved that fact. At this point it arises the question whether the news and the information circulating on Internet can be considered as a 'notorious fact'. At a first sight we could give a negative answer to our question for two reasons. First of all because not everyone actually has the possibility (economic and technical) to access the web and secondly because we cannot ensure the reliability of a news that is present on the web. As a matter of fact, in the Italian case law there is an attitude of closure to the possibility of considering as 'notorious fact' news obtainable from the Internet. In the jurisprudence of merit there is an important ordinance of the Court of Mantua of 16 May 2006 [10] that explicitly negates the possibility of considering as a notorious fact news circulating on the web. In particular in this ordinance the judge states that *'Le notizie reperibili in Internet non costituiscono di per se' nozioni di comune esperienza, secondo la definizione dellart. 115 ult. co. c.p.c., norma che, derogando al principio dispositivo, deve essere interpretata in senso restrittivo. Puo' infatti ritenersi 'notorio' solo il fatto che una persona di media cultura conosce in un dato tempo e in un dato luogo, mentre le informazioni pervenute da Internet, quandanche di facile diffusione ed accesso per la generalita' dei cittadini, non costituiscono dati incontestabili nelle conoscenze della collettivita"*. In other words the Court of Mantua states that the information present on the web cannot be considered as 'notorious fact', first because the second paragraph of the article 115 of the Italian civil procedure code represents an exception to the rule contained in the first paragraph and therefore it must be interpreted restrictively, and then because even if it is easy to spread and access the information contained in the web for most citizens, do not constitute indisputable data in the knowledge of the community. This is the answer of the jurisprudence of merit that is the Court of First Instance. But for a depth analysis, we have to consider the answer that the highest judicial organ in our legal system, the Court of Cassation, gives to our theme. The Court of Cassation in our legal system exercises the so-called 'funzione nomofilattica' that consists in ensuring the exact observance and uniform interpretation of the law [8]. So, thanks to the sentences of this Court, we have leading principles for the right comprehension of every norm and article of the Italian law. Through a general analysis of the sentences of the Court of Cassation, we can observe that even in the jurisprudence of this Court there is a limitative approach towards the second paragraph of the article 115, and therefore towards the notion of 'notorious fact'. Since 2005, in fact, the Court of Cassation is stating that the notion of 'notorious fact' must be interpreted restrictively. In particular, in the sentence [9] 19 November 2014, n. 24599, we can read that: *'Il ricorso alle nozioni di comune esperienza (fatto notorio), comportando una deroga al principio dispositivo ed al contraddittorio, in quanto introduce nel processo civile prove non fornite dalle parti e relative a fatti dalle stesse non vagliati n controllati, va inteso in senso rigoroso, e cioe' come fatto acquisito alle conoscenze della collettivita' con tale grado di certezza da apparire indubitabile ed incontestabile.'* Even if in this sentence the Court of Cassation doesn't mention explicitly the information and news that comes from Internet, we can deduce the hard tendency to reduce as far as possible the application ambit of the 'notorious fact', because the Court, insisting on the assumption that this notion is a derogation, an exception to the general rule and every exception can only be applied in limited cases, states that we can consider as 'notorious fact' just the fact that, in the knowledge of the community, presents a high level of certainty and indubitability. However, it seems necessary to overcome this restrictive approach towards the Internet news for two fundamental reasons. Firstly, because we cannot overlook the poignant datum that nowadays Internet is an essential tool of mass communication that contributes to spread countless facts, that can be classified as facts. Moreover, the ability to qualify an information obtained from the web as 'notorious fact' allows us to get a very valuable benefit from a procedural point of view that is reducing the process time. The introduction of a notorious fact in a process, as a matter of fact, exempts parties from proving that particular fact then it is not necessary an instruction on the same. Reducing the duration of the instruction phase, that is the phase of the process deputed to the collection of the proofs and which most dilates the duration of a process, we obtain a saving of procedural activities and thus of the times of the process [11]. This is very important, since it allows us to fulfill an important principle contained in the article 111 of Italian Constitution. The article 111 of the Italian Constitution is an exemplary norm of the Italian legal system because it expresses different important principles that must be applied in a trial. In particular, in the second paragraph, the article 111 states: '*Ogni processo si svolge nel contraddittorio tra le parti, in condizioni di parita', davanti ad un giudice terzo ed imparziale. La legge ne assicura la ragionevole durata.*' We can observe that this article requires each process a reasonable time because only a process that has a right duration can ensure the effective protection of individual rights. The importance of the 'reasonable duration' of every process, moreover, is confirmed by a norm that is present in the European Convention of Human Rights, the article 6. The first paragraph of this article states that: '*In the determination of his civil rights and obligations or of any criminal charge against him, everyone is entitled to a fair and public hearing within a reasonable time by an independent and impartial tribunal established by law*'. Then, this important perspective, that is the fulfillment of both a constitutional and European precept, makes us understand the need to overcome the restrictive approach of the Italian case law about the notion of 'notorious fact' in relation to the information present on the web. And just with this work we propose an important instrument to preventively examine the reliability of the information that comes from the web, the 'Notoriety system', in such a way as to select just the news that present a high degree of certainty and indisputability to be considered as 'notorious fact'. In this way we can achieve a saving of procedural activity, and therefore a reduction of the process duration, but at the same time, using just the information that presents a high level of certainty, we follow the instructions of the Court of Cassation about the notion of notorious fact. Ultimately it is important to note that a recent ordinance of the Court of Cassation (4 December 2015, n. 25707) gives strength to our work and underlines its importance because it states that it is possible to consider the

OMI quotes, that are the quotations useful to identify the market value of real estate for each homogeneous geographical area, as notorious fact since they are present on the site 'Agenzia delle Entrate' that is freely available to everyone.

Then, with this ordinance, the Court of Cassation admits that an information present on the web (in the present case, the OMI quotes) can be considered as notorious fact but only if the website that contains that information presents a high level of reliability, such as www.agenziaentrate.gov.

As showed afterwards, with the 'Notoriety system' we filter the results that the crawler gives us and we just use the websites that presents a high level of reliability such as the governments websites.

3   Database Notoriety Knowledge Base

An important aspect highlighted in this work regards the usage of an automatic model for websites classification. In fact, the focus of our notoriety system is the database where all the websites are classified by 3 different polarities: information, disinformation, misinformation. Having an objective and truthful classification of the websites is the base to obtain a reliable notoriety score. That is why it is important, for at least 2 aspects, to develop an automatic system for the web-sites classification:

(1) It is very difficult to cluster manually all the web-sites;
(2) The clustering is made through a subjective choice (automatically) and not by an objective one (manually).

Considering the huge amount of websites linked on the web, the classification is a problem of Big Data Analysis, and it is almost impossible to cluster manually lots of data as huge as the Websites existing on the web. Another aspect we need to take into consideration is that the score associated to every website is a dynamic data because it depends on the content inside the website that is dynamic; let us imagine, for example, the news that every days are added on a page.

Furthermore, a manual clustering could be not properly objective, because the database classification is made by users and could be influenced by a non-objective choice.

For these reasons it was necessary to adopt a system that automatically tracks a map of all sites on the Web labelled with a notoriety score (information, disinformation, misinformation). In order to reach this goal, the main prerogative is to create a system that, according to certain parameters, is able to search for the websites and to assign them the score, because we firstly need to find a relationship among the websites to cluster them. In our work, we solved this problem creating an algorithm that, starting from a labelled set of websites with a known notoriety, looks automatically for other websites with similar content, such as news, and in this way it tracks a maps of websites where the nodes of the graphs are represented by the websites and the edges are the relationships among them. In this way the relationship that exists between two websites, for example a similar news, allows the transfer learning of the notoriety feature of information, misinformation and disinformation from a website already clusters to another one not clustered.

3.1 Related Work

An interesting work related to transfer learning method applied for sentiment analysis of texts is presented in [2]. The aim of that work consists in creating a method for the classification of a set of text such as comments and opinions expressed on a social network in 3 different sentiment polarity: positive, negative and neutral. Also in this case, if we imagine the huge amount of data that we have regarding reviews and comments on the web, it is essential to find a system able to classify them automatically with a low computational cost. Therefore, it is desirable to use a system that starting from a set of labelled documents of a domain-such as book reviews-is capable of classifying the polarity of new document sets (no-labelled) whatever kind of topic they deal with, for instance electrical appliance reviews: this approach is called cross-domain sentiment classification. Since classification is typically driven by terms, the most critical issue in cross-domain setting is language heterogeneity, whereas only the classes are usually the same (i.e. positive, negative or neutral). The methods proposed by the authors in their work represents the possibility to transfer a polarity knowledge learned for a labelled text (positive, negative, neutral) to another text that is unlabeled even if it belongs to a different domain by using a set of terms that can create a correlation among the texts. Then, if two or even more texts treating different topics contain a set of words that are correlated trough terms that reflects the same polarity, (positive, negative or neutral,) the polarity knowledge learned for labelled text can be transfered.

For example, it is possible to compare 2 terms that belong to two different domains such as boring (book reviews) and noisy (electrical appliance) through a correlated term, in the example bad, and in this way transfer the negative polarity learned. The methods can be easily extended to an entire texts taking and weighting opportunity the terms to create a map where the graphs are the terms and the edge the relationship among them, and then applying the transfer learning to cluster un-labelled texts with a low computational cost.

4 Transfer Learning in Our Notoriety System

In the same way as in the Pagliarani and co-authors work, the aim of our system is to use the transfer learning of the notoriety feature such as (information, misinformation, disinformation) learned from a labelled set of websites to classify a set of unlabeled websites. In our case the relationship among websites is like a similar content of news in a given set of websites. So, the basic idea is to transfer the notoriety feature of information, misinformation and disinformation learned for a classified website to another one through the relationship expressed by a common content, such as a similar news. In this way, we can think to represent the web like a map where the graphs of this map are represented by the websites with their content and the edge by the relationship among them. A similar content of news means that exists a relationship in terms of polarity between two websites. The reason why we choose this method is the fact that news are shared by different sources; for example a news regarding a scientific discover launched by more websites like scientific papers, or a political news launched by different newspapers. In the same way a disinformation news is launched by different disinformation tools. Then, we can imagine to build up a map where the graphs are the websites and the relationship among them the common news. In this way taking a news by a labelled new paper it is possible to search for others websites treating the same news and automatically apply the transfer learning with the method below. First of all we used a strong hypothesis in classifying the websites with the grade of information websites. For this reason a websites is considered an information websites if all the news in it are classified like information. If a websites has one news classified like disinformation the grade automatically applied to the website is disinformation. On the other hand if a website shared a news with a disinformation websites is automatically classified with the grade of disinformation. Also, for being classified like information websites all the websites linked to its content have to be information websites. In other words a website un-labelled take the lowest grade among the labelled websites with whom it shares similar news.

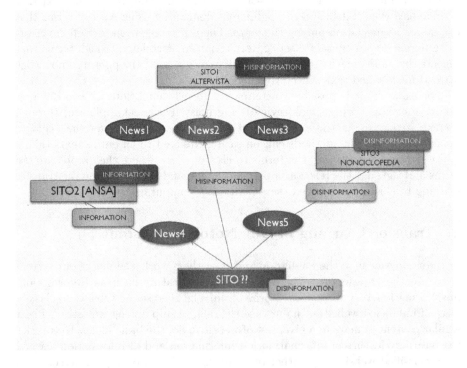

Fig. 1. Example of transfer learning applied to websites notoriety

5 Conclusion and Future Works

In this paper we debated about web data notoriety systems considering the only example retrieved in literature with the introduction of some improvements as Markovian model. There are still a lot of opportunities to improve the potentiality of our algorithm. An idea could be the addition to the text similarity [6] analyser also of a logic of sentiment analysis [7] for a better analysis of the content of a text and the study of the possibility to a user interaction to vote and improve the knowledge base. Furthermore, we would like to extend this idea to a Search Engine (e.g. using a plugin as Google Chrome Plugin) introducing a notoriety filter able to rank data considering their quality, with the target of making the web as a real repository of quality 'information'.

The authors would like to thank Luca Pala, for his support in the development of this work.

References

1. Santarcangelo, V., Buondonno, A., Oddo, G., Staffieri, F.P., Santarcangelo, N., Maragno, M., Trentinella, F.: WEB MISINFORMATION: A TEXT-MINING APPROACH FOR LEGAL ACCEPTED FACT., Choice and preference analysis for quality improvement and seminar on experimentation (2015)
2. Domeniconi, G., Moro, G., Pagliarani, A., Pasolini, R.: Markov chain based method for in-domain and cross-domain sentiment classification. In: 7th International Joint Conference on Knowledge Discovery, Knowledge Engineering and Knowledge Management (IC3K 2015) (2015)
3. Santarcangelo, V., Romano, A., et al.: Quality of web data: a statistical approach for forensics, Padova (2015)
4. Codice di Procedura Civile, REGIO DECRETO, n. 1443, 28 ottobre 1940
5. Balena, G.: Istituzioni di diritto processuale civile, vol. II. Cacucci Editore, Bari (2012)
6. Gomaa, W.H., Fahmy, A.A.: A survey of text similarity approaches. In: IJCA, vol. 68 (2013)
7. Santarcangelo, V., Oddo, G., Pilato, M., Valenti, F., Fornaro, C.: Social opinion mining: an approach for Italian language. In: SNAMS 2015 at FiCloud 2015 (2015)
8. Balena, G.: Istituzioni di diritto processuale civile, vol. I. Cacucci Editore, Bari (2009)
9. www.ilfiscoetasse.com/normativa-prassi
10. http://www.ilcaso.it/giurisprudenza/archivio/335m.htm
11. Giacalone, M.: Manuale di statistica giudiziaria. Bel-Ami Editore, Roma (2009)

Soft Computing and Applications

Rotation Clustering: A Consensus Clustering Approach to Cluster Gene Expression Data

Paola Galdi[(⊠)], Angela Serra, and Roberto Tagliaferri

NeuRoNe Lab, DISA-MIS, University of Salerno,
via Giovanni Paolo II 132, 84084 Fisciano, SA, Italy
{pgaldi,aserra,robtag}@unisa.it

Abstract. In this work we present Rotation clustering, a novel method for consensus clustering inspired by the classifier ensemble model Rotation Forest. We demonstrate the effectiveness of our method in a real world application, the identification of enriched gene sets in a TCGA dataset derived from a clinical study on Glioblastoma multiforme.

The proposed approach is compared with a classical clustering algorithm and with two other consensus methods. Our results show that this method has been effective in finding significant gene groups that show a common behaviour in terms of expression patterns.

Keywords: Clustering · Consensus clustering · Rotation Forest · Gene set enrichment · Pathways · Glioblastoma

1 Introduction

Technological advances lead to a huge increase in the number of technologies available to produce *omics* data such as gene expression, RNA expression (RNA), microRNA expression (miRNA), protein expression etc.

Nowadays, especially for microarray gene expression technology, the greatest effort no longer consists in the production of data, but in their interpretation to gain insights into biological mechanisms.

Microarray gene expression data allow to quantify the expression of thousands of genes across hundreds of samples under different conditions [4]. Here the main idea is that genes with similar expression patterns can have a relation in functional pathways or be part of a co-regulation system. This analysis is usually performed with exploratory techniques such as cluster analysis [7].

Clustering is an unsupervised technique used in data analysis to detect natural groups in data without making any assumption about their internal structure. There are two main reasons for choosing such an approach, that are (1) to try to confirm a hypothesis (e.g. about latent classes of objects) (2) to uncover previously unknown relationships among data points.

P. Galdi and A. Serra—Equal contribution.

© Springer International Publishing AG 2017
A. Petrosino et al. (Eds.): WILF 2016, LNAI 10147, pp. 229–238, 2017.
DOI: 10.1007/978-3-319-52962-2_20

Clustering has been successfully applied in bioinformatics in cancer subtyping [5,19,20,23] and in identifying groups of genes that show a similar behaviour [10,14,16].

However, one of the issues of full data-driven approaches (as opposed to hypothesis-driven approaches) is the risk of modelling noise or uninteresting properties w.r.t. the problem domain.

This is even more true for microarray gene expression data, since they are characterized by intrinsic noise, due to the high dynamics of the studied systems. Moreover, data have a background noise related to mechanical tools used to perform the analyses. For this reason, robust clustering techniques, such as consensus clustering, have been applied in gene expression clustering [9,17].

Consensus clustering has been devised as a method to deal with these issues by combining multiple clustering solutions in a new partition [22]. The main idea is to exploit the differences among base solutions to infer new information and discard results that might have been affected by the presence of noise or by intrinsic flaws of the chosen clustering algorithm. Being a consensus solution supported by the agreement of several base clusterings, not only it is more stable and robust to overfitting, but it also guarantees a higher degree of confidence in the results.

The ratio behind consensus clustering is analogous to that of classifier ensemble, where multiple "weak" classifiers are combined to obtain better performances [3]. Previous results have shown how diversity in the initial clustering solutions can lead to an improvement of the quality of the final consensus clustering [15]. For instance, in [8] the authors have investigated the relation between clustering accuracy (w.r.t. known classes of points) and the average normalized mutual information between pairs of partitions.

Since then many concepts have been borrowed from the classifier ensemble literature, such as subsampling or projections of the original data to promote diversity in the base partitions. Following this idea, here we propose a novel consensus clustering technique inspired by the Rotation Forest classifier [18] called Rotation Clustering.

2   Materials and Methods

2.1   Consensus Clustering

Consensus techniques are characterized by how the diversity among base solutions is generated and how the agreement among clusterings is quantified. In the following, two representative examples of consensus clustering methods are presented.

The approach by Monti et al. [17] generates multiple perturbed versions of the original data by computing random subsamples of the input matrix. Then a consensus (or co-association) matrix $M \in \mathcal{R}^{n \times n}$ (where n is the number of data objects) is built, where each entry $M(i,j)$ is the count of how many times items i and j were assigned to the same cluster across different partitions, normalized by the number of times that the two objects were present in the same subsample.

The approach by Bertoni and Valentini [1] uses random projections to build the perturbed versions of the input data. Random projections [2] are based on two main results. The first is the *Johnson-Lindenstrauss lemma* [13], that can be summarized as follows: if points in a vector space are projected onto a randomly selected subspace of suitably high dimension, then the distances between the points are approximately preserved. The second is the *Hecht-Nielsen lemma* [12], that states that "in a high-dimensional space, there exists a much larger number of almost orthogonal than orthogonal directions. Thus, vectors having random directions might be sufficiently close to orthogonal" [2]. Starting from these premises, the idea is to project the original d-dimensional data set to different k-dimensional ($k << d$) subspaces using random matrices whose elements are Gaussian distributed; a clustering is then executed on each subspace. Both rotation clustering and the approach based on random projections use the co-association matrix to measure the level of consensus. The final clustering can be obtained using the consensus matrix as a similarity matrix to be given as input to a hierarchical clustering algorithm.

In all the experiments the base solutions to be combined are generated with a single execution of the k-means algorithm with random initialization of the initial centroids.

2.2   Rotation Forest

Rotation Forest [18] is a classifier ensemble method that trains several base classifiers in the following way: starting from the training data matrix X (n samples $\times$ d features), the feature set is partitioned into K subsets, then from each of the submatrices X_i (for $i = 1, \ldots, K$) extracted from X selecting only one of the K subsets of features, a random subset of classes is eliminated and the remaining items are subsampled to obtain a sample size of, say, 75% of the original number of objects; PCA is applied on each one of the resulting matrices X_i and the computed coefficients are arranged in matrices C_i. All the components extracted by PCA are retained to not disrupt discriminatory information that might lie in the last components. The C_i matrices are then combined to build a sparse rotation matrix R which is arranged in such a way that its columns match the order of the original feature set; finally, the classifier is trained using XR as the training set.

This method has been proven to be able to outperform well established techniques in the classifier ensemble literature [18].

2.3   Rotation Clustering

Rotation Clustering follows the same steps of the Rotation Forest for building the input matrix except for what concerns the removal of a subset of classes since no prior information is available (see Algorithm 1 for the pseudo-code of the algorithm). Each of the input matrices for the base clustering is generated in the following way: first features are split randomly in subsets and for each subset a submatrix is built that contains only the features in the subset and a random subsample of the original data items; PCA is applied on each of the submatrices and a rotation matrix is built by combining the coefficients of all the principal components of each submatrix; finally the original data matrix is rotated using the obtained rotation matrix. Once the base clusterings are computed, a pair-wise co-association matrix is built counting how many times each pair of data objects was assigned to the same cluster across base solutions. The final consensus solution is the result of a hierarchical clustering algorithm applied to the co-association matrix.

Algorithm 1. Rotation clustering

Input:

- data matrix $X \in \mathcal{R}^{n \times d}$ where n is the no. of samples and d is the no. of features.
- the number M of base clusterings to generate
- the number K of feature subsets

for $i = 1, \ldots, M$ **do**
   Split the feature set in K subsets of size $\lfloor \frac{n}{K} \rfloor$
   for $j = 1, \ldots, K$ **do**
      $X_{i,j} \leftarrow$ select from X the $j - th$ subset of features j
      $X'_{i,j} \leftarrow$ select a subsample of items from matrix $X_{i,j}$
      $C_{i,j} \leftarrow$ apply PCA on matrix $X'_{i,j}$ ▷ PCA coefficients
   end for
   Arrange the $C_{i,j}$ in a rotation matrix R_i
   $X'_i \leftarrow X R_i$ ▷ Rotate input matrix
   $cls_i \leftarrow$ apply clustering algorithm on X'_i
end for
$final \leftarrow$ combine the cls_i in a consensus solution
return $final$

---

3   Experimental Setup and Validation

We compared the results obtained with the proposed method with those produced by the best of 100 runs of k-means with random initialization of the centroids and by two consensus techniques found in literature: the former builds a co-association matrix starting from subsamples of the original data; the latter is based on random projections (see Sect. 2.1).

The base solutions to be combined by each of the three consensus algorithms are generated by single runs of the k-means algorithm. The choice of using the same clustering algorithm across experiments is motivated by our dual goal of (1) comparing the performance of consensus methods versus a simple run of a clustering algorithm and (2) assessing the efficiency of our method compared to that of existing techniques. All clustering algorithms used Pearson's correlation coefficient as similarity measure between genes, since the goal is to identify gene sets that show a similar behaviour in terms of expression patterns.

The validation process considers two fundamental aspects of the gene clustering problems. Firstly, cluster analysis is a complex task and the results can change based on the number K of clusters selected as input parameter [11]. Therefore, each clustering algorithm was executed with different values of K (50, 100, 150, 200 and 250) and the performance was evaluated according to the index proposed in [19]: $ClVal = \frac{1}{4}\left(\frac{IC+1}{2} + 1 - \frac{EC+1}{2} + (1 - S) + CG\right)$. This index takes into account the average sample correlations inside each cluster (IC), the average sample correlation of the least similar objects for each pair of clusters (EC), the number of singletons (S) and the compression gain (CG). Its range is between 0 and 1, the higher the value, the better is the clustering result. Secondly, once the cluster analysis is performed on the gene expression dataset, the gene group needs to be interpreted from a biological perspective. Therefore, after selecting the best value of K, an enrichment set analysis was performed between all the clustering algorithms. The aim of the gene set enrichment is to evaluate microarray data at the level of gene sets [21]. Gene sets are defined based on *a priori* knowledge and, usually, they are gene sets with similar characteristics and behaviour. This is used to evaluate if the genes in a specific cluster have an homogeneous biological behaviour. This kind of analysis compares the clusters obtained with different methods by counting the number of gene sets enriched by each cluster and evaluating the gene ratio. For each pair of cluster and pathway, the gene ratio measures the proportion of genes in the cluster that are also included in the pathway. The best method is the one that, overall its clusters, has a higher number of enriched sets with the highest gene ratio. The analysis was performed by using the KEGG pathways from the Kyoto Encyclopedia of Genes and Genomes and the *compareCluster* function from the *clusterProfiler* R package. The last part of validation consists in verifying how many gene sets associated with specific diseases are identified by the used clustering algorithms. Known association between pathways and Glioblastoma were downloaded from the Comparative Toxicogenomics Database (CTD - http://ctdbase.org/) [6].

Dataset. Experiments have been performed on a real gene expression dataset, related to a Glioblastoma multiforme study. The dataset was accesses through the TCGA website (https://tcga-data.nci.nih.gov/tcga/ - Glioblastoma multiforme [GBM]) and publicly available gene expression data (level 3) were downloaded from 167 samples. As a further preprocessing step, features with low variance were eliminated and batch effect removal was performed with the *comBat* method in the *sva* R package).

4   Results and Discussion

We developed a new consensus clustering algorithm called Rotation Clustering. We applied this method to the problem of clustering gene expression data. Analyses were performed on a real gene expression dataset from TCGA repository with 2408 genes and 167 samples.

We clustered the genes with our method and we compared the results with other three classical clustering algorithms: the Kmeans clustering, the consensus clustering proposed by Monti et al. [17] and the random projection techniques proposed by Bertoni and Valentini [1].

Clustering algorithms give different results depending on the input parameters. To avoid this problem, the analyses were performed varying the number of clusters to be retrieved (K). The algorithm results were then evaluated according to the $ClVal$ measure proposed by [19]. Figure 1(a) shows that the best value for the parameter K is 50, in fact its score (light blue line) is the highest across the four clustering methods.

Moreover, in order to characterize the biological meaning of the obtained clusters, we performed an enrichment analysis with respect to the KEGG pathways. Figure 1(b) shows that the number of gene sets enriched by the clusters obtained from the Rotation clustering is the highest compared to the other methods when the parameter K assumes the values of 50, 100 and 150. On the other side, when the K value assumes values of 200 and 250, the best algorithm is the one proposed by Monti et al. To obtain a summary assessment of the algorithms we merged these 5 rankings with the Borda count method, implemented in the *TopKLists* R package. The final rating of the algorithm shows that the Rotation method is at the top, followed by the Random Projection approach, then the consensus clustering proposed by Monti et al. and in the last position the k-means algorithm. An important remark is that all the approaches based on consensus clustering give better results compared to the k-means. This justifies the higher computational effort that they require since they give more stable and less noisy results. We further investigated the obtained clusters with respect to the gene ratio. Here we report the results only for $K = 50$. Table 1 shows the quantiles of the distribution of the gene ratio for each clustering algorithm. The Rotation clustering, in addition to being the one with the highest count of gene sets, also reaches the highest value for the gene ratio. That means that there exists at least one cluster of genes in its solution that is completely included in a gene set.

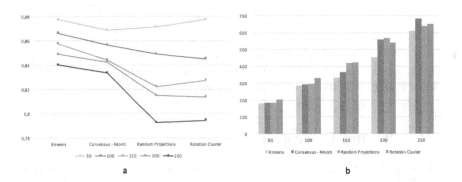

a

b

Fig. 1. (a) Evaluation or the $ClVal$ index when changing the parameter K. The figure shows the evaluation index for the parameter K for all the clustering algorithms and the different values of K. (b) Number of enriched KEGG sets. The figure shows the number of enriched KEGG sets for each algorithm and for each k. (Color figure online)

Table 1. Gene ratio

	k-means	Monti-consensus	Random projection	Rotation clustering
0%	0.05	0.04	0.04	0.03
25%	0.11	0.10	0.11	0.10
50%	0.14	0.14	0.17	0.14
75%	0.19	0.20	0.25	0.24
100%	0.42	1.00	0.57	1.00

Gene Ratio with K = 50

More details on the enriched pathway are shown in Fig. 2. The figure shows only the clusters that enrich at least one pathway. The clusters on the x-axis are ordered by gene-ratio. To have more insights into the biological meaning of the problem, we also checked if the pathways enriched from the four different

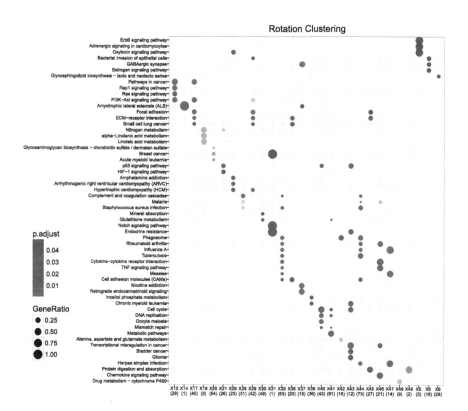

Fig. 2. Enrichment result for Rotation clustering (K = 50). The figure reports the enrichment analysis results for the clusters obtained with the Rotation clustering algorithm with $K = 50$ as input. Clusters with a significant p-value are reported on the x-axis, while pathways are reported on the y-axis. Colour and size of the bubbles indicate the p-value and gene ratio respectively. (Color figure online)

solutions are known to be, somehow, related to Glioblastoma. To do this, we downloaded a list of known pathways related to Glioblastoma from the CTD dataset [6]. This list contains 219 KEGG pathways. We then counted how many of these sets are in our solutions. Figure 3 shows the Venn diagram representing the number of pathways (related to glioblastoma) shared among the four algorithms. As we can see from the diagram, the Rotation Clustering solution is the one with the highest number of enriched pathways. Particularly, k-means is able to retrieve 46 pathways, the Monti's consensus method retrieves 48 pathways, the Random Projection method retrieves 49 of them and the Rotation clustering enriches 51 clusters. Most of the pathways of each of the three consensus methodologies are shared with the ones of the k-means. This can be due to the fact that the consensus algorithms are all based on the k-means, but the number of enriched sets increases when the variability imposed by the consensus methods to the data grows. In fact, while the Monti's method induces a relatively smaller perturbation by subsampling the data but preserving the original distribution of points in space, random projection and rotation also perturb the relative distance among points, probably allowing for previously unobserved relationships to emerge.

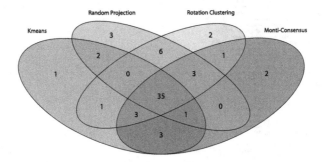

Fig. 3. Number of KEGG gene sets associated to Glioblastoma that give a significant p-value for each clustering technique.

5 Conclusion

In this work we presented a new consensus clustering method called Rotation clustering, that is based on the same idea of the Rotation Forest classifier. We successfully applied this method to a real gene expression clustering problem, related to a clinical study about patients affected by Glioblastoma. We validated our results with respect to both the structure of the clusters and the prior biological knowledge. We also compared the new method with a classical clustering and other consensus-based methodologies. We can conclude that this is an effective method for clustering noisy data because it gives stable and reliable results that resemble the known biological information. The reasons of the efficacy of this approach may be found in how diverse are the base clusterings that are combined in the final consensus solution, and this aspect will be further investigated in future work. However, diversity alone cannot guarantee the quality of the results, especially if we try to merge poor or incompatible clusterings. One possible way to overcome this problem might be merging only partitions that are sufficiently similar, thus adding a meta-clustering step to the consensus framework.

Acknowledgments. We would like to thank Teresa Savino and Luca Puglia for the helpful discussions.

References

1. Bertoni, A., Valentini, G.: Random projections for assessing gene expression cluster stability. In: Proceedings. 2005 IEEE International Joint Conference on Neural Networks, vol. 1, pp. 149–154. IEEE (2005)
2. Bingham, E., Mannila, H.: Random projection in dimensionality reduction: applications to image and text data. In: Proceedings of the Seventh ACM SIGKDD International Conference on Knowledge Discovery and Data Mining, pp. 245–250. ACM (2001)
3. Brown, G.: Ensemble learning. In: Sammut, C., Webb, G.I. (eds.) Encyclopedia of Machine Learning, pp. 312–320. Springer, Heidelberg (2011)
4. Brown, P.O., Botstein, D.: Exploring the new world of the genome with DNA microarrays. Nat. Genet. **21**, 33–37 (1999). http://www.nature.com/doifinder/10.1038/4462
5. Chang, H.Y., Nuyten, D.S., Sneddon, J.B., Hastie, T., Tibshirani, R., Sørlie, T., Dai, H., He, Y.D., van't Veer, L.J., Bartelink, H., et al.: Robustness, scalability, and integration of a wound-response gene expression signature in predicting breast cancer survival. Proc. Nat. Acad. Sci. US Am. **102**(10), 3738–3743 (2005)
6. Davis, A.P., King, B.L., Mockus, S., Murphy, C.G., Saraceni-Richards, C., Rosenstein, M., Wiegers, T., Mattingly, C.J.: The comparative toxicogenomics database: update 2011. Nucleic Acids Res. **39**(suppl 1), D1067–D1072 (2011)
7. D'haeseleer, P.: How does gene expression clustering work? Nat. Biotechnol. **23**(12), 1499–1501 (2005). http://www.nature.com/doifinder/10.1038/nbt1205-1499
8. Fern, X.Z., Brodley, C.E.: Random projection for high dimensional data clustering: a cluster ensemble approach. ICML **3**, 186–193 (2003)
9. Galdi, P., Napolitano, F., Tagliaferri, R.: Consensus clustering in gene expression. In: Serio, C., Liò, P., Nonis, A., Tagliaferri, R. (eds.) CIBB 2014. LNCS, vol. 8623, pp. 57–67. Springer, Heidelberg (2015). doi:10.1007/978-3-319-24462-4_5
10. Gautier, E.L., Shay, T., Miller, J., Greter, M., Jakubzick, C., Ivanov, S., Helft, J., Chow, A., Elpek, K.G., Gordonov, S., et al.: Gene-expression profiles and transcriptional regulatory pathways that underlie the identity and diversity of mouse tissue macrophages. Nat. Immunol. **13**(11), 1118–1128 (2012)
11. Handl, J., Knowles, J., Kell, D.B.: Computational cluster validation in postgenomic data analysis. Bioinformatics **21**(15), 3201–3212 (2005). (Oxford, England). http://www.ncbi.nlm.nih.gov/pubmed/15914541
12. Hecht-Nielsen, R.: Context vectors: general purpose approximate meaning representations self-organized from raw data. In: Computational Intelligence: Imitating Life, pp. 43–56 (1994)

13. Johnson, W.B., Lindenstrauss, J.: Extensions of lipschitz mappings into a hilbert space. Contemp. Math. **26**(189–206), 1 (1984)

14. Kimes, P.K., Cabanski, C.R., Wilkerson, M.D., Zhao, N., Johnson, A.R., Perou, C.M., Makowski, L., Maher, C.A., Liu, Y., Marron, J.S., et al.: SigFuge: single gene clustering of RNA-seq reveals differential isoform usage among cancer samples. Nucleic Acids Res. **42**(14), e113–e113 (2014)

15. Kuncheva, L.I., Hadjitodorov, S.T.: Using diversity in cluster ensembles. In: 2004 IEEE International Conference on Systems, Man and Cybernetics, vol. 2, pp. 1214–1219. IEEE (2004)

16. Lam, Y.K., Tsang, P.W.: eXploratory K-Means: a new simple and efficient algorithm for gene clustering. Appl. Soft Comput. **12**(3), 1149–1157 (2012)

17. Monti, S., Tamayo, P., Mesirov, J., Golub, T.: Consensus clustering: a resampling-based method for class discovery and visualization of gene expression microarray data. Mach. Learn. **52**(1/2), 91–118 (2003). http://link.springer.com/10.1023/A:1023949509487

18. Rodriguez, J., Kuncheva, L., Alonso, C.: Rotation forest: a new classifier ensemble method. IEEE Trans. Pattern Anal. Mach. Intell. **28**(10), 1619–1630 (2006). http://ieeexplore.ieee.org/document/1677518/

19. Serra, A., Fratello, M., Fortino, V., Raiconi, G., Tagliaferri, R., Greco, D.: MVDA: a multi-view genomic data integration methodology. BMC Bioinform. **16**(1), 1 (2015)

20. Shen, R., Mo, Q., Schultz, N., Seshan, V.E., Olshen, A.B., Huse, J., Ladanyi, M., Sander, C.: Integrative subtype discovery in glioblastoma using icluster. PLoS ONE **7**(4), e35236 (2012)

21. Subramanian, A., Tamayo, P., Mootha, V.K., Mukherjee, S., Ebert, B.L., Gillette, M.A., Paulovich, A., Pomeroy, S.L., Golub, T.R., Lander, E.S., Mesirov, J.P.: Gene set enrichment analysis: a knowledge-based approach for interpreting genome-wide expression profiles. Proc. Nat. Acad. Sci. US Am. **102**(43), 15545–15550 (2005). http://www.ncbi.nlm.nih.gov/pubmed/16199517, http://www.pubmedcentral.nih.gov/articlerender.fcgi?artid=PMC1239896

22. Vega-Pons, S., Ruiz-Shulcloper, J.: A survey of clustering ensemble algorithms. Int. J. Pattern Recogn. Artif. Intell. **25**(03), 337–372 (2011). http://www.worldscientific.com/doi/abs/10.1142/S0218001411008683

23. Wang, B., Mezlini, A.M., Demir, F., Fiume, M., Tu, Z., Brudno, M., Haibe-Kains, B., Goldenberg, A.: Similarity network fusion for aggregating data types on a genomic scale. Nat. Methods **11**(3), 333–337 (2014). http://www.nature.com/doifinder/10.1038/nmeth.2810

Efficient Data Mining Analysis of Genomics and Clinical Data for Pharmacogenomics Applications

Giuseppe Agapito, Pietro Hiram Guzzi, and Mario Cannataro[(✉)]

Department of Medical and Surgical Science,
University Magna Graecia of Catanzaro, 88100 Catanzaro, Italy
{agapito,hguzzi,cannataro}@unicz.it

Abstract. The identification of biomarkers for the estimation of cancer patients' survival is a crucial problem in oncology. The Affymetrix DMET microarray platform allows to determine the ADME gene variants of a patient and to correlate them with drug-dependent adverse events. We present a bioinformatics tool devoted to the discovery of gene variants correlated to a different response of cancer patients to drugs and able to compute the overall survival (OS) and progression-free survival (PFS) of cancer patients. The tool is based on the integration of DMET-Miner and OSAnalyzer. DMET-Miner is a data mining tool able to extract Association Rules from DMET datasets and OSAnalyzer is a software tool able to perform an automatic analysis of DMET data enriched with survival events. After presenting DMET-Miner and OSAnalyzer, we discuss a case study to highlight the usefulness of the pipeline constituted by DMET-Miner and OSAnalyzer when analyzing a large cohort of patients.

Keywords: Genotyping microarrays · ADME genes · Pharmacogenomics · Overall survival · Progression-free survival

1 Introduction

The discovery and identification of biomarkers of a particular disease state is a difficult and laborious task, but in the same time is a very important research topic in oncology field [1,2]. In fact, the identification of biomarkers for the estimation of cancer patients' survival is a crucial problem in modern oncology [3,4]. An aid to detect new biomarkers is represented by high-throughput technologies such as microarrays [5]. DNA microarrays are used to detect the presence of single nucleotide polymorphisms SNPs, to measure the expression levels of several genes simultaneously or to genotype multiple regions of a genome, producing a vast amount of data per single experiment. Doctors, researcher and clinicians can use the information produced by microarray, to evaluate in a more comprehensive way, the efficacy of the therapy in a cohort of patients, to draw more detailed conclusion on the benefits of the treatment, or to modify the drug dosage to reduce the side effects by following the genomic features of each patient and

© Springer International Publishing AG 2017
A. Petrosino et al. (Eds.): WILF 2016, LNAI 10147, pp. 239–248, 2017.
DOI: 10.1007/978-3-319-52962-2_21

improving the efficacy of the treatment. To pursue this goal, researchers need efficient algorithms and software platforms to analyze this enormous amount of data in the shortest time possible, since the manual analysis is infeasible. The Affymetrix DMET (Drug Metabolizing Enzymes and Transporters) microarray platform allows the possibility to determine the ADME (absorption, distribution, metabolism, and excretion) gene variants of a patient and to correlate them with drug-dependent adverse events. Annotating DMET data with the clinical information of each patient enrolled in the study would allow to researchers and clinicians to correlate the presence/absence of a particular SNP to the drug response, and to the different survival times of patients [6]. To enable the analysis of the DMET datasets annotated with clinical information of each enrolled patient, we present a new pipeline of software tools able to: (i) to discover gene variants that are correlated to a different response of cancer patients to drugs and (ii) able to compute the overall survival (OS) and progression-free survival (PFS) of cancer patients and (iii) able to evaluate their association with ADME gene variants. The software pipeline is based on the integration of DMET-Miner [7] and OSAnalyzer [8].

DMET-Miner is a software tool with data mining capabilities and correlates the presence of a set of allelic variants with the conditions of patient's samples by exploiting association rules. To face the high number of frequent itemsets generated when considering large clinical studies based on DMET data, DMET-Miner uses an efficient data structure based on FP-Growth algorithm [9] and implements an optimized search strategy that reduces the search space and the execution time. OSAnalyzer is a software tool able to compute the overall survival analysis (OS) and the progression-free survival (PFS) from a whole DMET dataset annotated by adding clinical information (temporal data).

In this work we focus on the correlation of the mined association rules from DMET datasets with the genomic characteristics and related OS and PFS of patients analyzed by DMET microarray technology. We provide an automatic analysis pipeline to evaluate the relation among the association rules mined by DMET-Miner and the overall survival analysis (OS) and the progression-free survival (PFS) computed by OSAnalyzer, from a whole DMET dataset produced by using the Affymetrix DMET PLUS platform and successively extended by adding temporal data. The main aim of this pipelines is to automatize and simplify the work of biomedical researchers and clinical researchers when interpreting and analyzing the data of observational studies. The rest of the paper is organized as follows. Section 2 describe genotyping data, Sect. 3 presents DMET-Miner, Sect. 4 presents the OSAnalyzer Sect. 5 discusses a case study and, finally, Sect. 6 concludes the paper.

2   Genotyping Data

The Affymetrix DMET platform is used in case-control association study, when a typical study involves the following steps:

- Sample collection: in this phase biological samples are collected and treated to perform microarray experiments; the Affymetrix DMET chip allows the investigation of 1936 different nucleotides that present possible variants as stored in SNP databases, each one representing a portion of the genome having a role in drug metabolism;
- DMET microarray analysis, this step produces the first raw microarray data (.CEL intensity data);
- DMET raw data preprocessing: CEL files are preprocessed by the DMET-Console software to produce a unique output file (usually arranged in a tabular format); the DMET Console software produces a table containing, for each nucleotide and for each sample, the detected SNP or a NoCall value (where NoCall means that the platform has not been able to detect the nucleotide).
- Data mining or Statistical analysis: in this phase data mining algorithms are used to analyse SNP data producing knowledge models (e.g. specific association rules able to discriminate if a patient belong to case or control group) or statistical models that help to find significant SNPs.

DMET-Console outcomes are arranged as a large $N \times M$ table of single nucleotide polymorphisms (SNPs), where N is the number of probes ($N = 1936$ for the current DMET chips) and M is the number of samples (patients), as depicted in Table 1.

Table 1. A simple SNP microarray dataset. S and P respectively refer to sample and probe identifiers.

Probes	Samples				
	s_1	s_2	s_3	$\cdots$	s_N
P_1	G/A	A/G	A/G	$\cdots$	T/T
P_2	G/A	A/G	A/G	$\cdots$	T/C
$\vdots$	$\vdots$	$\vdots$	$\vdots$	$\cdots$	$\vdots$
P_M	G/A	A/G	A/G	$\cdots$	T/C

In Table 1 the first column contains the probes identifier, while the other columns contain the identifier of the samples. A generic element of the table contains the *i-th* identified SNP in the *j-th* sample, represented as X/Y, where $X,Y \in \{A,T,C,G,-\}$ or by the symbol *NoCall*.

To extract relevant rules from the input SNP dataset by using DMET-Miner, it is necessary to pre-process the input dataset, to make it suitable to the association rules mining task. Input SNP dataset (e.g. Table 1) is initially loaded, filtered by iteratively applying the Fisher's Test Filtering (FTF) that removes the pointless probes and finally, the filtered dataset is transposed obtaining a $M \times N$ matrix of alleles called T. Thus, each row of the table T represents a transaction, where all SNPs detected in samples, on the various probes, are the items of the transactions. Table 2 shows the transformed dataset T for the input dataset of Table 1.

Table 2. Transposed DMET microarray dataset. S and P respectively refer to sample and probe identifiers. Some probes may be deleted after the Fisher's Test filtering.

Samples	Probes			
	P_1	P_2	$\cdots$	P_M
S_1	G/A	G/A	$\cdots$	G/A
S_2	A/G	A/G	$\cdots$	A/G
S_3	A/G	A/G	$\cdots$	A/G
$\vdots$	$\vdots$	$\vdots$	$\vdots$	$\vdots$
S_N	T/T	T/C	$\cdots$	T/C

To make the automatic computation of OS and PFS possible for OSAnalyzer, it was necessary to extend the output produced by the DMET platform by adding clinical information after the header row. For each sample (patient), the supplementary information are, temporal information i.e., the period between the start and end of the clinical observation along with a status, indicating whether or not each patient had a clinical event of interest, e.g., death or metastasis. The extended dataset, hereinafter called OS-dataset, is arranged as conveyed in Table 3.

Table 3. A simple DMET-OS dataSet. S and P respectively refer to sample and probe identifiers. OS refers to the collected time for each sample, Status-OS is a boolean variable where 1 means that the event was observed whereas, 0 refers to censored data. PFS is a measure of the activity of a treatment on a disease. Status-PfS is a boolean variable where 1 means that the event was observed whereas, 0 refers to censored data. PFS can only be measured in patients in which a tumor is present. Thus, we need to add the metastatic variable. Metastatic is a boolean variable, where 1 means that the $i - th$ sample presents cancer, whereas 0 the absence of metastasis.

Probes and OS-Data	Samples				
	S_1	S_2	S_3	$\cdots$	S_N
OS	**20.6**	**19.7**	**12.2**	$\cdots$	**21.3**
Status-OS	1	0	0	$\cdots$	1
PFS	**16.6**	**4.7**	**3.8**	$\cdots$	**27.3**
Status-PFS	1	0	1	$\cdots$	1
Metastatic	1	1	0	$\cdots$	1
P_1	A/A	G/A	G/A	$\cdots$	G/G
P_2	G/A	A/G	A/G	$\cdots$	NoCall
$\vdots$	$\vdots$	$\vdots$	$\vdots$	$\cdots$	$\vdots$
P_M	G/G	C/G	C/C	$\cdots$	C/C

3 DMET-Miner

DMET-Miner is a software tool developed in Java to mine association rules from DMET datasets. DMET-Miner is available under Creative Commons License, is freely downloadable for academic and not-for-profit institutions at: https:// sites.google.com/site/dmetminer/get-software. DMET-Miner automatically filters useless rows by iteratively applying the Fisher's Test Filter allowing to reduce the search space, improving the performance. The efficacy of using FTF is demonstrated in [10]. To start DMET data analysis with DMET-Miner, as first step the user has to load the input dataset by using the "Frequent Item Sets extraction" command from the Menu Bar DataMining (see Fig. 1a and 1b). Then, the user has to select the patients to assign the class (e.g. not respond class) as conveyed in Fig. 1c and can start the Association Rules Mining analysis by clicking on Start Preprocessing button on the bottom of the window (see Fig. 1c). The DMET input table is then filtered and transposed after user entered the FTT significance value (see Fig. 1c). Finally, DMET-Miner asks to the user to enter the value of minimum support and confidence with which to extract association rules from the data and shows the results in a new window (see Fig. 1e). Clicking with the right button of the mouse on the rule panel it is possible save the rule on disk by using the pop-up menu "Save" as conveyed from Fig. 1e.

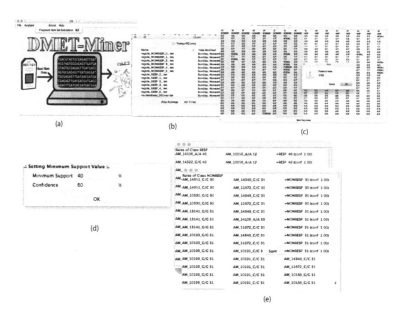

Fig. 1. DMET-Miner GUI. (a) Input loading menu; (b) File system navigation; (c) FTT significance value setting; Minimum Support and Confidence values setting (d); and (e) Association rules visualizer.

Further information about interesting SNPs (i.e. those contained in the extracted association rules) are visualized to the user by providing SNP annotations provided by the Affymetrix libraries embedded in DMET-Miner, and by providing links to dbSNP and PharmaGKB databases respectively.

4   OSAnalyzer

OSAnalyzer is a platform-independent application and it is developed by using the Java 6 programming language, making it available for Windows, Linux, and MacOSX operating systems. OSAnalyzer provides a simple and essential graphical user interface (GUI) allowing the users easy access to the tool's functionalities. OSAnalyzer is very simple to use, the analysis of a whole OS-dataset requires only some clicks with the mouse, which are: (i) load the input OS-dataset by using the command "File" located in the menu bar as shown in Fig. 2a; as result, OSAnalyzer shows to the user a file system navigation windows; (ii) browse the file system, to select the file to analyze (see Fig. 2b); and (iii) wait that OSAnalyzer finishes the computation and shows the results sorted in descending order according to the statistical significance of the log-rank test, and conveyed in two separate navigation panel results, one for OS, and one for PFS.

The navigation panel results are designed to simplify the analysis of the results. In fact, each meaningful probe is related with three curves obtained by comparing the detected alleles in pairs, which present different values of statistic

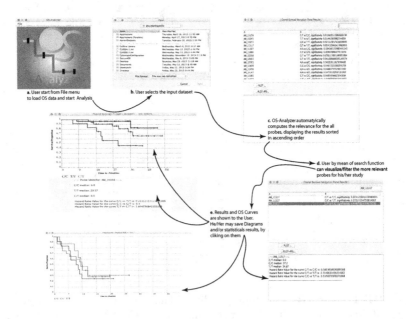

Fig. 2. OSAnalyzer GUI. (a) Input loading menu; (b) File system navigation; (c) OS and PFS Navigation Panel Results; and (d) OS and PFS curves visualizer.

relevance. Thus, using the search function (locate in the top right corner of the window), it is possible to see the value of each curve (log-rank test) obtained comparing the three groups among them, Fig. 2c. Selecting one probe per time and using the buttons "plot-os" or "plot-pfs" it is possible to see OS or PFS curves as depicted in Fig. 2d. Moreover, at the bottom of the window, OSAnalyzer provides a quick summary of the most important measures for survival curves comparison such medians and the hazard ratio for each curve, as shown in see Fig. 2d. The user can save the results and the curves that he/she considers relevant on file, by clicking the right button of the mouse on the chart. Finally, OSAnalyzer can automatically annotate the relevant SNPs related to the overall survival by using annotation libraries provided by Affimetrix, or retrieving further information from dbSNP and PharmaGKB databases. OS-Analyzer is distributed under Creative Commons License, is freely downloadable for academic and not-for-profit institutions at: https://sites.google.com/site/overallsurvivalanalyzer/. OSAnalyzer avoids wasting time on the manual analysis of all probes in order to figure out which probes are relevant from an overall survival or PFS point of view. Users who are exploiting this feature of OSAnalyzer can automatically analyze a whole DMET microarray dataset in one go without further effort, as opposed to other available tools where the user is forced to manually organize the analysis of the whole dataset each time, increasing the possibility of introducing mistakes. This way, OSAnalyzer allows the users to focus only on the analysis of the results.

5 Case Study

In this section, we will discuss how the mined association rule by DMET-Miner can be used in the context of overall survival, and how to evaluate how SNPs affect the overall survival of the two classes. We generated a synthetic DMET dataset composed of 1936 rows (probes) and 100 columns (samples), by adding OS and PFS data randomly generated, to obtain the OS-dataset. By using

Table 4. Two associative rules mined by DMET-Miner from the synthetic dataset, that correlate the presence of certain SNPs with a negative response to the drug.

Rules of class NONRESP
IF(AM_14351_C/C:30 **AND** AM_ 14348_ C/C:35) = NONRESP: 30 (conf: 1.00)
IF(AM_14351_C/C:30 **AND** AM_11872_C/C:35) =NONRESP: 30 (conf: 1.00)

Table 5. Two associative rules mined by DMET-Miner from the synthetic dataset, that correlate the presence of certain SNPs with a positive response to the drug.

Rules of class RESP
IF(AM_14106_A/A:46 **AND** AM_10053_A/A:52)=RESP: 46 (conf: 1.00)
IF(AM_14322_G/G:48 **AND** AM_10053_A/A:52)=RESP: 48 (conf: 1.00)

DMET-Miner following the methodology described in the previous section, we mined the following association rules (see Tables 4 and 5).

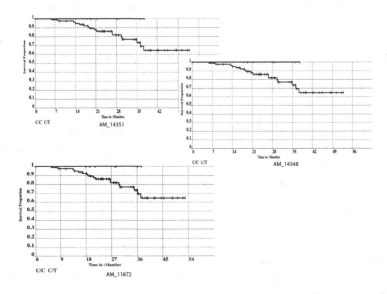

Fig. 3. The overall survival curves related to the SNP detected in the patients belong to the NORESP class.

Association rules can be used to assess how to change the survival of patients related to the presence or the absence of these particular SNPs. In Fig. 3 are reported the survival curves of the SNPs related to the NONRESP class, while in Fig. 4 are reported the survival curves of the SNPs related to the RESP class.

Analyzing the overall survival of the two classes, it is worthy to note that the subject belonging to the RESP class present a slightly better overall survival than subject belonging to NORESP class. Subject belonging to NORESP class show an average life expectancy less than 60 percent, whereas the subject belonging to RESP class present a better life expectancy. This results highlight the advantage of using association rules to evaluate which are the factors (SNPs) that influence positively or negatively the overall survival.

6   Conclusion

We presented a software pipeline composed by DMET-Miner and OSAnalyzer for the automatic generation of survival curves starting from association rules mined from DMET datasets annotated with temporal data. The automatic analysis of a whole microarray avoids to waste time in manual configuration and preprocessing step reducing the probability to make mistakes, since these steps have to be done manually without the tools. The software pipeline overcomes the problem related

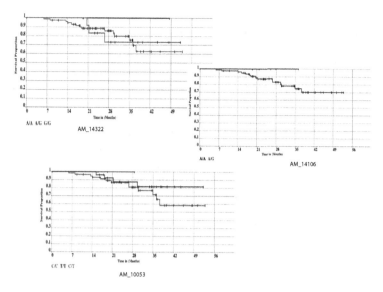

Fig. 4. The overall survival curves related to the SNP detected in the patients belong to the RESP class.

with data filtering and preprocessing by introducing an automatic step of data filtering based on the FTF in the DMET-Miner tool, while the analysis of a whole OS-dataset avoids that user has to manually investigate each probe of the dataset to discover which rows are significant from a survival or PFS point of view.

Acknowledgments. This work has been partially funded by the following research project funded by the Italian Ministry of Education and Research (MIUR): "BA2Know-Business Analytics to Know" (PON03PE_00001_1).

References

1. Abeel, T., Helleputte, T., Van de Peer, Y., Dupont, P., Saeys, Y.: Robust biomarker identification for cancer diagnosis with ensemble feature selection methods. Bioinformatics **26**(3), 392–398 (2010)
2. Hu, S., Arellano, M., Boontheung, P., Wang, J., Zhou, H., Jiang, J., Elashoff, D., Wei, R., Loo, J.A., Wong, D.T.: Salivary proteomics for oral cancer biomarker discovery. Clin. Cancer Res. **14**(19), 6246–6252 (2008)
3. Phillips, M., Altorki, N., Austin, J.H., Cameron, R.B., Cataneo, R.N., Greenberg, J., Kloss, R., Maxfield, R.A., Munawar, M.I., Pass, H.I., Rashid, A.: Prediction of lung cancer using volatile biomarkers in breath. Cancer Biomark. **3**(2), 95–109 (2007)
4. Hoyt, K., Castaneda, B., Zhang, M., Nigwekar, P., di Sant'Agnese, P.A., Joseph, J.V., Strang, J., Rubens, D.J., Parker, K.J.: Tissue elasticity properties as biomarkers for prostate cancer. Cancer Biomark. **4**(4,5), 213–225 (2008)

5. Chu, W., Ghahramani, Z., Falciani, F., Wild, D.L.: Biomarker discovery in microarray gene expression data with Gaussian processes. Bioinformatics **21**(16), 3385–3393 (2005)
6. Arbitrio, M., Di Martino, M.T., Scionti, F., Agapito, G., Guzzi, P.H., Cannataro, M., Tassone, P., Tagliaferri, P.: DMET TM (Drug Metabolism Enzymes and Transporters): a pharmacogenomic platform for precision medicine. Oncotarget **7**(33), 54028–54050 (2016)
7. Agapito, G., Guzzi, P.H., Cannataro, M.: DMET-miner: efficient discovery of association rules from pharmacogenomic data. J. Biomed. Inform. **56**, 273–283 (2015)
8. Agapito, G., Botta, C., Guzzi, P.H., Arbitrio, M., Di Martino, M.T., Tassone, P., Tagliaferri, P., Cannataro, M.: OSAnalyzer: a bioinformatics tool for the analysis of gene polymorphisms enriched with clinical outcomes. Microarrays **5**(4), 24 (2016)
9. Borgelt, C.: An implementation of the FP-growth algorithm. In: Proceedings of the 1st International Workshop on Open Source Data Mining: Frequent Pattern Mining Implementations, pp. 1–5. ACM, August 2005
10. Guzzi, P.H., Agapito, G., Di Martino, M.T., Arbitrio, M., Tassone, P., Tagliaferri, P., Cannataro, M.: DMET-analyzer: automatic analysis of affymetrix DMET data. BMC Bioinform. **13**(1), 258 (2012)

Extraction of High Level Visual Features for the Automatic Recognition of UTIs

Paolo Andreini[1], Simone Bonechi[1(✉)], Monica Bianchini[1], Andrea Baghini[1],
Giovanni Bianchi[1], Francesco Guerri[1], Angelo Galano[2], Alessandro Mecocci[1],
and Guendalina Vaggelli[2]

[1] Department of Information Engineering and Mathematics,
University of Siena, Via Roma 56, Siena, Italy
simo_bone@alice.it
[2] Department of Medical Biotechnologies,
University of Siena, Strada delle Scotte 4, Siena, Italy
http://www.diism.unisi.it,
http://www.dbm.unisi.it

Abstract. Urinary Tract Infections (UTIs) are a severe public health problem, accounting for more than eight million visits to health care providers each year. High recurrence rates and increasing antimicrobial resistance among uropathogens threaten to greatly increase the economic burden of these infections. Normally, UTIs are diagnosed by traditional methods, based on cultivation of bacteria on Petri dishes, followed by a visual evaluation by human experts. The need of achieving faster and more accurate results, in order to set a targeted and sudden therapy, motivates the design of an automatic solution in place of the standard procedure. In this paper, we propose an algorithm that combines a "bag–of–words" approach with machine learning techniques to recognize infected plates and provide the automatic classification of the bacterial species. Preliminary experimental results are promising and motivate the introduction of a visual word dictionary with respect to using low level visual features.

Keywords: Color image processing · Clustering techniques · Bag–of–words · Artificial neural networks · Support vector machines · Urinoculture screening

1 Introduction

Urinoculture represents a screening test in the case of hospitalized patients and pregnant women. For women, the lifetime risk of having a UTI is greater than 50%. Pregnant women seem no more prone to UTIs than other women. However, when the UTI occurs during pregnancy, it is more likely that the infection extends to the kidneys, giving rise to more serious pathologies. For this reason, health care providers routinely screen pregnant women for UTIs during the first 3 months of pregnancy. On the other hand, nosocomial urinary tract infections

© Springer International Publishing AG 2017
A. Petrosino et al. (Eds.): WILF 2016, LNAI 10147, pp. 249–259, 2017.
DOI: 10.1007/978-3-319-52962-2_22

account for up to 40% of all hospital–acquired infections and, most importantly, nosocomial pathogens causing UTIs tend to have a higher antibiotic resistance than simple UTIs [1].

From the operational point on view, for the urinoculture test, the urine sample is seeded on a Petri plate that holds a culture substrate, used to artificially recreate the environment required for the bacterial growth. There exist many different culture media which allow to perform different kinds of analysis, from isolating specific types of bacteria, to promoting a wide range of microbial growth. The seeding procedure (streaking) consists on spreading the urine sample over the whole plate and can be performed both manually or automatically, with an *ad hoc* device. Then, the plate is incubated in a controlled environment for a fixed period of time (16–24 h). After the incubation phase, each plate is visually examined by a microbiologist with the aim of recognizing the possible presence of bacterial colonies and eventually their species and number, adding some more time to the medical report emission. This common situation significantly departs from the requirement to have results in quick time, to set a targeted therapy, avoiding the use of broad–spectrum antibiotics and improving the patient management.

In recent years, significant improvements in biology and medicine applications and decision support systems [2] have been obtained by using hybrid approaches, based on the combination of advanced image processing techniques [3,4], and artificial intelligence methods [5–8]. In fact, the development of automated tools for results assessment (screening systems) has attracted increasing research interest during the last decade, due to their higher repeatability, accuracy, reduced staff time, and lower costs [9]. In particular, in [10–14], different methods have been proposed to automate the uriculture screening, based on image processing and machine learning techniques. Actually, in [10], after segmentation and background subtraction, the classification of the infection type is performed using support vector machines (SVMs) and multilayer perceptron (MLPs), trained with low level visual features, such as the Cie–Lab color components, and the average colony dimension.

In this paper, we propose an algorithm that combines a "bag–of–words" approach with machine learning techniques to recognize infected plates and provide the automatic classification of the bacterial species. A dataset of 753 images has been collected in partnership with the Microbiology and Virology Laboratory of the Careggi Hospital (Florence). The images represent Petri plates that have been automatically seeded on a chromogenic substrate (Chromagar Orientation) by the Copan WASPLab specimen processor. From the dataset, a visual word dictionary based on shape and color features has been extracted. Preliminary experimental results are promising and motivate the introduction of "visual words" with respect to using low level features (such as color and texture).

The paper is organized as follows. In the next section, we briefly describe how the image dataset has been collected, before defining the codebook generation procedure. In Sects. 3 and 4, the automatic infected plate recognition and infection classification methods are, respectively, presented, also reporting experimental results. Finally, Sect. 5 collects some conclusions.

2 Dataset Collection and Codebook Generation

The Microbiology and Virology Laboratory of the Careggi Hospital in Florence is the Tuscany reference center for microbiological and virological tests, with more than 250 urinoculture tests performed per day. In the MV–Lab, urine samples are automatically seeded on a chromogenic substrate (Chromagar Orientation) by the Copan WASPLab specimen processor. A chromogenic culture ground exploits the presence of specific enzymes, which are common in bacterial cells, to produce different colors, depending on the bacterial species; in particular, Chromagar Orientation allows to distinguish between the following pathogens:

– Escherichia Coli – produces dark pink to reddish colonies;
– Enterococcus – produces turquoise blue colonies;
– Proteus – produces a brown halo;
– Klebsiella, Enterobacater, Serratia, Citrobacter (KESC) – produces metallic blue colonies;
– Staphylococcus Aureus – produces golden opaque small colonies;
– Staphylococcus Saprophitycus – produces pink opaque small colonies;
– Candida Albicans – produces colorless colonies;
– Streptococcus Agalactiae – produces light blue colonies;
– Pseudomanas Aeuroginos – produces translucent, cream to blue colonies.

A dataset of 753 images of Petri dishes has been collected, gained via the WASPLab, which are first segmented, in order to detect and remove the culture ground.

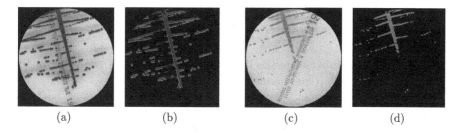

(a)   (b)   (c)   (d)

Fig. 1. Results of the background removal procedure: in (a) and (c) the original images; in (b) and (d) the result of the background subtraction.

After the segmentation phase, a bag–of–words approach is then applied to generate a visual codebook, composed by color and shape descriptors. Bag–of–words is a common way to represent documents in natural language processing and information retrieval applications. In this model, a sentence or a document is represented as the collection of its words; only the word frequencies in the text are considered, disregarding grammatical rules and even the word flow. The bag–of–words approach is also widely used for image classification, when visual features are treated as visual words. This involves three main steps;

firstly, visual features are extracted from the image. Then descriptors – arrays collecting such features – are built. After this step, images are represented by a collection of vectors of the same dimension. Finally, descriptors are grouped and converted to codewords, the analogous of words in text documents. Clustering algorithms are usually employed to this aim. The set of codewords, belonging to the same cluster, produces a codebook, i.e. a "word dictionary". For the present study, different kind of features (describing color and shape) have been evaluated, to extract meaningful information from the images of Petri dishes. Based on the background removal procedure described in [10] (see Fig. 1), a set of foreground segments have been extracted from each of the 162 images that compose the training set. Each segment is then represented by a descriptor (feature vector). The selected features are described in the following.

- Color moments characterize the color distribution of a segment, and they are scale and rotation invariant. Among the possible color spaces, experimental results have shown that the use of HSV provides better performance. Three color moments are used, mean (Eq. (1)), standard deviation (Eq. (2)) and skewness (Eq. (3)); they are computed separately for each HSV channel:

$$m_i = \sum_{j=1}^{N} \frac{1}{N} v_{ij} \tag{1}$$

$$\sigma_i = \sqrt{\frac{1}{N} \sum_{j=1}^{N} (v_{ij} - m_i)^2} \tag{2}$$

$$s_i = \sqrt[3]{\frac{1}{N} \sum_{j=1}^{N} (v_{ij} - m_i)^3} \tag{3}$$

where N is the number of pixels in the segment and v_{ij} is the value of the j–th pixel in the i–th color channel.
- The shape of the segments is also a useful characteristic, since bacterial colonies produced by different types of infections may have different size; moreover, the shape is important to evaluate the infection severity, because it allows to distinguish between single and overlapping colonies. Two shape features have been used, the segment area (**A**) and the elongation (Eq. (4)):

$$E = \sqrt{\frac{i_2}{i_1}} \tag{4}$$

where i_1 and i_2 are, respectively, the minor and the major axis of the smallest bounding ellipse of the segment.

Therefore, for each image segment in the training set, a descriptor has been computed, in the form:

$$D = [m_H, m_S, m_V, \sigma_H, \sigma_S, \sigma_V, s_H, s_S, s_V, A, E]$$

Once the descriptors were extracted, the codebook generation relies on an unsupervised clustering procedure. The set of codewords obtained in this way forms the final codebook. In particular, the k–means clustering algorithm[1] has been used to extract the codewords (each word corresponds to a cluster centroid). Let us first introduce some notations.

Let X be the dataset to be clusterized and let $x_i \in X$, for $i \in \{1, ..., N\}$, with $N = \mid X \mid$. Let $C = \{c_1, c_2, ..., c_K\}$ be the obtained K disjoint sets, with $\bar{c}_k = \frac{1}{|c_k|} \sum_{x_i \in c_k} x_i$, the centroid of the k–th cluster and $\bar{X} = \frac{1}{N} \sum_{x_i \in X} x_i$ the dataset centroid. The optimal number of clusters K has been selected by evaluating the best partition among data, according to the four different measures, described in the following.

– Silhouette (SL) [15]

$$SL(C) = \frac{1}{N} \sum_{i=1}^{N} s(x_i) \qquad s(x_i) = \frac{(b(x_i) - a(x_i, c_j))}{\max\{b(x_i), a(x_i, c_j)\}} \qquad (5)$$

where, $\forall x_i \in X$, $a(x_i, c_j)$ is the average distance of the object x_i from all the points belonging to a different cluster c_j and $b(x_i)$ is the minimum average distance of x_i from all the points belonging to the other clusters.

– Calinski–Harabasz (CH) [16]

$$CH(C) = \frac{(N- \mid C \mid) \sum\limits_{c_k \in C} \mid c_k \mid d(\bar{c}_k, \bar{X})}{(\mid C \mid -1) \sum\limits_{c_k \in C} \sum\limits_{x_i \in c_k} d(x_i, \bar{c}_k)} \qquad (6)$$

where $d(x_i, c_j)$ is the average distance of the object x_i from all the points belonging to a different cluster c_j.

– Davies–Bouldin (DB) [17]

$$DB(C) = \frac{1}{\mid C \mid} \sum_{c_k \in C} \max_{c_l \in C \backslash c_k} \left\{ \frac{S(c_k) + S(c_l)}{d(\bar{c}_k, \bar{c}_l))} \right\} \qquad S(c_k) = \frac{1}{\mid c_k \mid} \sum_{x_i \in c_k} d(x_i, \bar{c}_k)$$

$$(7)$$

– Gap (G) [18]

$$G(C) = E_N^*\{\log(W_k)\} - \log(W_k) \qquad W_k = \sum_{i=1}^{k} \frac{1}{2 \mid c_i \mid} D_i \qquad (8)$$

where D_i is the sum of the pairwise distance for all the points belonging to cluster c_i and the expected value $E_N^*\{\log(W_k)\}$ is determined by Monte Carlo sampling from a reference distribution.

The optimal number of clusters, computed in the range $[1, 100]$, is reported in Table 1.

---

[1] Other clustering methods—such as DBSCAN, OPTICS, SOM—have been tested. k–means was chosen since it offers the best trade off between simplicity and performance.

Table 1. The best value of K selected by the four different validity measures, SL, CH, DB, and G.

	SL	CH	DB	G
K^*	3	3	3	96

We have also tested different values for K within the same range (selecting $K \in \{16, 28, 45, 68\}$), in order to check the ability of the chosen metrics to produce a correct dictionary dimension. The overall codebook generation procedure is summarized in Fig. 2, whereas, in Fig. 3, an example of the obtained set of words with respect to a test image is reported.

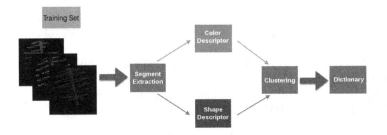

Fig. 2. Codebook generation schema.

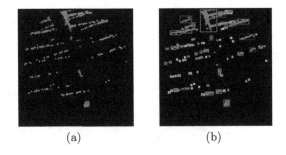

(a)   (b)

Fig. 3. In (a), the original image without the background and, in (b), words found within the image.

3   Infected Plate Recognition

Automatic recognition of infected plates aims at distinguishing between positive and negative tests, where a positive plate is characterized by the presence of bacterial colonies, whereas on negative plates no bacteria are grown. Nevertheless, sometimes, plates can be considered as negative even if bacterial colonies are actually present on the culture medium, especially in the case of atypical infections that, in specific circumstances, can be considered as a contamination of the urine sample. The rate between infected and not infected samples is imbalanced, with a higher prevalence of the latter. Actually, about 70% of the samples are negative[2]. Hence, automatically recognizing infected plates with a high accuracy is fundamental, since it can significantly reduce the biologist workload.

Table 2. Dataset composition (infected plate classification).

Dataset	Positive	Negative	Total
Training set	162	162	324
Test set	84	345	429

In order to detect infected plates, using the codebook generation procedure previously described, six codebooks, with a different number of words (3, 16, 28, 45, 68, and 96, respectively), have been extracted from images belonging to the dataset (Table 2). Then, two different classifiers, namely, MLPs and SVMs, have been trained[3], obtaining the results summarized in Table 3.

Table 3. Accuracy gained by SVMs/MLPs based on the different dictionaries.

K	3	16	28	45	68	96
MLP parameters	3–10–2	16–50–2	28–30–2	45–25–2	68–70–2	96–200–2
MLP accuracy	94.63%	94.41%	96.73%	96.2%	96%	95.1%
MLP TP Rate	0.92	0.79	0.91	0.88	0.8	0.82
SVM parameters	Poly2	RBF	Poly2	Poly2	RBF	RBF
(Kernel/C/Gamma)	1/0.3	1/0.1	1/0.2	1/0.3	1/0.1	15/0.4
SVM accuracy	95.57%	93.94%	94.87%	95.33%	95.8%	96.5%
SVM TP rate	0.92	1	0.88	0.88	0.91	0.9

As we can observe from the results in Table 3, the performance of the two classification models are very similar, with the best accuracy obtained by MLPs using the codebook with 28 words. However, it is worth noting that, in medical applications, not only the accuracy of the system is important, but also its capacity of avoiding false negatives since, actually, a false negative could lead to ignore the infection and to expose the patient to possible risks. Therefore,

---

[2] For this reason, the training set dimension has been reduced to balance the number of positive and negative patterns (see Table 2).

[3] The MLP structures are described in Table 3. Both hidden and output neurons are sigmoidal. Two neurons constitute the ouput layer in order to improve the network *flexibility* in modeling complicated relationships. All the architectural parameters (for MLPs and SVMs) were chosen via a trial–and–error procedure and crossvalidation.

Table 4. Accuracy (a) and confusion matrix (b) obtained by an RBF kernel SVM, with $C = 1$ and $\gamma = 0.1$, on the test set composed by 429 images.

<table>
<tr><td colspan="3" align="center">(a) Accuracy</td><td colspan="2" align="center">(b) Confusion Matrix</td></tr>
<tr><td></td><td>Number of Images</td><td>Percentage</td><td>Negative</td><td>Positive</td></tr>
<tr><td>Incorrectly Classified</td><td>26</td><td>6.06 %</td><td>319</td><td>26</td></tr>
<tr><td>Correctly Classified</td><td>403</td><td>93.94 %</td><td>0</td><td>84</td></tr>
</table>

observing the True Positive (TP) rate, the SVM trained using the codebook with 16 words must be preferred with respect to other alternatives. Table 4 collects detailed results obtained in this case.

4 Infection Species Recognition

Automatic infection classification aims at recognizing the infection strain(s) present on the infected plates. There is a huge number of different bacterial strains, which can possibly be present in an urine sample and, as discussed in Sect. 2, using the Chromagar Orientation medium, nine different bacterial types can be distinguished. The most common infection is E. Coli, with about 60–70% of occurrence, whereas all other species are much less frequent. As a consequence, our image dataset is too small to represent, with statistical significance, all the nine classes. Actually, for example, only four images of Pseudomanas Aeuroginos and Proteus, and two of Staphylococcus Saprophitycus are present. Therefore, the number of classes to be recognized has been reduced, grouping together some underrepresented classes with similar (color) properties (see Table 5).

Table 5. Dataset composition (infection classification).

Infection classes	Training set	Test set	Total
E. Coli	94	48	142
Enterococcus Spp	26	14	40
KESC	22	12	34
Other (Proteus, S. Aureus, Pseudomonas, Candida)	20	10	30

As it can be observed in Table 5, the dataset is imbalanced; as expected, E. Coli is highly prevalent, whereas the other classes contain a small number of samples. From the classification point of view, this is an undesirable situation and, to address this problem, the training set has been pre–processed and artificially balanced[4]. Then, the codebook has been generated using the positive

---

[4] The Weka Class Balancer function has been used to balance the data. This function reweights the instances in the data so that each class has the same total weight. The total sum of weights across all instances will be maintained. Only the weights in the first batch of data received by this filter are changed.

Table 6. Accuracy gained by SVMs/MLPs based on the different dictionaries.

K	3	16	28	45	68	96
MLP parameters	3–10–4	16–25–4	28–20–4	45–30–4	68–60–4	96–100–4
MLP accuracy	23.8%	88%	88%	90.47%	92.85%	85.71%
SVM parameters	RBF	Poly2	Poly2	Poly2	Poly2	RBF
(Kernel/C/Gamma)	10/0.2	1/0.6	1/0.1	112/0.1	1/0.3	1/0.2
SVM accuracy	54.7%	73.8%	80.95%	77.38%	82.15%	78.57%

sample images. Actually, six codebooks with a different number of words (3, 16, 28, 45, 68, and 96, respectively) have been extracted from the images of the dataset (Table 5). Two different classification architectures, MLPs and SVMs, have been trained, producing the results reported in Table 6.

In this case, the best accuracy has been obtained using the codebook with 68 words. It is worth noting that, differently from the infected plate detection, a greater number of words seems to be necessary to conveniently characterize the infection type. This may reflect the more complex nature of this problem (see Table 7).

Table 7. System overall accuracy and confusion matrix obtained by an SVM with a polynomial kernel of degree 3, $C = 1$, $\gamma = 0.1$, and by an MLP architecture, with 68–60–4 units.

(a) Accuracy

	Number of Images	Percentage
Total Number of Images	84	
Incorrectly Classified	15 / 6	17.85 / 7.14 %
CorrectlyClassified	**59 / 78**	**82.15 / 92.85 %**

(b) Confusion Matrix

E.Coli	Enterococcus	KESC	Other
45 / 47	0 / 0	1 / 0	2 / 1
1 / 1	10 / 13	0 / 0	3 / 0
0 / 0	1 / 1	10 / 11	1 / 0
5 / 0	1 / 3	0 / 0	4 / 7

The accuracy for each single class gained here is very similar to the results obtained in [10] even if, unfortunately, a precise comparison is not possible due to the differences in the training dataset and in the classification approach (different number of classes in the two problems).

Finally, briefly considering the computational complexity of the proposed approach, we can notice that the image segmentation module (which is out of the scope of this article) represents the most demanding task, taking 6 to 10 s for each image, whereas the whole procedure constituted by the two phases of word frequency histogram extraction and image classification took 3 s, on average, for each image[5]. All the experiment were carried out using an Intel i5 CPU.

5   Conclusions

Urinary tract infections can be caused by many different microbes, including fungi, viruses, and bacteria. Bacteria are actually the most common cause of UTIs. The body is usually able to rapidly removing bacteria that enter the urinary tract before they cause symptoms. However, sometimes bacteria overcome the body natural defenses and, in fact, roughly 150 million of infections occur annually worldwide. In this paper, an automatic method capable of detecting the presence of UTIs and to establish their type, was described. The system shows a good accuracy in distinguishing positive and negative samples, and also a very good sensitivity. Moreover, the proposed procedure is able to recognize different infection types with a high accuracy. Unfortunately, some classes are under-represented, and this lead to group different infection types together (based on similar colors) in order to obtain a meaningful representation. A larger dataset (hopefully available soon) could avoid this issue and allow to distinguish among a higher number of different infections, even with very similar colors.

Aknowledgements. The authors would like to thank Prof. Rossolini and the whole staff of the MV–Lab of the Careggi Hospital for their willingness to provide real data, and for their invaluable experience in interpreting their microbiological meaning.

References

1. National Institute of Diabetes and Digestive and Kidney Diseases, Urinary Tract Infections in Adults. https://www.niddk.nih.gov/health-information/health-topics/urologic-disease/urinary-tract-infections-in-adults/Pages/facts.aspx
2. Berlin, A., Sorani, M., Sim, I.: A taxonomic description of computer-based clinical decision support systems. J. Biomed. Inform. **39**, 656–667 (2006). Elsevier
3. Deserno, T.M.: Biomedical Image Processing. Springer-Verlag, New York (2011)
4. Belazzi, R., Diomidous, M., Sarkar, I.N., Takabayashi, K., Ziegler, A., McCray, A.T., Sim, I.: Data analysis, data mining: current issues in biomedical informatics. Methods Inf. Med. **50**(6), 536–544 (2011). Schattauer Publishers
5. Agah, A.: Artificial Intelligence in Healthcare. CRC Press, Boca Raton (2014)
6. Heckerling, P.S., Canaris, G.J., Flach, S.D., Tape, T.G., Wigton, R.S., Gerber, B.S.: Predictors of urinary tract infection based on artificial neural networks and genetic algorithms. Int. J. Med. Inform. **76**(4), 289–296 (2007)
7. Bianchini, M., Maggini, M., Jain, L.C.: Handbook on Neural Information Processing. Intelligent Systems Reference Library, vol. 49. Springer-Verlag, Heidelberg (2013)
8. Bandinelli, N., Bianchini, M., Scarselli, F.: Learning long-term dependencies using layered graph neural networks. In: Proceedings of IJCNN-WCCI 2012, pp. 1–8 (2012)

---

[5] Instead, the codebook generation required about 15 min, using a training set of pre–segmented images.

9. Bourbeau, P.P., Ledeboer, N.A.: Automation in clinical microbiology. J. Clin. Microbiol. **51**(6), 1658–1665 (2013)
10. Andreini, P., Bonechi, S., Bianchini, M., Garzelli, A., Mecocci, A.: Automatic image classification for the urinoculture screaning. Comput. Biol. Med. **70**, 12–22 (2016). Elsevier
11. Andreini, P., Bonechi, S., Bianchini, M., Garzelli, A., Mecocci, A.: ABLE: an automated bacterial load estimator for the urinoculture screening. In: ICPRAM, pp. 573–580. Springer (2016)
12. Andreini, P., Bonechi, S., Bianchini, M., Mecocci, A., Massa, V.: Automatic image analysis and classification for urinary bacteria infection screening. In: Murino, V., Puppo, E. (eds.) ICIAP 2015. LNCS, vol. 9279, pp. 635–646. Springer, Heidelberg (2015). doi:10.1007/978-3-319-23231-7_57
13. Andreini, P., Bonechi, S., Bianchini, M., Mecocci, A., Massa, V.: Automatic image classification for the urinoculture screening. In: Neves-Silva, R., Jain, L.C., Howlett, R.J. (eds.) Intelligent Decision Technologies. SIST, vol. 39, pp. 31–42. Springer, Heidelberg (2015). doi:10.1007/978-3-319-19857-6_4
14. Ferrari, A., Signoroni, A.: Multistage classification for bacterial colonies recognition on solid agar images. In: Proceeding of IEEE IST 2014, pp. 101–106 (2014)
15. Rousseeuw, P.J.: Silhouettes: a graphical aid to the interpretation and the validation of cluster analysis. J. Comput. Appl. Math. **20**, 53–65 (1987)
16. Calinski, T., Harabasz, J.: A dendrite method for cluster analysis. Commun. Stat. **3**(1), 1–27 (1974)
17. Davies, D.L., Bouldin, D.W.: A cluster separation measure. IEEE Trans. Pattern Anal. Mach. Intell. **1**(2), 224–227 (1979)
18. Tibshirani, R., Walther, G., Hastie, T.: Estimating the number of clusters in a data set via the gap statistic. J. Royal Stat. Soc. Ser. B **63**(2), 411–423 (2001)

Two-Tier Image Features Clustering for Iris Recognition on Mobile

Andrea F. Abate, Silvio Barra$^{(\boxtimes)}$, Francesco D'Aniello, and Fabio Narducci

BipLab, University of Salerno, Fisciano, Salerno, Italy
{abate,fnarducci}@unisa.it, barra.silvio@gmail.com,
fdaniello@studenti.unisa.it
http://www.biplap.unisa.it

Abstract. Nowadays, many smartphones are provided with built-in sensors for the acquisition and the recognition of specific biometric traits of the user. This policy has been adopted since the massive use of such devices brought the user to store sensible data in them as well as effectuate sensitive transactions on-the-move. As a consequence, many biometric systems have been migrated from stand alone to mobile environments. The methodology proposed in the following presents an approach to the iris recognition in visible spectrum. Iris images are first enhanced by a fuzzy color/contrast preserving technique and then passed to a two-tier clustering: the first is based on the linear decomposition of the iris into superpixels; the second one exploits an unsupervised learning network model to built a feature vector of the iris. According to the performance obtained in terms of time and recognition rate, the method is compliant with the needs of real-time and in-movement environments.

Keywords: Fuzzy image enhancement · SLIC · SOM network · Iris recognition · Biometrics · Mobile devices

1 Introduction

The widespread diffusion of mobile devices and the ever increasing processing power available pose the proper conditions for an exclusive usage of them for all daily and extraordinary operations by the owners (e.g. mail control, bank transfer, storage of personal data and so on). Consequently, the privacy and safety issues are progressively becoming more critical but also very hard to address. For such reasons, very recent mobile phones integrated fingerprint scanners or infrared sensors for face/iris/periocular recognition with the aim to ensure a controlled access to mobile functions, applications and stored data [4,6]. With the ever increasing use of such devices amongst the population, many biometric recognition systems have been migrated to mobile environments [1,3,18]. The ubiquity of this kind of devices represents a very challenging issue to face. The variation of illumination introduced in indoor versus outdoor conditions has a big impact on achievable performances [9]. When involving the use of mobiles,

© Springer International Publishing AG 2017
A. Petrosino et al. (Eds.): WILF 2016, LNAI 10147, pp. 260–269, 2017.
DOI: 10.1007/978-3-319-52962-2_23

the design of the applications has to deal with memory consumption, battery discharge, near real-time response and working with limited quality images and sensors. Even though this limitations, the research community is particularly active on these topics, a trend that is driven by the promising improvements on mobile devices quality in terms of processors, gpu processing, sensors, long-life batteries and so on. Iris has been confirmed as one of the strongest and most reliable biometric traits for human identification. Although in controlled and laboratory conditions it has been demonstrated to be very robust, in uncontrolled environments and without users' cooperation there is still a significant gap to be filled to definitely state the solely use of iris as a reliable biometric trait for owner's recognition on mobile. In [12], the authors presented one of the very first algorithms specifically developed for mobile devices: an Adaptive Gabor Filter was used to extract the iris code from iris images. In [4], the authors exploited the spatial histograms (*spatiograms*) to verify owner identity during any kind of transaction involving the exchange of sensible data. On the other side, the work in [9] is another of the first attempts to recognize humans' eyes with the usage of information surrounding the iris, that is known as periocular region.

In this work, we focus the attention on iris images usage only acquired by mobile camera in the visible spectrum (RGB images) in noisy conditions. The goal is that of designing a fast and reliable approach to iris recognition on mobiles, based on a two-steps clustering of iris pixels achieved by a combination of the *simple linear iterative clutering* (SLIC) [2] and Self Organizing Maps (SOM) [15]. SLIC generates superpixels by clustering pixels based on their color similarity and proximity in the image plane. This is done in the five-dimensional [*labxy*] space, where [*lab*] is the pixel color vector in CIELAB color space and [*xy*] is the color position. A special-purpose distance measure, tailored on 5-dimension representation of the problem, is applied to achieve the clustering, that is the localization of the superpixels (see Fig. 1).

Fig. 1. The figure presents the SLIC decomposition applied to sample color images. The images are iteratively decomposed in groups of pixels until the residual errors of centroids of two successive iterations does not exceed a given threshold. (Color figure online)

In this paper we propose the usage of SLIC decomposition as a preliminary step of clustering iris pixels in an image. To improve the work of superpixels localization, the iris image is pre-processed by a fuzzy equalization of the image histogram. The advantage is that it preserves the brightness and enhances the contrast in the image while reducing the computational complexity (short detailed description can be found in Sect. 2.2). Once a partitioning of the iris is obtained, the proposed approach computes three measures at block level, and it is done for all blocks of the partitioning. The result is then used as input for a SOM network which further analyses the clusters to achieve the biometric recognition. Self-organizing map (SOM) (or self-organising feature map (SOFM)) represents a class of artificial neural networks (ANN) that implement unsupervised learning to produce a low-dimensional representation of the input space, called *map*. In the recent past, SOM networks have been efficiently used for detection and localization in color images [9, 19]. Jiang and Zhou [13] showed that clustering of image pixels by many SOM neural networks performed better than some other existing clustering-based image segmentation methods. Also known as Kohonen maps, SOM networks differ from error-correction learning networks (e.g., backpropagation with gradient descent [7]) since they use a neighborhood function that preserves the topological properties of the input space [14]. This is a desirable property since most of the information characterizing a human iris is in its minutiae. The localization of such characteristics is a key aspect for biometric recognition purposes. In the field of biometrics, SOM networks have been demonstrated useful for the recognition of faces from a single image [23] while Lawrence et al. [16] exploited the output of a SOM network as input for a Convolutional Neural Network. Tan et al. [22] used SOM maps as a way of representing the face subspace. On the contrary, a very limited research have been conducted on iris recognition by SOM network. The work by Liam et al. [17] represents one of the few attempts of proving the benefits of using SOM network for recognition of human's irises in the visible spectrum. The major difficulty arises from the high uncertainty of clustering by SOM when working at pixel level. By pre-clustering the iris image by SLIC decomposition, we lighten the work of the SOM which is able to compute faster and provide more reliable results.

The content of the paper is organized as follows: Sect. 2 describes in detail the processing pipeline and the operators used; Sect. 3 presents the preliminary results conducted on iris datasets and Sect. 4 draws some conclusions and future improvements of the proposed work.

2   The Proposed Approach

The approach starts with the segmentation of the iris region and the selection of the ROI that has to be processed; the Sect. 2.1 details the steps involved. Section 2.2 briefly introduces the color/contrast fuzzy enhancement applied over the iris data before pre-clustering. The first tier clustering is achieved by the application of the *SLIC* over the region of interest (hereinafter ROI) (Sect. 2.3); then, for each block identified by the method, three measures, namely *entropy*, *magnitude* and *directions* are computed and arranged in 2D matrices (Sect. 2.4). The application of a SOM represents the second tier of the clustering (Sect. 2.5). The matching between the feature vectors is carried out by the application of a simple euclidean distance.

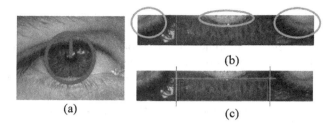

Fig. 2. The ROI selection process. (a) The Hough transform detects the circular shape of the iris; (b) the iris area is normalized to be in cartesian coordinates; the eyelids are located on the upper zone of such as image; (c) the green region represents the ROI selected for the further processing operations. (Color figure online)

2.1   The Segmentation and the ROI Selection

The first step involves the selection of the ROI of the iris. One of the most sensible and challenging issues to be faced in the iris recognition fields regards the presence of external factors in the images, mainly eyelids and eyelashes, which introduce noise, shadows and occlusions, so getting the extraction of the discriminative areas of the iris more difficult. In the current state-of-the-art, many segmentation methods are capable of isolating these elements, so returning a quite good "denoised" image, see [11]. Unfortunately, the main drawback of such method lies on the high computing time of the segmentation mask, mainly due to the detection of these defects. Since in mobile environments any feedback to the user has to be provided in the lowest time possible, such solutions to iris segmentation is hard to implement and objectively unsuitable. Therefore, we preferred exploiting a more rough approach needing lower computing time, rather than using one more precise but heavier in terms of time consumption: the Hough transform method for the detection of circles into an image suited our needs (Fig. 2(a)). The circle extracted is then polarized by means of the process of linearization described in [10] and resized to 100×600. Usually, in such normalized images, eyelids are located in the upper part of the image, due to the linearization process, how is depicted in Fig. 2(b). In order to overcome the noise brought by the presence of such external elements, the central-lower part of each normalized iris is selected as ROI (the green region in Fig. 2(c)) for the further processing operations.

2.2   The Fuzzy Enhancement

As mentioned in the introduction, we used an image enhancement technique based on BPDFHE (Brightness Preserving Dynamic Fuzzy Histogram Enhancement). It manipulates the image histogram in such a way that no remapping of the histogram peaks takes place, while only redistribution of the gray-level values in the valley portions between two consecutive peaks. Briefly, the algorithm consists in four steps: (i) computation of fuzzy histogram; (ii) partitioning of the histogram; (iii) dynamic histogram equalization of the partitions; (iv) normalization of the image brightness.

The reader can find a comprehensive description of the algorithm and how it works in [21]. It is mainly built around the concept of a *fuzzy histogram* which is meant as a sequence of real numbers $h(i), i \in 0, 1, ..., L - 1$. $h(i)$ represents the frequency of the occurrence of gray levels that are "*around i*". By considering the gray value $I(x, y)$ as a fuzzy number $\tilde{I}(x, y)$, the fuzzy histogram is computed as:

$$h(i) \leftarrow h(i) + \sum_x \sum_y \mu_{\tilde{I}(x,y)_i}, k \in [a, b] \tag{1}$$

where $\mu_{\tilde{I}(x,y)_i}$ is the triangular fuzzy membership function defined as $\mu_{\tilde{I}(x,y)_i} = max\left(0, 1 - \dfrac{|I(x, y) - i|}{4}\right)$ and $[a, b]$ is the support of the membership function. Fuzzy statistics is able to handle the inexactness of gray values in a much better way compared to classical histograms thus producing a smoother histogram. We therefore applied such a filtering on iris data in each RGB channel separately and then combined the three streams to get the original color image with fuzzy enhancement. A sample iris with color enhancement is shown in Fig. 3 (first row).

2.3   The Image Decomposition

The first tier of clustering is achieved by means of the application of the *SLIC* over the extracted ROI. SLIC is a simple and efficient method to decompose an image in visually homogeneous regions. It starts by dividing the image domain into a regular grid with M equally-sized tiles such that each contains about K pixels. For each superpixel, the centroid $C_i, i = 1, ...M$ is computed. At each iteration, the new regions are obtained by running k-means clustering, started from the computed centers. For each one of the regions the new centroid coordinates are obtained. The superpixel is no longer expanded when the residual error of the Manhattan distance on new and previously-computed centroids is less than a threshold. The result of the superpixel decomposition on a sample iris image is shown in Fig. 3 (second row). By enhancing the contrast while preserving the color properties of the image, the SLIC algorithm can more efficiently locate superpixels. As can be seen in Fig. 3, the images in the right column present a more detailed description of the iris and, consequently, a finer decomposition of the image in smaller superpixels compared to the results obtained on original resolution (on the left column in Fig. 3).

2.4   The Block Filling

Once the decomposition of the image into uniform superpixels has been obtained, simple mathematical operators are computed on each of them. The operators are: (i) the entropy, (ii) the gradient magnitude and (iii) the gradient direction. Entropy is a key measure of information, which quantifies the uncertainty involved in predicting the value of a random variable. In the field of the images,

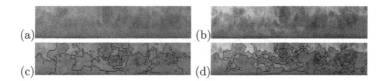

Fig. 3. BPDFH Enhancement of color images and SLIC. (a) The original aspect of normalised iris pixels. (b) The color/contrast enhanced iris image. (c)(d) The SLIC on (a) and (b) respectively. (Color figure online)

it has been widely used as a statistical measure of randomness that can be used to characterize the texture [5]. It is computed as $\sum p * log_2(p)$ where p contains the histogram counts of the pixels inside the block. To a more reliable usage of the entropy, the computed value has been normalised in the range $[0, 1]$. The second and third operators are the magnitude and the direction of the gradient computed on the superpixel. Mathematically, for an image function $f(x, y)$, representing the pixels of the block, the gradient magnitude $g(x, y)$ and the gradient direction $\theta(x, y)$ are computed as:

$$g(x, y) \cong \left(\Delta x^2 + \Delta y^2 \right)^{\frac{1}{2}} \quad \text{and} \quad \theta(x, y) \cong atan\left(\frac{\Delta y}{\Delta x}\right) \tag{2}$$

where $\Delta x = f(x + n, y) - f(x - n, y)$ and $\Delta x = f(x, y + n) - f(x, y - n)$ and n matches with the size of the block.

The three computed components per each superpixel are then used to till the block itself. This means that all pixels inside the block are assigned to the same triple of entropy, magnitude and direction. The obtained results, which replace the RGB values of each pixel can also been visually presented as shown in Fig. 4.

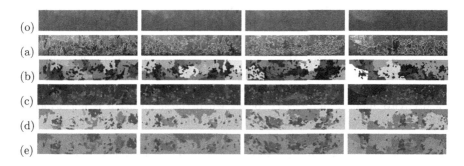

Fig. 4. The discriminative capability of the features extracted. Although the irises may look like similar to each other (o), the extraction of the features emphasizes the differences. (a) The superpixel decomposition by SLIC. (b) The entropy matrices; the darker the spot, the lower the entropy value. (c)(d) Magnitudes and directions of the features of the superpixels. (e) The final composition of the features in an RGB image to be passed as input to the SOM. (Color figure online)

2.5   The Clustering by SOM

Self-Organizing Map networks belong to a class of neural networks that implements unsupervised learning, which is used to draw inferences from datasets consisting of unlabelled input data. A detailed description of the principles and math behind them is out of the scope of this work. Briefly, the principal goal of an SOM is to transform an incoming signal pattern of arbitrary dimension into a one or *two dimensional discrete map*, and to perform this transformation adaptively in a topologically ordered fashion. Neurons are placed at the nodes of a one or two dimensional lattice. The neurons become selectively tuned to various input patterns during the course of the competitive learning. The neurons change their locations to meaningful coordinate system for the input features, thus forming the required topographic map of the input patterns [8]. We used a standard implementation of SOM available in the Neural Network Toolbox of Matlab. The pre-clustering of the iris image explained above is therefore given as input to an SOM network of small size which processes the input pixels and produces the map. Maps of two irises are then compared in terms of Euclidean distance of images as in [24].

3   Preliminary Results

Preliminary results have been obtained on a simulation of the algorithm with MATLAB 2013a on commodity hardware, a Microsoft Windows machine equipped with an Intel(R) Core(TM) Duo CPU P8600 @ 2.40 GHz. The selected hardware architecture is comparable with the average CPU performances of modern mobile devices, thus providing a good estimate of the time consumption of the algorithm when running on mobile hardware. More aimed investigations on time and processing consumption of the proposed algorithm will be carried out on a collection of different mobile devices as future improvements of this study. We used the dataset UBIRISv1 [20] as a reference for the experiments. Iris images of 241 different subjects have been used. The acquisition policy adopted in UBIRISv1 suits the goal of this study since it provides two different acquisition sessions of subjects' irises. The first one minimizes the noisy factors, especially those related to the illumination and reflections. The second session introduces typical environmental factors that occur when the acquisition is performed in another time, e.g. variation of the illumination and non-ideal visibility of the iris. On the other hand, images are almost planar to the camera in both sessions, which introduces the assumption of cooperative users. Results obtained on the chosen dataset are depicted in Fig. 5. Pairs of segmented and normalized iris images of 241 available subjects have been used in the closed-set experiment. Of each pair of irises, one as been used as probe and the second one as gallery. To reduce the computational demand, we applied a strong resizing of the images at a quarter of the original resolution and used a SOM network of size 5×5. As shown in Fig. 5, the preliminary level of performance achieved by using very simple and light operators is encouraging. Results have been also reported in terms of Recognition Rate (RR), Equal Error Rate (EER), Area Under Curve

(AUC) and Decidability (DEC). Particularly significant is the rough index of decidability achieved. The separation amongst genuine and impostors is a key aspect to make assumptions on the reliability of the system. In terms of computing time, the single recognition process needs about 1 second: (i) 0.62 s for the preparation of the input (including fuzzy enhancement, SLIC, computation of entropy, magnitude and direction values); (ii) 0.25 s for the training of the SOM on the single image; (iii) 0.318 s for the matching among the feature vectors of the images. The times shown have been computed over the execution of the whole test and then averaged on the single samples. Such performance suggest that the method is appropriated at working in real time environments. Since the results come from a very simple algorithm at an early-stage of its design, we believe that it can be reasonably improved following three main directions: (1) adaptive SLIC decomposition (2) fine-tuning of mathematical operators at block level, (3) other forms of unsupervised learning.

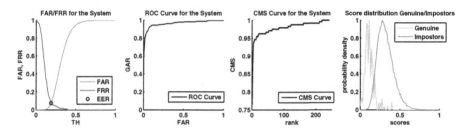

$$RR = 80.7\%, \ EER = 7.53\% \ AUC = 96.53\% \ DEC = 2.38$$

Fig. 5. Performance curve plots achieved on UBIRISv1 dataset. Particularly relevant is the initial good separation among genuine and impostors using very simple and light operators and functions.

4   Conclusions

Although iris recognition applications have been demonstrated particularly powerful in laboratory/controlled conditions, the level of performance achieved decays significantly when having to deal with environmental noise factors, e.g., illumination, reflections and shadows, out-of-focus and others. The amount and variability of uncontrolled factors affecting the quality of iris data when acquired on mobiles is sensibly high. This is intrinsically linked to the ubiquity of such devices which exposes the camera sensors to a wide range of adverse conditions. The research seems to suggest that the approaches based on iris only can not achieve desirable levels of performance, nor scalable solutions, if not relaxing some constraints (e.g., availability of a good quality and computationally inexpensive segmentation of the iris from the image, user's cooperation and awareness as well as the control of prohibitive conditions). In this work we aim at exploring if light and simple fuzzy image enhancement together with image descriptors and power of unsupervised learning, can effectively support the feasibility of a reliable application of iris recognition on mobile devices. We involved the usage of a simple clustering of pixels in images by SLIC to achieve a light and easy-to-manage representation of iris features. To achieve a more reliable and detailed decomposition of the image, a fuzzy enhancement of color and contrast is globally applied to the iris data. The obtained decomposition is then used to apply some operators at superpixel level with the aim of building a descriptor of each block which acts as local descriptor of the iris features. The pre-clustering is used as input for a SOM network which is responsible for the building of a *map* of the iris. Such a map, which is a low-dimensional representation of the iris, consists in a biometric identifier of the users used for the biometric identification. The linear processing power required to compute the biometric identifier on the iris image represents an encouraging result which makes its execution on mobile architectures feasible. Moreover, preliminary results obtained on noisy iris datasets achieved promising level of performance which leads to further confirm the applicability of the proposed approach.

Future improvements of the method proposed in this paper can be split into two directions: (1) thoroughly assess the strengths and weaknesses of the approach and investigate on additional descriptors to improve the identification on a larger set of iris datasets and (2) facing with the issues related to iris segmentation which still represents a prominent bottleneck of the processing pipeline for mobile devices.

References

1. Abate, A.F., Nappi, M., Narducci, F., Ricciardi, S.: Fast iris recognition on smartphone by means of spatial histograms. In: Cantoni, V., Dimov, D., Tistarelli, M. (eds.) BIOMET 2014. LNCS, vol. 8897, pp. 66–74. Springer, Heidelberg (2014)
2. Achanta, R., Shaji, A., Smith, K., Lucchi, A., Fua, P., Süsstrunk, S.: SLIC superpixels compared to state-of-the-art superpixel methods. IEEE Trans. Pattern Anal. Mach. Intell. **34**(11), 2274–2282 (2012)
3. Barra, S., Casanova, A., De Marsico, M., Riccio, D.: Babies: biometric authentication of newborn identities by means of ear signatures. In: Proceedings of the 2014 IEEE Workshop on Biometric Measurements and Systems for Security and Medical Applications (BIOMS), pp. 1–7. IEEE (2014)
4. Barra, S., Casanova, A., Narducci, F., Ricciardi, S.: Ubiquitous iris recognition by means of mobile devices. Pattern Recogn. Lett. **57**, 66–73 (2015). http://www.sciencedirect.com/science/article/pii/S0167865514003286, mobile Iris CHallenge Evaluation part I (MICHE I)
5. Barra, S., De Marsico, M., Cantoni, V., Riccio, D.: Using mutual information for multi-anchor tracking of human beings. In: Cantoni, V., Dimov, D., Tistarelli, M. (eds.) BIOMET 2014. LNCS, vol. 8897, pp. 28–39. Springer, Heidelberg (2014)
6. Barra, S., De Marsico, M., Galdi, C., Riccio, D., Wechsler, H.: Fame: face authentication for mobile encounter. In: 2013 IEEE Workshop on Biometric Measurements and Systems for Security and Medical Applications (BIOMS), pp. 1–7. IEEE (2013)

7. Bengio, Y., Simard, P., Frasconi, P.: Learning long-term dependencies with gradient descent is difficult. IEEE Trans. Neural Netw. **5**(2), 157–166 (1994)
8. Bullinaria, J.A.: Self organizing maps: fundamentals. In: Introduction to Neural (2004)
9. Cho, D.H., Park, K.R., Rhee, D.W., Kim, Y., Yang, J.: Pupil and iris localization for iris recognition in mobile phones. In: Seventh ACIS International Conference on Software Engineering, Artificial Intelligence, Networking, and Parallel/Distributed Computing (SNPD 2006), pp. 197–201, June 2006
10. De Marsico, M., Nappi, M., Daniel, R.: Isis: iris segmentation for identification systems. In: Proceedings of 20th International Conference on Pattern Recognition, pp. 2857–2860 (2010)
11. Haindl, M., Krupika, M.: Unsupervised detection of non-iris occlusions. Pattern Recogn. Lett. **57**, 60–65 (2015). http://www.sciencedirect.com/science/article/pii/S0167865515000604, mobile Iris CHallenge Evaluation part I (MICHE I)
12. Jeong, D.S., Park, H.-A., Park, K.R., Kim, J.: Iris recognition in mobile phone based on adaptive gabor filter. In: Zhang, D., Jain, A.K. (eds.) ICB 2006. LNCS, vol. 3832, pp. 457–463. Springer, Heidelberg (2005). doi:10.1007/11608288_61
13. Jiang, Y., Zhou, Z.H.: Som ensemble-based image segmentation. Neural Process. Lett. **20**(3), 171–178 (2004). doi:10.1007/s11063-004-2022-8
14. Kiviluoto, K.: Topology preservation in self-organizing maps. Helsinki University of Technology (1995)
15. Kohonen, T.: The self-organizing map. Proc. IEEE **78**(9), 1464–1480 (1990)
16. Lawrence, S., Giles, C.L., Tsoi, A.C., Back, A.D.: Face recognition: a convolutional neural-network approach. IEEE Trans. Neural Netw. **8**(1), 98–113 (1997)
17. Liam, L.W., Chekima, A., Fan, L.C., Dargham, J.A.: Iris recognition using self-organizing neural network. In: Student Conference on Research and Development, SCOReD 2002, pp. 169–172. IEEE (2002)
18. Marsico, M.D., Galdi, C., Nappi, M., Riccio, D.: Firme: face and iris recognition for mobile engagement. Image Vis. Comput. **32**(12), 1161–1172 (2014). http://www.sciencedirect.com/science/article/pii/S0262885614000055
19. Ong, S., Yeo, N., Lee, K., Venkatesh, Y., Cao, D.: Segmentation of color images using a two-stage self-organizing network. Image Vis. Comput. **20**(4), 279–289 (2002). http://www.sciencedirect.com/science/article/pii/S0262885602000215
20. Proença, H., Alexandre, L.A.: UBIRIS: a noisy iris image database. In: Roli, F., Vitulano, S. (eds.) ICIAP 2005. LNCS, vol. 3617, pp. 970–977. Springer, Heidelberg (2005). doi:10.1007/11553595_119
21. Sheet, D., Garud, H., Suveer, A., Mahadevappa, M., Chatterjee, J.: Brightness preserving dynamic fuzzy histogram equalization. IEEE Trans. Consum. Electron. **56**(4), 2475–2480 (2010)
22. Tan, X., Chen, S., Zhou, Z.H., Zhang, F.: Recognizing partially occluded, expression variant faces from single training image per person with som and soft k-NN ensemble. IEEE Trans. Neural Netw. **16**(4), 875–886 (2005)
23. Tan, X., Chen, S., Zhou, Z.H., Zhang, F.: Face recognition from a single image per person: a survey. Pattern Recogn. **39**(9), 1725–1745 (2006)
24. Wang, L., Zhang, Y., Feng, J.: On the euclidean distance of images. IEEE Trans. Pattern Anal. Mach. Intell. **27**(8), 1334–1339 (2005)

A Decision Support System for Non Profit Organizations

Luca Barzanti[1](✉), Silvio Giove[2], and Alessandro Pezzi[3]

[1] Department of Mathematics, University of Bologna,
Piazza di Porta San Donato 5, 40126 Bologna, Italy
luca.barzanti@unibo.it
[2] Department of Economics, Ca' Foscari University of Venice,
Cannaregio 873, 30121 Venice, Italy
sgiove@unive.it
[3] School of Economics, Management and Statistics, University of Bologna,
P.le della Vittoria 15, 47121 Forli, Italy
a.pezzi@unibo.it

Abstract. A crucial activity of Non Profit Organizations (NPO's) is the fund raising management, by which Organizations sustain the achievement of their mission. In this context, the development of a structured Decision Support Systems (DSS) is becoming increasingly important. The process of fund raising is very complex and in part different in function of the characteristics of the considered Organization. Recently for the medium-sized Associations a model has been developed, which explicitly considers the specificities of this kind of NPO's in order to optimize the fund raising process and the related algorithm. In this contribution, we enhance and complete this model by considering a fuzzy approach for the donor's ranking and a simple cost function with the aim to evaluate if the preset campaign target is reached. Moreover, we implement an effective DSS (FS) and we show the results obtained with a properly simulated large Data Base (DB), by analyzing the achieved computational results.

Keywords: Social economics · Fund raising management · Medium-sized organizations · Decision support systems

1 Introduction

The Fund Raising management allows NPO's achieving their mission, by the collection of the resources for this aim [1,19]. In this context, a great importance has the role of the *donor*, see *e.g.* [11,14,21], and his efficient management [15,18]. For this reason, econometric literature dealt with (potential) donors profiles that match some specific gift inclination, see *e.g.* [9,10], in order to support the effectiveness of the process. The use of a rudimentary Data Base management is documented in the classical operational literature [12,13]. Recently an innovative approach has been performed and developed in this field in [2], by

© Springer International Publishing AG 2017
A. Petrosino et al. (Eds.): WILF 2016, LNAI 10147, pp. 270–280, 2017.
DOI: 10.1007/978-3-319-52962-2_24

using mathematical modeling and Decision Support Systems (DSS) techniques, in order to help Associations both to decide the kind of campaign they have to organize, and the donors of the DB list which must be contacted to maximize the expected return of the campaign, satisfying time and cost constraints. This approach has been specialized for different kind of Organizations (see [3,4,6]) and it has been validated both in the operative world by Associations that test it (as documented in [3,4,6]) and in the pertaining literature, see [23].

In this contribution we improve the model described in [5], by introducing a donors fuzzy ranking algorithm. Based on past data, the method computes first an estimation of the expected gift and of the gifting probability for each donor, then a donors' ranking is performed using two fuzzy thresholds for gift and probability. The obtained DSS is based on a robust model, but at the same time it is usable and targeted for the management of a medium-sized Organization. The goal is obtained including inside the computational process the principal instances of fund raising management (like *e.g.* the fund raising pyramid), simplifying in this way the structure of the system, that is completely enclosed in the DB management context and not includes sophisticated and laborious to manage information techniques, like *artificial intelligence* tools.

Moreover an actual DSS has been performed and a numerical analysis has been developed, showing in details the effectiveness both of the proposed approach and of their implementation.

2   The Donor's Gift Forecasting Model

Organizing a fund raising campaign, the Management (DM for brevity) has to decide the donors to contact, from a large Data Base of donors, and has also to specify both the campaign budget *Budget* and the net estimated minimum return G, that is the minimum amount that the NPO requires to gain, with cost not greater than *Budget* (contact a donor requires a cost).

The purpose of the system consists in: (i) the production at time t of an ordered list of donors in function of probability and expected gift (estimated using past data); (ii): the choice of a subset (coalition) of donors to maximize the Expected Value of total gift, given the budget constraints.

To this aim, the donor's gift forecasting model proposed [5] is extended in what follows using two fuzzy memberships with the aim to implement a DSS useful for the NPO management. As a result, the donors will be listed in decreasing order from the most promising to the less. The donors ranking is based on the gift probability and value, estimated from the donors Data Base, a large set of data formed by the sampled historical data of each donors, that are the number and date of contacts, and the offered gift, if any. The Data Base $DB = \{d_1, d_2, ..., d_M\}$ includes all the stored donors information, where M is the number of stored donors. For each $d_i \in DB$, the Data Base lists the sampled time series of past gifts, i.e. the sequence of the request dates and the corresponding gift (zero if the case of no gift). Fixing the time origin in t_0, the series conventionally starts from $t - n_i$ up to t for the $i - th$ donor, being t the current time. Thus each donor is characterized by the *profile* d_i:

$$d_i = \{(t_i(k), D_i(k))\}, k = 1, ..., n_i$$

being:

$$\{t_i(1), t_i(2), ..., t_i(n_i)\}$$

the data of the request (or contact), n_i the number of contacts, and:

$$\{D_i(1), D_i(2), ..., D_i(n_i)\}$$

the corresponding gift sequence, $D_i(j) \geq 0$. Moreover, we suppose that the donor gives the gift, if he does, at the same time of the contact. In the line of principle, starting from the sampled data, it could be possible to infer a probability distribution for each stored donor, and subsequently compute a probability distribution for each subset of donors in the Data Base, and select the subset whose net expected value is maximum. Finally, the expected value can be used to check if the target could be reached or not. The optimization problem can be stated as follows:

$$max_{S \subset DB} EV(S) - Cost(S))$$

where $EV(S)$ is the Expected Value of the probability distribution of the coalition S, which depends on the probability distribution of the donors belonging to S, while $C(S)$ is the cost of contacting the donors[1] in S. A first difficult consists in the determination of the distribution probability of each donors, given that the gift tendency varies in time thus the phenomenon is not stationary. Moreover in some cases too few data are collected. Anywise, a second more critical drawback is the computation of the cumulative probability distribution of the coalition S. Apparently it should be clear that the optimization problem above cannot be analytically solved. Namely it is an NP-hard problem, given the huge number of coalitions that can be obtained by the Data Base.

For all these reasons, the unique possibility relays on some heuristic to obtain a sub-optimal but satisfactory solution. A natural and rational idea consists into privileging the donors with high frequency and gifts in the past. In the sequel, this rough idea is better formalized with the help of the expert *way of reasoning* inferred by some experts manager of NPO.

The way by which an expert NPO manager bases his/her judgement about a donor, is based on some simple heuristic rules acting on a limited set of parameters. Usually, an expert decision judgement about a donor is based on the frequency of gifts and on the average gift amount. A donor is as more considered as more both frequency and gift amount are high. Again, for what it concerns the frequency, an important item is the frequency variation, usually estimated observing the *short-term* frequency, that is the frequency in the recent past. Based on this three variables, the DM infers a binomial distribution, formed by the gift probability and the gift amount he can aspect[2]. In a second phase,

---

[1] This cost depends on the type of contact, like e-mail, normal mail, direct contact, etc.

[2] This simplified approach is partially justified by the fact that usually a limited gift variability characterized a single donor.

depending on the expected target, the availability budget and other non observable variables, she can subjectively decide if a donor has to be considered a promising donor or not. Trying to implement this type of expert knowledge, at first the expected gift amount together with the gift probability are estimated for each donors, as done in [5]. Subsequently, using two fuzzy thresholds, whose parameters will be tuned by the DM, the donors will be ranked from the most promising to the less, while donors with totally unsatisfactory probability or expected gift will be discharged. Finally starting from the most promising donor and taking the contact cost into account, using the Expected Value criterion, suitably modified, the expected Total Revenue of the campaign can be estimated, thus the DM can evaluate if the pre-defined target can be reached or not. The next Subsections describes the proposed method for gift and probability estimation, and the fuzzy approach for the donors selection and the total revenue estimation.

2.1   Gift Value and Probability Estimation

At first the DM has to specify the minimum value for the global expected gift, the target of the campaign, G, and the available budget $Budget$, used to contact the donors[3]. Then the DB is *filtered*, eliminating all the donors which do not satisfies a minimum value for the number of requests (a proxy of the *robustness*), r_{min}, for the sampled gift frequency, f_{min}, and for the average past gift, V_{min}. Moreover the DM specifies a penalization function $\mu(ET)$ for the *elapsed time* in function of the frequency and of the variance of the past gift sequence, a time window w to compute the *empirical* frequency of the last gifts, to include the *coldness* effect, see below.

The algorithm will compute, for the $i-th$ donor, the estimated gift probability $PD_i(t)$ at time $(t+1)$, computed at time t, and between τ_1 and τ_2, the gift's frequency, and the average gift value, taking only non zero gift into account. Gift probability and value can be estimated as follows, see [5]:

$$f_i(\tau_1, \tau_2) = \frac{\sum_{\tau_1 \leq t_i(j) \leq \tau_2} I(D_i(j))}{\sum_{\tau_1 \leq t_i(j) \leq \tau_2} j}$$

$$V_i(\tau_1, \tau_2) = \frac{\sum_{\tau_1 \leq t_i(j) \leq \tau_2} D_i(j)}{\sum_{\tau_1 \leq t_i(j) \leq \tau_2} j}$$

being $I(x)$ the indicator function of x: $I(x) = 1$ if $x > 0$, 0 otherwise.

It has been observed that the gift attitude can change in time for some donors. Thus it can be argued that the probability $PD_i(t)$ monotonically depends both on the *global* empirical frequency, $f_i(t_0, t)$, and the empirical frequency in the last time window of width w, $f_i(t - w, t)$, through a suitable function $F(f_i(t_0, t), f_i(t - w, t))$[4].

---

[3] The contact cost depends on the type of contact, see [2], but in this contribution we suppose only one type of contact.

[4] If $f_i(t - w, t) < f_i(t_0, t)$ we affirm the presence of a *coldness* effect, conversely the donor becomes more *generous*.

Moreover, the time passed since the last donation, the *elapsed time* $ET_i(t) = t - t_i(n_i)$, plays a critical role (*memory effect*). We suppose that a donor is less prone to give a gift if he has gifted too much recently; the shorter ET_i, the lower the gift probability, and the same for the average gift. Then the algorithm estimates $PD_i(t)$ and $V_i(t)$ by a linear combination of $f_i(t_0, t)$ and $f_i(t - w, t)$ with coefficients ω_L and $\omega_S = 1 - \omega_L$ previously assigned by the DM.

2.2   Donors Selection and Total Revenue Estimation

Up to now, each donor is then characterized by the ordered couple $PD_i(t), V_i(t)$, the probability and the expected amount of the gift. The ordered list of donors can then be obtained through a two arguments utility function, which in our context is made by the aggregation of two separated *fuzzy* thresholds. Namely, the DM has to express his subjective preference about the probability and the expected gift, discharging at first the ones which do not satisfy minimum requirement for both the two criteria. This can be done introducing two thresholds, but, given the usual uncertainty about the border point in the thresholds, a fuzzy function can be introduced. The simplest form of the fuzzy threshold can be a piecewise linear defined by two parameters: the *minimum* requirement, m_p, m_V and the *complete satisfactory* points, M_p, M_V, which are to be asked to the DM. Values below m_p, m_V are assigned zero, and values above M_p, M_V will be assigned one. Both the two thresholds can be represented by the following memberships $f(p), g(V)$:

$$f(p) = \begin{cases} 0, & p < m_p \\ \frac{p - m_p}{M_p - m_p}, & m_p \leq p \leq M_p \\ 1, & p \geq M_p \end{cases}$$

$$g(V) = \begin{cases} 0, & V < m_V \\ \frac{V - m_V}{M_V - m_V}, & m_V \leq V \leq M_V \\ 1, & V \geq M_V \end{cases}$$

The first fuzzy threshold can be interpreted as a probability *distortion*, see [17,22] for the relationship among distorted probabilities and fuzzy measures. Namely, the DM modifies the probability estimation computed by the system applying his subjective trust, thus, for instance, neglecting too low probabilities (inferior to m_p). A similar interpretation holds for the expected gift threshold. Finally, let us observe that a membership function for a fuzzy threshold can also be intended as a particular *utility* function; [7] showed the equivalence between an utility function and an uncertain target described by a smoothed threshold. The two values $f(p)$ and $g(V)$ are subsequently aggregated using a suitable *aggregation operator*, usually a *triangular* norm, see [8] is used in fuzzy logic. In this case we prefer to be as less conservative as possible, thus we adopt the MIN operator, which is the upper bound in the family of triangular norm, the only which satisfies idempotency, a rational requirement. Summarizing, for each donor in the data base, represented by the ordered couple $\{PD_i(t), V_i(t)\}$, a scoring is computed as follows:

$$Score_i = MIN\{f(PD_i), g(V_i(t)\}$$

The donors are ordered using the values $Score_i$, and let $D = \{d_{(1)}, d_{(2)}, ...,$ $d_{(N)}\}$ be the ordered list of all the donors with $Score_i > 0$ $(N \leq M)$, i.e. $d_{(1)} \leq d_{(2)} \leq ... \leq d_{(N)}\}$. Moreover, let $Ucost$ be the cost for a contact (mail, direct contact or other, here supposed equal for all the donors) and $K_{MAX} = INT(\frac{Budget}{Ucost})$ be the maximum number of contacts given the available budget $Budget$. The system proposes to the DM the first K^* Donors in the ordered list, with $K^* = MIN\{K_{MAX}, N\}$, and supposing all the donors to be statistically independent together[5], a rough estimation of the total gain TG can be easily done through the expected value of the first K^* expected gifts[6]:

$$TG(K^*) = \sum_i^{K^*} PD_{(i)}(t) \times V_{(i)}(t)$$

Moreover the system computes the total average score:

$$TScore(K^*) = \frac{1}{K^*} \sum_i^{K^*} Score_{(i)}$$

If $TG - Budget \geq G$ the target will be reached with good probability, and at the same time, the DM is advised about the average score $TScore$ of the involved contacts. Anywise if the average score is too low but the expected total gain is satisfactory, the DM can decide to decrease the value of K^* and the system will compute again $TG(K)$ and $TScore(K)$ until a good compromise is reached.

As pointed out in [5] the proposed algorithm produces only an *empirical* estimation of the probability of giving and of the average gift, [16], trying to simulate a real DM way of reasoning, thus capturing expert knowledge. It does not require any hypothesis about the statistical distribution of the data, in this sense, it belongs to non parametric statistic approach and machine learning see [20].

3   The Tool Architecture

The tool is designed in Visual Basic, using SQL language on a Ms Access DB. The graphical interface allows the DM to set up the Campaign Parameters (*Target, Budget* and *Unit Cost*), the Algorithm Parameters (*omega L, omega S, Time*

---

[5] In the line of principle, given that all the donors attitude to gift is represented by binomial distribution, the probability to reach the target could be computed. But with different probabilities and gift values, this problem is NP-hard.

[6] As for the ordered score it holds: $P_{(1)}(t) \leq P_{(2)}(t) \leq ... \leq P_{(N)}(t)\}$ and $V_{(1)}(t) \leq V_{(2)}(t) \leq ... \leq V_{(N)}(t)$.

Horizon and *Time Window*), the Preliminary Selection Parameters (*Robustness,
Frequency* and *Average Past Gift*) so as the Elapsed Time Thresholds (*Min
Elapsed Time, Max Elapsed Time*) already present in the base algorithm. Fuzzy
parameters of aggregation are also set, by the indication of the bound of both the
thresholds (*Min Probability, Max Probability, Min Expected Value, Max Expected
Value*). Figure 1 shows the whole interface.

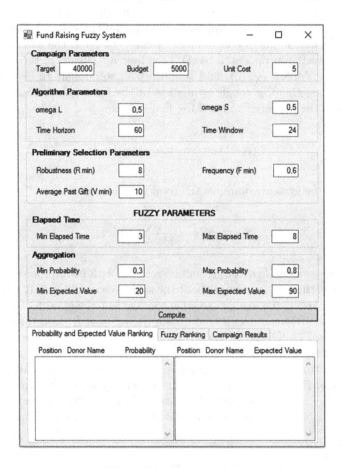

Fig. 1. The user interface

Notice that the results are widely described. First of all, it is presented a
synthesis of the ranking both by probability and by expected value (the *Probability and Expected Value Ranking* Tab). Then, the ranking obtained by the
fuzzy algorithm is showed (the *Fuzzy Ranking* Tab) and finally the synthetic
indicators for the results of the campaign (*Campaign Results*) are presented.

4   Computational Results

A DB with 30.000 donors is considered, a medium to high number for a medium-sized Organization, where about 403.000 gifts requests are collected. The parameters are set like in Fig. 1. The Campaign Parameters are typical of a medium-sized Organization; the Technical Ones are set in function of the characteristics of the DB and the preliminary choices of the DM. The results of the first ranking phase (*Gift value and probability estimation*) are presented in Fig. 2.

Probability and Expected Value Ranking			Fuzzy Ranking	Campaign Results	
Position	Donor Name	Probability	Position	Donor Name	Expected Value
1	D 19897	95,00%	1	D 2073	121
301	D 8631	72,50%	601	D 11744	96
601	D 9469	63,33%	1201	D 21409	89
901	D 28938	55,00%	1801	D 27044	86
1201	D 15463	44,32%	2401	D 22234	82
1501	D 26357	35,00%	3001	D 24209	79
1801	D 5628	26,79%	3601	D 19807	77
2101	D 27163	17,50%	4201	D 16882	74
2401	D 23715	13,94%	4801	D 26472	72

Fig. 2. The first ranking phase

In this case, after the preliminary selection, the considered donors decrease from 30000 down to about 9500. Notice that the DM can choose how deep to explore each single ranking. This gives a general idea of the features of data with respect of the ranking criteria. In this case, the selected step is 300 for the Probability and 600 for the Expected Value. The fuzzy aggregation implementing the first new extension of the algorithm (*Donors selection*) is showed in Fig. 3.

Probability and Expected Value Ranking		Fuzzy Ranking		Campaign Results	
Position	Donor Name	$f(p)$	$g(V)$	**Score**	
1	D 2834	1,00	1,00	1,00	
51	D 9605	1,00	0,95	0,95	
101	D 18362	0,87	0,96	0,87	
151	D 5678	1,00	0,82	0,82	
201	D 26650	1,00	0,79	0,79	
251	D 6007	1,00	0,76	0,76	
301	D 10401	1,00	0,73	0,73	
351	D 20076	0,70	0,89	0,70	
401	D 3802	1,00	0,68	0,68	

Fig. 3. The fuzzy aggregation

The step of visualization here selected for the results is 50 (in any case, chosen by the DM). The global results of the Campaign (*Total revenue estimation*), with the Average Score, are presented in Fig. 4.

Notice that the target is well achieved (indeed abundantly exceeded), with a satisfactory average score. The whole budget is used, as specified by the algorithm. If anyway the DM wants to decrease the considered donors number

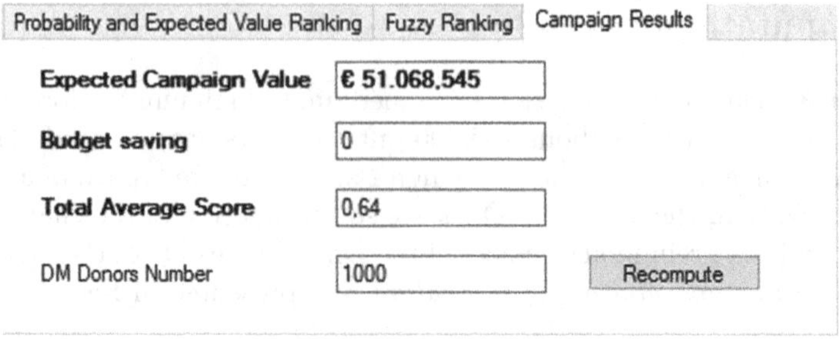

| Probability and Expected Value Ranking | Fuzzy Ranking | Campaign Results |

Expected Campaign Value	€ 51.068,545
Budget saving	0
Total Average Score	0,64
DM Donors Number	1000	Recompute

Fig. 4. Campaign results

(like proposed at the end of Sect. 2), in order to increase the Average Score, maintaining the goal of reaching the Target, there is the possibility to recompute the Campaign Results with the new donors number, by using the parameter *DM Num Donors* and the *Recompute* function. In this way it is also obtained a *Budget saving*, which is to be in principle added to the Expected Campaign Value. Figure 5 shows the results with 750 donors.

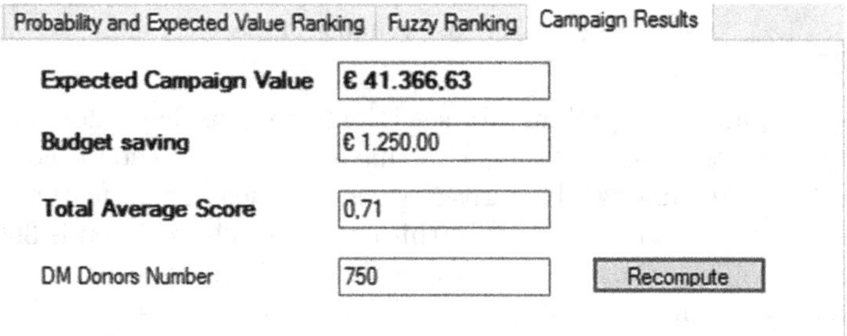

| Probability and Expected Value Ranking | Fuzzy Ranking | Campaign Results |

Expected Campaign Value	€ 41.366,63
Budget saving	€ 1.250,00
Total Average Score	0,71
DM Donors Number	750	Recompute

Fig. 5. Campaign results with budget saving

An intermediate choice of the DM (800 donors) gives the results showed in Fig. 6.

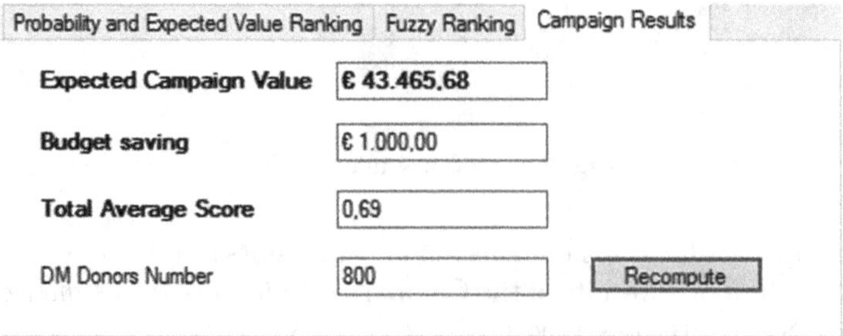

| Probability and Expected Value Ranking | Fuzzy Ranking | Campaign Results |

Expected Campaign Value	€ 43.465,68
Budget saving	€ 1.000,00
Total Average Score	0,69
DM Donors Number	800	Recompute

Fig. 6. Campaign results with an alternative choice of donors number

5   Conclusions

In this contribution, we enhance and complete a model for fund raising management focused specifically for medium-sized Organizations. This is achieved both by developing a fuzzy approach for the donors' ranking and by the implementation of an effective DSS, which allows also the consideration of a scenario approach (performed by the Recompute function). Future work includes the complex problem of determining an estimation of the probability of reaching the target.

References

1. Andreoni, J.: Philantropy. In: Gerard-Varet, L.A., Kolm, S.C., Ythier, J. (eds.) Handbook of Giving, Reciprocity and Altruism. North Holland, Amsterdam (2005)
2. Barzanti, L., Dragoni, N., Degli Esposti, N., Gaspari, M.: Decision making in fund raising management: a knowledge based approach. In: Ellis, R., Allen, T., Petridis, M. (eds.) Applications and Innovations in Intelligent Systems XV, pp. 189–201. Springer, Heidelberg (2007)
3. Barzanti, L., Gaspari, M., Saletti, D.: Modelling decision making in fund raising management by a fuzzy knowledge system. Expert Syst. Appl. **36**, 9466–9478 (2009)
4. Barzanti, L., Giove, S.: A decision support system for fund raising management based on the Choquet integral methodology. Expert Syst. **29**(4), 359–373 (2012)
5. Barzanti, L., Giove, S.: A decision support system for fund raising management in medium-sized organizations. M2EF - Math. Methods Econ. Finance, vol. 9, no. 10 (forthcoming)
6. Barzanti, L., Mastroleo, M.: An enhanced approach for developing an expert system for fund raising management. In: Expert System Software, pp. 131–156. NYC Nova Science Publishers (2012)
7. Bordley, R., LiCalzi, M.: Decision analysis using targets instead of utility functions. Decis. Econ. Finan. **23**, 53–74 (2000)
8. Calvo, T., Major, G., Mesiar, R. (eds.): Aggregation Operators; New Trends and Applications. Physica-Verlag, Heidelberg (2012)
9. Cappellari, L., Ghinetti, P., Turati, G.: On time and money donations. J. Socio-Economics **40**(6), 853–867 (2011)
10. Duffy, J., Ochs, J., Vesterlund, L.: Giving little by little: dynamic voluntary contribution game. J. Public Economics **91**(9), 1708–1730 (2007)
11. Duncan, B.: Modeling charitable contributions of time and money. J. Public Economics **72**, 213–242 (1999)
12. Flory, P.: Fundraising Databases. DSC, London (2001)
13. Kercheville, J., Kercheville, J.: The effective use of technology in nonprofits. In: Tempel, E. (ed.) Hank Rosso's Achieving Excellence in Fund Raising, pp. 366–379. Wiley, Hoboken (2003)
14. Lee, L., Piliavin, J.A., Call, V.R.: Giving time, blood and money: similarities and differences. Soc. Psychol. Q. **62**(3), 276–290 (1999)
15. Melandri, V.: Fundraising Course Materials. D.U Press, Bologna (2004). (in Italian)
16. Mood, A.M., Graybill, F.A., Boes, D.C.: Introduction to the Theory of Statistics. McGraw-Hill, New York (1974)

17. Narukawa, Y., Torra, V.: On distorted probabilities and m-separable fuzzy measures. Int. J. Approx. Reason. **42**(9), 1325–1336 (2011)
18. Nudd, S.P.: Thinking strategically about information. In: Tempel, E. (ed.) Hank Rosso's Achieving Excellence in Fund Raising, pp. 349–365. Wiley, Hoboken (2003)
19. Rosso, H., Tempel, R., Melandri, V.: The Fund Raising Book. ETAS, Bologna (2004). (in Italian)
20. Russell, S., Norvig, P.: Artificial Intelligence: A Modern Approach, 2nd edn. Prentice Hall, Upper Saddle River (2003)
21. Sargeant, A.: Using donor lifetime value to inform fundraising strategy. Nonprofit Manag. Leadersh. **12**(1), 25–38 (2001). John Wiley and Sons Inc
22. Torra, V., Narukawa, Y.: On distorted probabilities and other fuzzy measures. In: EUROFUSE 2007 (2007)
23. Verhaert, G.A., Van den Poel, D.: The role of seed money and threshold size in optimizing fundraising campaigns: past behavior matters! Expert Syst. Appl. **39**, 13075–13084 (2012)

Author Index

Printed in the United States
By Bookmasters